Marin Mersenne and the Study of Harmony

From Sound to Music

Leendert van der Miesen

AF872692

Routledge
Taylor & Francis Group
LONDON AND NEW YORK

Publication has been generously supported by the Research Group "Visualizing Science in Media Revolutions" at the Bibliotheca Hertziana – Max Planck Institute for Art History, Rome (project number BH-P-22-17).

BIBLIOTHECA HERTZIANA MAX PLANCK INSTITUTE FOR ART HISTORY

VISUALIZING SCIENCE IN MEDIA REVOLUTIONS

This work has been submitted and accepted as a doctoral thesis at the Humboldt University of Berlin at Faculty of Humanities and Social Sciences in 2022.

First published in 2025 by Amsterdam University Press Ltd.

Published 2025 by Routledge
4 Park Square, Milton Park, Abingdon, Oxon OX14 4RN
605 Third Avenue, New York, NY 10158

Routledge is an imprint of the Taylor & Francis Group, an informa business

© L. van der Miesen / Taylor & Francis Group 2025

The Open Access version of this book, available at www.taylorfrancis.com, has been made available under a Creative Commons Attribution-NonCommercial-NoDerivatives (CC-BY-NC-ND) 4.0 International license.

Trademark notice: Product or corporate names may be trademarks or registered trademarks, and are used only for identification and explanation without intent to infringe.

ISBN: 9789048564132 (hbk)
ISBN: 9781041182474 (pbk)
ISBN: 9781003699323 (ebk)
NUR 662

Cover illustration: Mario Bettini. *Apiaria universæ philosophiæ mathematicæ. In quibus paradoxa et nova pleraque Machinamenta ad usus eximios traducta, & facillimis demonstrationibus confirmata. Tomus Secundus*. Vol. 2 of 2 vols. Bologna: Ferroni, 1642, 8 ("Musica"). Augsburg, Staats- und Stadtbibliothek – 2 Math 11 -2 / Münchener DigitalisierungsZentrum.

Cover design: Coördesign, Leiden

DOI: 10.5117/9789048564132

Every effort has been made to obtain permission to use all copyrighted illustrations reproduced in this book. Nonetheless, whosoever believes to have rights to this material is advised to contact the publisher.

For Product Safety Concerns and Information please contact our EU representative:
GPSR@taylorandfrancis.com
Taylor & Francis Verlag GmbH, Kaufingerstraße 24, 80331 München, Germany

Marin Mersenne and the Study of Harmony

Table of Contents

Note on Typography and Citation

When quoting early modern sources, I have maintained the original spelling, apart from small changes where there is no ambiguity, such as the change from *i* to *j* and *v* to *u*. For Mersenne's *Harmonie universelle*, I have used the facsimile edition published by François Lesure in 1963. The *Harmonie universelle* is notoriously difficult to cite because of its incongruities and the many overlapping page numbers. In order to make clear which section I am citing, I indicate the title of the book section in the footnotes. I have included these book titles and the overall composition of *Harmonie universelle* and *Harmonicorum libri XII* in the two appendices at the end of this book. For works such as *Traité de l'harmonie universelle Questions harmoniques*, *Questions theologiques*, *Questions inouyes*, and *Les preludes de l'harmonie*, I have followed the page numbers of the original publications, rather than the edition published by *Corpus des Oeuvres de Philosophie en Langue Française* (Fayard) unless otherwise noted. When I refer to Mersenne's *Quaestiones celeberrimae*, I include the number at the top of each column of text (abbreviated with col.). In citing Mersenne's letters, I use the edition published by Paul Tannery et al. between 1932 and 1988, abbreviated with CM (*Correspondance du P. Marin Mersenne*), followed by the number of the volume and the page number. In my footnotes, I refer to the last names of the correspondents unless ambiguity could arise (for example with Constantijn and Christiaan Huygens). Unless otherwise stated, translations are my own.

List of Figures

List of Music Examples

Introduction: Harmonies at Work

Abstract: This introductory chapter presents the work of Marin Mersenne and the main themes of this book. Mersenne's *Harmonie universelle* (Paris, 1636–37) can be counted among the central scholarly works on music of the early modern period, positioning music at the forefront of the empirical sciences while at the same time being steeped in notions of divine harmony. For Mersenne, music's ability to be both quantified and experienced aesthetically made it the central science through which the divine harmonies present in the universe could be demonstrated. Interrogating how music obtained this central position, this chapter introduces Mersenne's concept of harmony, his intellectual and institutional surroundings, and his defense of music as a science.

Keywords: Marin Mersenne, music, early modern science, sound, universal harmony

Throughout the 1620s and early 1630s, the French polymath, Minim theologian, and music theorist Marin Mersenne roamed throughout Paris with open ears. The city was full of sounds, then as much as now, and provided a test site for many of Mersenne's experiments. From the balconies of his own convent, Mersenne made sounds and observed the echoes of the building. While others strolled and rested on the benches in the gardens of the Louvre, Mersenne was measuring an echo in a small alley. In the nearby galleries of the palace, he envisioned himself performing an experiment with an enormous string of a thousand feet. At the nearby Minim convent in Vincennes, he measured that it took twenty seconds for the sound of a cannon of the Parisian arsenal to reach him. He performed similar ballistic experiments on the banks on the Seine toward Saint-Cloud. On the outskirts of Paris, he observed men and women twisting animal guts to make strings for lutes and tennis racquets. Returning to his convent, he walked through the alleys surrounding the Rue St. Jacques, where the many instrument makers of Paris were located. Or he went through to

Van der Miesen, Leendert. *Marin Mersenne and the Study of Harmony: From Sound to Music.* Taylor & Francis Group, 2025.
DOI 10.5117/9789048564132_INTRO

the university district to visit the royal music printer of the king, Pierre Ballard, to acquire the latest editions of *airs de cour*, the most popular form of vocal music at the time. Among the many musical meetings in the city, Mersenne joined the gatherings of the harpsichordist Jacques Champion de Chambonnières, whom he described as the instrument's "final master."

Back in his room at a Minim convent next to the Place Royal (now Place des Vosges), Mersenne gathered music books, instruments, measuring devices, and other scientific instruments such as telescopes. In the convent, he could retreat from "the hassle and mud" [*de tracas et de boüe*] of the French capital.[1] In his room, he also likely kept the hundreds of letters he received, which his later student Hilarion de Coste collected in three large volumes. Mersenne had correspondents all over France and western Europe, who kept him up to date with any strange sounds or music they heard in their towns. He did not seem to have to worry about the costs of his postal exchange, as his convent covered all the charges.[2] The Minim convent had one of the largest libraries of seventeenth-century Paris, and this is where Mersenne was able to collect music editions from Italian and French composers, theoretical treatises, and even books the censors had deemed dangerous. Mersenne also collected prints and drawings of musical instruments, partially collected through his correspondence but also partially ordered from Parisian engravers, perhaps after his own sketches. With all these activities, his room must have been surrounded by papers and experimental objects, which was not unusual for the Minims. As one historian of the order writes, some members of the order made "their cells appear more like laboratories and libraries than traditional cells."[3]

All this was to lead up to a book on music entitled *Harmonie universelle* (Paris, 1636–37), which assembled more than ten years of research and has been described as "unlike any other work of music theory that had ever been published."[4] The book brings together music theory, experiments on sound, the construction of instruments, and the use of music for architects,

1 Mersenne to Rivet, 25 November 1640, CM 10, 293. I abbreviate the *Correspondance du P. Marin Mersenne* with CM and the number of the volume.

2 Samuel Hartlib stated: "Mersennus the whole Cloister maintaining the charges." Ephemerides 1639, Part 1, Hartlib 30/4/7A. Available online through Greengrass, Leslie, and Hannon, *The Hartlib Papers*.

3 Whitmore, *The Order of Minims*, 122. For more on the intellectual culture of the Minim convent, see Krakovitch, "La vie intellectuelle. On the Minim library and Mersenne's role in shaping it, see Guillo, "Sous la main du père Mersenne."

4 Christensen, "Mersenne's Sound World," 60.

preachers, and engineers. As such, it contains far more than what is now considered within the remit of music theory and addresses not only musical composition, counterpoint, and instrument-making but also mechanics, the anatomy of the human body, and combinatorics. These various elements were used to elevate the status of music and direct the attention of Mersenne's contemporaries to the true goal of music: demonstrating the order and harmony of God's creation. Rather than a cultural practice or custom, music was an instrument for universal knowledge, closely tied to morality and the search for truth. The study of music was the study of universal harmonies, dealing with the immutable laws of nature. True music, Mersenne argued, was spiritual, not mediated through the senses. Pointing to its divine origins and the always imperfect nature of human music, he wrote, "music is in God and in the angels, for God has the knowledge of all things in the highest degree of perfection."[5] Although Mersenne's ears were well attuned to the sounds and music of his contemporaries, he never lost sight of what, in his eyes, was the origin and ultimate goal of music, namely the divine.

Mersenne was an early modern polymath like many of his contemporaries and published on optics, mechanics, and mathematics, but over the course of his life, he found himself particularly drawn to music. Music was his field of study by choice, he wrote more than once, and although he was drawn to melodies, compositions, and various singers and composers, the intellectual aspect of music was especially important for him. Throughout his work, Mersenne emphasized that the intellectual and theoretical engagement with music was the highest goal. In his *Les preludes*, he formulates it as follows:

> Indeed, if music is to serve any purpose, and if its practice has any end, it can have none more excellent—after the glory of God, which is the last end of all possible things—than the recreation of the wise, who consume their time and their minds in meditation on the mysteries of religion, and in the search for reasons, which serve to combat all those who oppose the infallible truth of our faith and to persuade everyone of this truth and of the virtues which depend on it.[6]

5 "la musique est en Dieu, & aux Anges, car Dieu a la science de toutes choses en souverain degré de perfection." Mersenne, *Traité*, 19.

6 "En effect si la Musique doit servir à quelque usage, & si sa pratique a quelque fin, elle n'en peut avoir de plus excellente, apres la gloire de Dieu, qui est la derniere fin de toutes les choses possibles, que la recreation des sçavans, qui consomment leur temps, & leur esprit à la meditation des mysteres de la Religion, & à la recherche des raisons, qui servent pour combatre tous ceux qui s'opposent à la verité infaillible de nostre Foy, & pour persuader cette verité, & les vertus qui en dependent, à tout le monde." Mersenne, *Les preludes*, 183–84.

The ultimate purpose of music should be theoretical, the "recreation of the wise" in Mersenne's words, as the ultimate goal of science was theological. Music could thus be used to convince those who questioned the principles of the Catholic Church, such as heretics and skeptics. For Mersenne, music was an instrument of truth and virtue, simultaneously moral and epistemological. In these respects, music was superior to other arts. Mersenne described the art of painting as "dead," in contrast to music, as "music is life-giving, and in some way conveys the life, soul, spirit and affection of the singer, or musician, to the ears and souls of the listeners."[7] Since sound is measurable, it is more related to knowledge than the other senses, or in Mersenne's words: "sounds approach closer to the mind and to that which is intellectual than the objects of the other senses."[8]

This relation between sound and harmony was crucial for Mersenne. As the titles of his books demonstrate, harmony played a central role in his approach to the study of music, sound, and science at large. One of his correspondents, Christophe de Villiers, summarized Mersenne's achievements thusly: "You have investigated the source of harmonies."[9] As Villiers' letter describes, Mersenne was investigating, or even searching, for harmonies. Harmony was an overarching concept that could benefit all fields and sciences. As he writes in *Harmonie universelle*, "There is hardly any art, science or profession that harmony ... is not able to serve."[10] Mersenne's notion of harmony was especially theological; musical harmony gives "a foretaste of divine pleasures."[11] Whereas previous generations described harmony as something abstract—between the four elements, the human body, or the heavenly spheres—Mersenne found harmony in the world of materials: in the measurement of an organ pipe, the notes of a trumpet, an echo reflecting on a wall, and the physical composition of sound. Thomas

7 "morte, mais la Musique est vivante, & transporte en quelque façon la vie, l'ame, l'esprit & l'affection du Chantre, ou du Musicien, aux oreilles & dans l'ame des auditeurs." Mersenne, *Harmonie universelle*, "Livre second, des chants," 92. The book is notoriously difficult to cite because of the many treatises or "books" and the incongruent page numbers. In my citations, I therefore always include the part of the book from which the citation derives alongside the page number. I have included the composition of *Harmonie universelle* and *Harmonicorum libri* in the appendices at the end of this book.

8 "les sons approchent plus pres de l'esprit & de l'intellectuel, que l'objet des autres sens." Mersenne, *Harmonie universelle*, "Livre premier des consonances," 88.

9 "vous avez recherché la source des harmonies." Villiers to Mersenne, 13 May 1634, CM 4, 119.

10 "Il n'y a quasi nul art, nulle science, ou profession, à qui l'harmonie, & les livres precedens ne puissent servir." Mersenne, *Harmonie universelle*, "Livre de l'utilité de l'harmonie," 1.

11 "un avant-goust des plaisirs divins." Mersenne, *Questions harmoniques*, 61.

Christensen aptly writes that for Mersenne, "the music of the spheres has been yoked and brought crashing down to earth resoundingly."[12]

A central paradox or complexity in Mersenne's work is this tension between the abstract and the material. Although he sought an "intellectual music" (*musique intellectuelle*) that spoke more to the mind than to the ear, his science of music was firmly grounded in the sounds, instruments, and courtly songs of his day. Rather than retreating to his convent, his books, notes, and letters show someone who was very much engaged with the sounds in the world around him. He went far beyond some of his contemporaries to obtain materials of hitherto unknown music. He sent out numerous requests and images to the eastern Mediterranean to find materials on music played there, questioned and observed instrument makers, looked in the cabinets of curiosities of his age, and sought out the roaring of cannons. How Mersenne was able to unite all these different sounds, forms of music, and materials is the subject of this study. This book investigates how musical and acoustic experiences were turned into evidence in Mersenne's project on universal harmony by looking at the changing roles of experience, techniques of observation, and use of material instruments. Although others have noted the central role of harmony for Mersenne, there is a lack of scholarship at the interrelation of different forms of knowledge and the epistemic role of music in his work. My key questions are: How could the world of experiences and materials provide evidence for Mersenne of the harmonious world that lies beyond human perception? What types of music and musical materials could provide this evidence? And why was it music precisely that became the central instrument for this knowledge? To answer these questions, I situate Mersenne's endeavor within the world of instrument makers in Paris and correspondence networks across Europe and take into account the increased mobility and transcultural contacts of the period. I here show that Mersenne developed new methods and experiments to cement the image of a harmonious world, knowable through sound and music. To approximate the divine oscillations and cadences around him, the scholar centralized sound among the shifting landscape of the early modern sciences, imploring his contemporaries to perceive harmony in the noise and clamor of the early modern age.

Marin Mersenne and the Study of Harmony explores two larger arguments, one historical argument and a more historiographical one. First, I argue that Mersenne's work demonstrates the new epistemic promise of sound in the early seventeenth century, as materials, instruments, and sound in general

12 Christensen, "Mersenne's Sound World," 81.

gained a central role in the study of music. For Mersenne, it was sound that provided a privileged point of access to truth, knowledge, and the divine, and subsequently the sense of hearing took on a central role in his scientific praxis, albeit carefully deployed. And secondly, I hold that the transformation of the discipline of music involved a wide variety of methods and forms of knowing that cannot be explained solely by an increasing empiricism or application of mathematical methods. As this study will demonstrate, various forms of knowledge and ways of knowing that could be described as aesthetic, theological, antiquarian, artisanal, and ethnographic were all mobilized in the framework of music as the study of harmony. Music was not solely a subalternate science of arithmetic in the quadrivium or belonging to mixed-mathematics; rather, it was a cultural practice that involved various sciences, techniques, methods, and forms of knowledge. This wider understanding of music permeated Mersenne's scientific praxis, as it was closely tied to cultures of collecting, antiquarianism, French colonial discourse, and contemporary musical practices.

By investigating specific themes within Mersenne's output and that of his contemporaries, this book tries to understand how the study of music meandered between these different realms, methods, and practices. Through a consideration of his investigations into musical instruments discussed in chapter 2, the role of curiosity and collecting in chapter 3, his interest in music from outside Europe in chapter 4, and the role of beauty and compositional techniques in chapter 5, each of my case studies shines a different light on how and under which conditions music could provide evidence of the universal harmonies. Rather than assigning Mersenne to a particular epistemological position, my study demonstrates how new forms of knowledge were piloted and older orthodoxies were maintained. As this study shows, musical knowledge moved through many domains, many of which fall outside the purview of the borders of the present-day discipline of musicology. Mersenne's work demonstrates that early modern knowledge cultures were complex and entangled, combining different fields, methods, and objects of study.

By interrogating the role of music in the field of early modern sciences, this study builds on the long-standing interest in the relationship between music and science. For several decades, musicologists and historians of science have pointed to the role of music within the so-called "Scientific Revolution."[13] The foundational studies on this subject have unearthed

13 For a critical investigation of this concept, see Shapin, *The Scientific Revolution*. For several of the earlier groundbreaking studies in this regard, see Palisca, "Scientific Empiricism"; Drake,

close connections between musicians and scholars, such as the deep bonds between the experimental methods of father and son Vincenzo Galilei and Galileo Galilei. Often these works have singled out the development of specific theories or the introduction of empirical methods. A central theme has been the emancipation of acoustics as a separate field of study as scientific and aesthetic concerns grew apart. But as others have shown, concepts such as "scientific" and "aesthetic" did not carry the same meaning for early modern scholars, who dealt with aesthetic questions throughout their scholarly writings. Terms like "science" and "music" were not stable entities, in constant flux, and were not understood in the seventeenth century in the same way as they are conceived by the modern reader.[14] And the emphasis on empirical methods of acoustic research have obscured the degree to which older traditions such as Pythagoreanism or occult philosophies were upheld.

In viewing Mersenne's work as a conduit to observe several trajectories, I build here on a longer-standing interest in the shifting roles of music among the sciences around 1600. In his study on early modern discovery and imagination, Timothy Reiss concludes that "music came to traverse the arts in new ways and to new purposes."[15] Focusing on early modern France, Kate van Orden has similarly highlights the role of music as an important framework for measurement and bodily discipline. Van Orden writes, "Music was a practical and *practicable* mathematics that occupied both body and intellect, bridging the sensible realm of sound and the speculative realm of number."[16] In her work on early modern England, Katie Bank shows how "music not only reflected changes in episteme but also actively participated in its generation."[17] Bank shows that there were not only many relations between composers, performers, and natural philosophers but that music was fundamental in the reorientation of what counted as knowledge around 1600.

Inspired by such research, this study emphasizes that Mersenne's work is especially fruitful for looking at the interchange of the arts, sciences, and

"Renaissance Music"; Dostrovsky, "Early Vibration Theory"; Walker, *Studies in Musical Science*; Cohen, *Quantifying Music*, Palisca, *Music and Ideas*; Coelho, *Music and Science*; Kassler, *Inner Music*; Gozza, *Number to Sound*.

14 Gouk, *Music, Science, and Natural Magic*; Austern, *Both from the Ears and Mind*. The need for such histories is already expressed by Jackson, "Music and Science"; Coelho, "Preface." For the changing meanings of "science" in the early modern period, see Garber, "Philosophia, Historia, Mathematica."

15 Reiss, *Knowledge, Discovery and Imagination*, 158.

16 Van Orden, *Music, Discipline, and Arms*, 53.

17 Bank, *Knowledge Building*, 3.

practices in the early modern period. For Mersenne, music's very nature stimulated questions about what could count as knowledge and what not. How did music function as evidence for something beyond the immediate? Could a concert function as a scientific proof? What is the relation between mathematics and physics in explaining the sensory world? Here Mersenne was not alone. Many of his contemporaries already noted music's special epistemic status. In an introduction to his works, a contemporary of the composer Claude le Jeune writes of Le Jeune's contrapuntal techniques: "One perceives in his music / the secrets of mathematics / well observed from point to point."[18] An introduction to an edition of Eustache Du Caurroy's *Preces ecclesiasticae* (Paris, 1609) describes music as belonging to mathematics and as a "demonstrative and very certain science," able to engender "marvelous effects."[19] Du Caurroy's nephew Andre Pitard drew the connection even further in a posthumous publication of Du Caurroy's *Fantasies* (Paris, 1610). Du Caurroy's music in particular, Pitard declares, has as its object "the heavens and the holy image of truth," taking part in a desire to find truth, much as Mersenne did.[20] Listening to music, Pitard writes, one has the opportunity to contemplate "yourself as a resonant soul."[21] Playing and listening to music trains the mind and the senses, much as scientific advancement was considered a step closer to the divine. This tradition was reevaluated in the early seventeenth century, as musical discourse moved increasingly away from a strict Pythagoreanism and commentators situated music between individual taste and custom, closer to the sense of taste. And according to later historians, the seventeenth century witnessed an "untuning of the sky," as the idea that each of the planetary spheres brings forth a tone became more and more a metaphor, and even the "destruction" of world harmony.[22] Music as an art based on aesthetic principles and music as a science of immutable principles were increasingly seen as two different domains.

For Mersenne and many of his contemporaries, music was an especially interesting if challenging body of knowledge that did not always fit into clear-cut

18 "On apperçoit en sa Musique / les secrets de Mathematique / Bien observez de poinct en poinct." Le Jeune, *Le Printemps*, n.p. ("Ode sur la musique").

19 "Et d'autant que ceste science faisant partie des Mathematiques, est demonstrative & tres-certaine, il est impossible qu'elle ne produise admirables effects." Du Caurroy, *Preces ecclesiasticae*, n.p. ("Au roy").

20 "esprit vif, qui n'ayant que le ciel & la saincte image de verité pour object, sera sonner par tout soubs les figures de ces notes." Du Caurroy, *Fantasies*, n.p. ("A monseigneur").

21 "contemplant vous mesme en ame resonante." Du Caurroy, *Fantasies*, n.p. ("A monseigneur").

22 Hollander, *Untuning of the Sky*; Cohen, "Music as Science and as Art."

disciplinary and epistemic boundaries. Music was something theoretical and practical, divine and worldly, artistic and scientific all at the same time. Music making depended on individual skill, and preferences in music could be highly personal. At the same time, music had claims to be something much broader, theological, and even universal. Each chapter of this book tries to highlight these paradoxes and offers a particular case study in how Mersenne oscillated and mediated between these realms, drawing out a different constellation and answer each time. In what follows in this introduction, I will introduce the role of harmony for Mersenne more extensively, before considering Mersenne's relationship to music and his Minim order and his defense of science.

Unbreakable Sway

Mersenne never seemed to doubt the existence of harmony in the world. Already in his earliest works, he defended the certainty of knowledge and the existence of order in the world. Before he published the books on music and mechanics that would cement his reputation as an empirical scientist, he published a commentary on the Bible and multiple attacks on several groups he considered dangerous, such as skeptics, atheists, Protestants, and libertines. One example is *L'impiété des deistes* (Paris, 1624), dedicated to Cardinal Richelieu who became chief minister to Louis XIII that year. The book is an 800-page refutation of an anonymous poem that denounced the Christian God. In the book, Mersenne stages a conversation between a theologian and a deist, who is eventually led to conversion. Although seemingly removed from Mersenne's work on music and sound, the theologian in the book expresses the same convictions on the existence of a universal harmony as the author's musical writings. Writing on the order evident in the world, he expounds,

> Truly, when we contemplate the beautiful order in the world and sees that each thing maintains its rank and place, in spite of all the disorders that seem to arrive, one needs to conclude that there is someone who governs the whole world and maintains all things in good order; for the world could not keep the unbreakable sway [*branle*] and regular cadence that we perceive in it, if it were not for the divine Orpheus who touches the strings of the great lute of the universe and who takes care of all the activity and motions that appear in the heavens and in the elements.[23]

23 "Veritablement quand on contemple le bel ordre qui est au monde, & qu'on voit que chaque chose retient son rang, & son lieu, nonobstant tous les desordres, qui semblent arriver, il faut

The image of the universe as a musical instrument encapsulates the order and correspondence that existed in the world for Mersenne. God is a musician who plucks the strings of the universe. When commenting upon the world's order, Mersenne centered the role of contemplation. Turning inward into one's mind and meditating on the existence of God, one could find true knowledge. But this process was just as much an act of observation, in Mersenne's words, of perceiving "the unbreakable sway [*branle*] and regular cadence" in the world around us. The world moves, shakes, and is seemingly chaotic, but through careful and guided observation, one can discover an order to all things.

To describe this cosmic rhythm, Mersenne used the word *branle*, meaning wobble, shake, or swaying. The branle was also a type of round dance popular in France in the late sixteenth century (fig. 0.1). The dancers move by two or four larger steps to the left and the same number of smaller steps to the right, creating a line or circular motion.[24] The image of the dance is revealing of Mersenne's concept of order. Just as a single dancer might not be able to perceive the movement of the entire group, the perception of order was a question of perspective. It is possible that Mersenne took this idea from his favorite author, Augustine, bishop of Hippo. Writing on music, Augustine compared the position of a person perceiving the universe to a statue in the corner of an enormous, beautiful building. The statue cannot perceive the beauty of the construction it is part of, nor is a soldier "able to see the order of the whole army," nor the syllable in a poem able to perceive the whole poem.[25] Humans, from their limited point of perception, could hardly observe the harmony connecting the creation of which they are a part. Only a position like that of Er in Plato's *Politeia* or Scipio Africanus in Cicero's *Somnium Scipionis* was able to perceive the entire harmony of the world.[26]

For Mersenne, the limited perspective from which humans could perceive the harmony of the world did not mean that one should reject the idea of harmony itself. Rather, through different methods, some of this harmony

conclure qu'il y a quelqu'un qui gouverne tout le monde, & qui maintient toutes choses en bon ordre, car le monde ne pourroit pas garder le branle inviolable, & la cadence reguliere, que nous y appercevons, n'estoit l'Orphee divin, qui touche les cordes du grand luth de l'Univers, & qui a soing de tous les ressorts, & mouvemens, qui paroissent dans les Cieux, & dans les elemens." Mersenne, *L'impiété des deistes*, 81.

24 Mersenne was well acquainted with the dance form, referencing it numerous times in his works and giving six different types with each a different melody in his *Harmonie universelle*. See Mersenne, *Harmonie universelle*, "Livre second, des chants," 168–69; Mersenne, *Harmonie universelle*, "Livre sixiesme de l'art de bien chanter," 391, 396, 405.

25 Augustinus, *De musica liber VI*, 67–69. Translation by Martin Jacobsson.

26 Plato, *Republic*, 319–26 (10.614–10.621); Cicero, *On the Republic*, 116–24 (VI, 12–33).

Figure 0.1: A group dancing the branle (detail). Guérard (attributed), "Le branle du Mirliton," seventeenth century. Paris, Bibliothèque nationale de France, département Musique, VM PHOT MIRI-17 (101).

could become evident. In his scholarly activities, he emphasized that despite the chaos in the world, a harmony can be observed. Through the investigation of musical objects like strings, bells, and organ pipes (but also falling bodies, echoes, and the parabolic motion of a cannonball), something of this harmony could be approximated. Throughout his experiments, Mersenne hoped to find order, expressible in numbers. This Augustinian (sometimes described as Pythagorean) conviction guided all his experiments. Mersenne would always be convinced of the presence of harmony in all sensory experiences, writing that if humans could understand the songs of birds, sounds of animals, and noises of thunder or wind, then "we would find an admirable harmony."[27] Discussing the mechanics of a music string, Mersenne assumed that the whole world moves in a similar way: "The earth, the walls, and everything we see and touch, moves by similar turns and returns [*tours et retours*], we cannot perceive any of it; so that all the bodies of the world could make a perpetual harmony."[28] The world shakes or oscillates in perpetual harmony, but the human mind and senses are too weak to comprehend this all the time. A full perception of this divine harmony is reserved for humans in the afterlife.

As studies by Melanie Wald-Fuhrmann, Geoffrey Baker, Kate van Orden, Aviva Rothman, and others have shown, the ideal of harmony was especially useful in a period of religious strife and wars raging across the European

27 "nous y trouverions une admirable harmonie." Mersenne, *Traité*, 63.

28 "par exemple la terre, les murailles, & tout ce que nous voyons, & ce que nous touchons se mouvoir par de semblables tours & retours, nous ne pourrions l'appercevoir en aucune façon; de sorte que tous les corps du monde peuvent faire une perpetuelle harmonie." Mersenne, *Harmonie universelle*, "Livre troisiesme du mouvement," 168.

continent and beyond.[29] The notion of (musical) harmony was widely used to speak about a variety of topics, such as architecture, the human body, and nature itself. The sixteenth-century poet Pontus de Tyard wrote that "nature is tuned like music," and the political theorist Jean Bodin relied on musical harmony to explain different state forms.[30] In sixteenth-century France, the Academie des poesie et musique, a group of poets and composers who were directly linked to Mersenne, projected music as a moral and societal force: wherever music is undisciplined, the morals are depraved, and where it is well ordered, the people are disciplined. Harmony had the power to affect civil affairs.[31] In his dedication of *Harmonice mundi* (Linz, 1619), the astronomer Johannes Kepler explicitly connected his own study of celestial and musical harmony with the harmony of the continent, writing, "What else is a kingdom but a harmony?"[32] Mersenne himself also used harmonic terminology to discuss political events. In a letter from 1642 to André Rivet that mentions the possible threat of the Ottoman Empire, he wrote, "It would be good for God that twenty or so of the most knowledgeable and good people among you can come together or communicate by letters filled with the charity of God, to shake the world and make it return to concord and union."[33] Mersenne saw his scientific work as beneficial to this unity, since it demonstrated the central role of order and harmony. Science could elucidate what is natural and subsequently what is supernatural. As he wrote in another letter to Rivet, "We must use reason to embrace religion."[34] Simultaneously, Mersenne maintained that this view was veiled, and there continued to be mysteries of the faith: a full view on creation in its essence was impossible.[35] The idea that advancement of knowledge would be a service of the faith was especially pronounced in his work on music. Already in his 1623 *Quaestiones*, he hoped for a music academy that would instruct young people in harmony so that all of France and the entire world would

29 Peil, "Concordia discors"; Baker, *Imposing Harmony*; Van Orden, *Music, Discipline, and Arms*; Wald-Fuhrmann, "'Discors Concordia'"; Rothman, *The Pursuit of Harmony*; Honisch, "Encounters with Music."

30 "Nature a accordé comme une Musique." Tyard, *Les discours philosophiques*, 283v. Bodin, *The Six Bookes of a Common-Weale*, 790.

31 Yates, *The French Academies*, 319.

32 Kepler, *The Harmony of the World*, 4.

33 "Plust à Dieu que vintaine des plus sçavans et des plus gens de bien d'entre vous peussent s'assembler, ou se communiquer par lettres remplies de la charité de Dieu, pour ébransler le monde et luy faire r'ambrasser la concorde et l'union." Mersenne to Rivet, 12 October 1642, CM 11, 294.

34 "Il faut donc user de la raison pour embrasser la religion." Mersenne to Rivet, 14 December 1640, CM 10, 335.

35 Lenoble, *Mersenne*, 311–12; Phillips, *Church and Culture*, 147–51; Fabbri, "Hérésie et orthodoxie."

have the most skilled singers and so that their hearts would "be inflamed with divine love."[36] His notion of harmony was fundamentally social, moral, and political, with the explicit aim of establishing a musical re-education program designed to praise God.

The question of the extent to which these harmonies could be measured and approximated had been the subject of intense discussion in the early seventeenth century. Mersenne fought his way into the discussion between the English physician Robert Fludd and Johannes Kepler, which started when Kepler included a three-page attack of Fludd in his *Harmonice mundi*. Kepler found that the English physician obscured harmony more than illuminating it by discussing harmonies that could not be measured, such as the harmonic proportions between light and darkness.[37] Instead, Kepler emphasized that he only sought harmonies in physical movements that can be measured. Mersenne embroiled himself in this debate with his 1623 *Quaestiones*, where he quoted Kepler's statement that Fludd's representation is poetical and oratorical, not mathematical or philosophical. Fludd's universal harmony had no foundation, Mersenne found.[38] "The compass," not "the furnace and the fire," should be the foundation of any harmonic investigations.[39] Gassendi offered a similar critique of Fludd, namely that the proportions that Fludd gave "are nowhere to be observed."[40] Although he sided with Kepler in 1627, Mersenne would later also criticize Kepler's approach and describe his figures as "symbolic relations and analogies."[41]

It would be a mistake to limit the role of harmony to the Pythagorean theme of the music of the spheres. This was in fact not Mersenne's primary interest. Some of his correspondents even found that he spoke too little of the harmonies of the planets in his book to merit the title *Harmonie universelle*.[42] Rather, harmony was a much broader concept that united the

36 "tota Gallia, imo totus mundus ad majorem Dei gloriam, & omnium corda amore divino inflammanda personaret." Mersenne, *Quaestiones*, col. 1687.

37 Kepler, *The Harmony of the World*, 507. On Fludd's theory of music, see Ammann, "The Musical Theory"; Fludd, *The Temple of Music*. On the debat itself, see especially De Buzon, "Harmonie et métaphysique"; Fabbri, "Questioning Fludd"; Mehl, "Gassendi in the Philosophical Debate."

38 Mersenne, *Traité*, 446–47.

39 "l'un se sert du compas, & l'autre des fourneaux & du feu." Mersenne, *Traité*, 352.

40 "hoc Mundo proportiones illas nusquam observari." Gassendi, *Epistolica exercitatio*, 104–5. Fludd replied to Mersenne in two publications, and Mersenne asked the natural philosopher Gassendi to defend him. For the sources surrounding the controversy, see Taussig, *L'examen de la philosophie de Fludd*.

41 "rapports symboliques, & à ces analogies." Mersenne, *Harmonie universelle*, "Livre troisiesme des genres," 188.

42 Villiers to Mersenne, 13 May 1634, CM 4, 120.

world of sensations and materials with the larger universe. In the preface to the second book of his *Traité*, Mersenne writes that his book shows that "harmony is found in everything in the world."[43] For example, as chapter 4 of this book demonstrates, Mersenne's correspondence includes an extensive discussion of whether other cultures outside Europe have "harmony." Whereas Mersenne's correspondent Peiresc wrote to the theorist that the Ottomans were "great lovers of harmony," Mersenne was more skeptical.[44] Although Mersenne rejected much of the evidence that was presented to him, I show how the question of what harmonies are used beyond Europe became part of his investigation, in which he went beyond the efforts of many of his contemporaries. A different orientation of harmony is explored in chapter 2, which considers musical instruments. Mersenne was especially concerned with the question of whether instruments can operate as evidence of harmony. I show that the trumpet occupied a special role for Mersenne, as an instrument that "naturally" brings forth the first notes of the diatonic scale. At the same time, he rejected the evidence of the trumpet tones because he considered the number seven an enemy of harmony. Although the notion of harmony was an a priori fact for Mersenne, it had to be newly approximated time and again, as the ultimate disclosure of harmony would only be in the afterlife. Harmony was thus an incredibly mobile concept, uniting many different subjects and forms of knowledge. Going beyond the image of the universal monochord found in his *L'impiété des deistes*, Mersenne searched for different methods and experiments to foreground the idea of a harmonious world knowable through music.

Minim and Musicus

Few figures within early modern science have received such disparate treatment as Mersenne. This can be partially explained by the many paradoxical qualities his work demonstrates. On the one hand, his work seemed to initiate a stronger role for empiricism and observation. He criticized his contemporaries for upholding the same views that had been espoused by music theorists for centuries, even when faced with evidence to the contrary. Simultaneously, his concept of universal harmony was deeply traditional, derived from Platonism and Augustine. Consequently, both musicologists

43 "l'Harmonie se trouve en toute ce qui est au monde." Mersenne, *Traité*, n.p. ("Preface du second livre").

44 "grandement friands de l'harmonie." Peiresc to Mersenne, 18 June 1634, CM 4, 179.

and historians have drawn conflicting conclusions about him. For some, Mersenne represents a modern, initiating a new era of scientific method. According to Leslie Blasius, Mersenne's work is dominated by "rationalism and empiricism" and "anticipates the new science of the eighteenth century," in contrast to the Jesuit Athanasius Kircher.[45] Others emphasize the conservative aspects of his work, such as his adherence to universal harmony and theologically motivated science.[46] Jamie Croy Kassler probably best sums up the different view by describing Mersenne's work as "Janus-faced," bearing contrasting attitudes.[47]

There are also widely different claims regarding Mersenne's scholarly career. Some commentators have suggested that Mersenne's work from 1634 onward represents an entirely different approach, a break or rupture with his previous works. In his foundational study of the Minim, Robert Lenoble singles out 1634 as the year in which Mersenne's approach saw a shift from Aristotelianism to mechanistic philosophy.[48] Others, too, have pointed to the different characters of his books from this period onward, arguing that "universal harmony" slowly became an empty signifier for Mersenne, or that he ultimately completely separated the discipline of acoustics from the broader investigations of music.[49] Several other scholars such as Peter Dear, Natacha Fabbri, Thomas Christensen, Daniel Garber, and Claudio Buccolini have all stressed the more enduring aspects of his work, showing how Mersenne continued to have similar concerns before and after 1634.[50] This study does not aim to answer this question but rather traces the longevity of Mersenne's musical project, which occupied him for at least several decades. Building on the wide-ranging previous scholarship on the Minim, I emphasize that his work complicates clear-cut epistemic boundaries, as he mobilized different traditions and forms of knowledge to convince his readers and gather as many arguments as he could find to retain music's status as a science. Throughout Mersenne appears as a highly flexible thinker who approached old questions with new methods, and new questions with old methods.

By showing how Mersenne's aesthetic and scientific interests collided, this study argues that the two domains cannot be separated, as Mersenne

45 Blasius, "Mapping the Terrain," 34.

46 For example Copenhaver, "The Occultist Tradition," 469.

47 Kassler, *The Beginnings of the Modern Philosophy*, 2.

48 Lenoble, *Mersenne*, 38.

49 Wald-Fuhrmann, "Musik und Universalwissenschaft," 321; Cohen, *Quantifying Music*, 101.

50 Dear, *Mersenne*; Garber, "On the Frontlines"; Fabbri, "Genesis"; Buccolini, "L'exégèse du dernier Mersenne"; Christensen, "Mersenne."

continued to struggle with how the aesthetic appreciation of music and the physical study of sound relate. The central argument of the book is captured in the subtitle of this study, *From Sound to Music*, as I aim to reverse the traditional narrative of the trajectory of Mersenne's musical interests. Rather than an emancipation of the scientific study of sound out of the discipline of music, I show that this process was far from complete, as Mersenne argued that knowledge on the mechanics and physics of sound should loop back to musical practitioners, composers, and listeners. Although he entertained the idea of a separate science of sound, as I show in chapter 1, he nonetheless published his acoustic investigations firmly within *Harmonie universelle*, which bears the subtitle "the theory and practice of music." This subtitle should not be seen as a designation devoid of meaning, but Mersenne always hoped that the knowledge of instrument building, the physics of sound, and the mechanics of movement would flow back to the musical practitioners and listeners so that they might use this knowledge to elevate themselves and help to position music in a better position than it was in before.

Most information about Mersenne's life is documented by his fellow Minim Hilarion de Coste, who published a short biography a year after his death.[51] After receiving his elementary education in Le Mans, Mersenne was sent to the Jesuit college in La Flèche. The college at La Flèche had just been opened in 1604 by Henri IV and would become one of the most important schools for the French cultural, political, and ecclesiastical elite. The Jesuits were an important arm of the Counter-Reformation and one of the most organized religious institutions of the early modern period. Beginning in the second half of the sixteenth century, they were active in France and under the protection of the king from 1561.[52] When Mersenne left the college in 1609, he had enjoyed an education at one of the most elite institutions in France. From here he went to Paris, where he continued his philosophical and theological studies at the Sorbonne and the Collège de France. But he left the university in 1611 to enter the order of Minims, a mendicant order established in Italy in the fifteenth century.[53] After he was fully initiated, Mersenne taught for some years at the Minim college in

51 De Coste, *La vie*. This biography can be supplemented by the biography of Cornelis de Waard, in CM 1, xix–lv.

52 The educational principles in La Flèche during Mersenne's time were founded on the society's rule book *Constitutiones* and the *Ratio Studiorum* of 1599. See Farrell, *The Jesuit Ratio Studiorum*. More information on the principles at La Flèche during this period can be found in Rochemonteix, *Un college de jésuites*; Dear, *Mersenne*, 15–22; Gaukroger, *Descartes*, 38–67.

53 Whitmore, *The Order of Minims*; Pierre and Vauchez, *Saint François de Paule et les Minimes*.

Nevers before settling at the convent near the Place Royale in Paris in 1619. His cell became a central visiting place for many travelers and the meeting place of his own "Academia Parisiensis." It was renowned beyond France, and according to the English philosopher Thomas Hobbes, it outrivaled any school in the city.[54]

The Minims followed the teachings of their patron saint, Francis of Paola. First a Franciscan friar, Francis used to refer to himself as *il minimo dei minimi*, the least of the least.[55] From this designation, his followers were known as the Minims, emphasizing their austerity. The life of the friars in this order was characterized by the four vows of poverty, chastity, obedience, and a perpetual lent, meaning abstinence from meat and all other animal products except fish. The order had a strong eremitic heritage and daily life was directed toward penitence and charity. Although originally from Italy, it held a special place in seventeenth-century France. At the end of his life, Francis of Paola went to France to heal Louis XI, initiating a long connection to the French monarchy. By the seventeenth century, the order had grown to be an important institution, with close associations with the French court, as is shown by the location of the Paris convent. It was located right in the heart of Paris next to the square built by Henri IV, which was conceived as a social and artisanal epicenter of the city, signaling the Minims' importance to the crown and city at large.

The Place Royale convent had one of the richest libraries in the city, with large collections on theology, navigation, history, and chronology.[56] As Odile Krakovitch has shown, it is clear from their library that the friars were interested in almost all scientific, theological, and historical domains. The building itself provided a space for exploration and reflection, for example through two anamorphic murals painted by the Minim Jean-François Niceron, who returned to the Paris convent from Rome in 1640. In the library, Niceron also staged other optical games.[57] Like most other Minim convents, the Parisian convent had a garden, and several members developed an interest in botany and healing. Mersenne himself experimented with the echo in the courtyard of the convent, finding that the echo responded an

54 "Cujus Cella Scholis erat omnibus anteserenda." Hobbes, *Angli Malmesburiensis*, 61.

55 Whitmore, *The Order of Minims*, 3–4.

56 Whitmore, *The Order of Minims*, 124; Krakovitch, "La vie intellectuelle"; Guillo, "Sous la main du père Mersenne"; Kent, "'Minimizing.'"

57 "present en la Bibliotheque de nostre Convent de la place Royale à Paris." Niceron, *La perspective curieuse*, 115; De Rosa, Bortot, and D'Elia, "The Digital Reconstruction."

octave higher than his original voice after he moved around and shouted words at different pitches.[58]

The important position of the order in France was also evident in the elaborate architecture of some of their convents, leading Krakovitch to write that "everything at the Minims is seemingly contradictory."[59] One of these contradictions could be found between their spiritual, almost mystic, philosophy and their deep interest in the arts and sciences. Although Mersenne was certainly among the most famous, he was not the only Minim with a scholarly career, as the order seemed to have stimulated its brothers to pay attention to nature, crafts, and the arts. As Emily Kent has shown, such intellectual concerns were "especially common" at the Place Royal convent.[60] Toward the end of his life, Mersenne was praised for his scholarly activities by the head of the order.[61] Although the collective intellectual culture of the Minims might have been less institutionalized than, for example, the Jesuit order, the order did have its own intellectual and social agenda, associated with the Counter-Reformation and the resurgence of Catholic movements in late sixteenth-century France.[62] The Minims thus shaped Mersenne's output, provided him with a network, and structured his day-to-day life. Here he could read and write, seemingly undisturbed by daily obligations at the convent, possibly distracted by the freezing cold of the rooms, even in the month of May.[63] Mersenne himself was outspoken on how his religiosity informed his scholarly writings. In the preface to *Harmonie universelle*, he writes, "I have also taken the liberty of making a few small elevations of spirit to God in a few corollaries, because I did not think I should neglect them, since they came to me without seeking them out."[64] Even when seeking out scientific truths, spiritual inspiration came

58 "J'en remarqué un au haut de nostre 2me et 3me degré du 2e estage qui respond une octave plus haut que ma voix plus creuse." Marginal note at Mersenne, *Harmonie universelle*, "Livre premier de la nature," 50. Transcribed in CM 8, 69.

59 "Tous, chez les Minimes, semble apparemment contradictoire." Krakovitch, "L'architecture des trois couvents minimes," 245.

60 Kent, "'Minimizing,'" 63. See also Whitmore, *The Order of Minims*, 143. For more on the intellectual contributions of some of the Minims, see Krakovitch, "La vie intellectuelle"; Beltrán, "Nature *au natural*"; Kellman, "Mendicants, Minimalism, and Method"; De Rosa and Bortot, "Anamorphosis."

61 Spinozza to Mersenne, 14 September 1647, CM 15, 425.

62 Whitmore, *The Order of Minims*, 6.

63 Mersenne for example in May 1635 wrote to Peiresc that "it was so cold today that the rooms were freezing" ("Le froid a esté si grand ajourd'hui qu'on geloit dans les chambres"). Mersenne to Peiresc, 26 May 1635, CM 5, 216.

64 "J'ay aussi pris la liberté de faire quelques petites eslevations d'esprit à Dieu dans quelques Corollaires, parce que je n'ay pas creu les devoir negliger, puis qu'elles me sont venuës sans les

to him during his scholarly work, which could be used by preachers "to draw morals from all the sciences."[65] His scholarly work was thus a deeply spiritual, with wide religious and social ramifications.

The Minims were not a particularly musical order and did not develop extensive musical traditions like the Franciscans and Oratorians. In terms of music, as in many other fields, they were characterized by austerity, with a recited rather than sung office. In his history of the Order of the Minims in France, Whitmore writes, "It seems that any aesthetic pleasure to be derived from adornment of the ritual was proscribed so that the fullest attention could be paid to the meaning of the liturgical language."[66] The rule books of the order would usually state that singing with notes was prohibited.[67] Any more elaborate forms of music were not allowed, and as Thuillier puts it, the brothers "shall perform the divine offices with great respect and observing all the ceremonies according to the usage of the Roman Church, nevertheless without making use of any notes."[68] The same rule book held for feast days: "They [the brothers] will celebrate it in the same way, without notes, and simply by counting or by developing the syllables."[69] The rule books states that one should "sing with *squared* voice, as if counting, according to our rules."[70] In French this is translated from Latin as "voix quarrée, ou egale."[71] Mersenne would himself note that "we form no intervals, and observe no other measure than that of syllables."[72]

rechercher." Mersenne, *Harmonie universelle*, n.p. ("Preface au lecteur"). Frédéric de Buzon has already shown that in Mersenne's view to understand the harmonious order of creation, both experimental investigations and theological discourses were necessary; see De Buzon, "Harmonie et métaphysique."

65 "de tirer des Moralitez de toutes les sciences." Mersenne, *Harmonie universelle*, n.p. ("Preface au lecteur").

66 Whitmore, *The Order of Minims*, 65.

67 Whitmore, *The Order of Minims*, 65; Shaw, "The Minims," 362. For more on music within the Minim order, see Pons, *Droit ecclésiastique*; Vaast, "L'esthétique musicale" 258–62; Kent, "'Minimizing,'" 70.

68 "s'acquitteront des divins Offices avec grand respét, & en observant toutes les Cerémonies selon l'usage de l'Eglise Romaine, sans néanmoins se servir d'aucune notes; mais simplement comme en expliquant, & comptant les mots." Thuillier, *Traduction nouvelle des regles*, 69.

69 "ils la celebreront de même, sans notes, & simplement en comptant, ou en développant les syllables." Thuillier, *Traduction nouvelle*, 69.

70 "cantari voce quadrata, quasi computando, juxta Regulam nostram, cum pausis competentibus." De Paule, *Regulae fratrum*, 262.

71 Thuillier, *Traduction nouvelle des regles*, 114; Kent, "'Minimizing,'" 70n48. There were traditions that performed plainchant either "measured" in equal notes or without measure; see Sherr, Plainsong in the Age of Polyphony."

72 "nous ne faisons aucuns intervalles, & que nous n'observons point d'autre mesure que celle des syllabes." Mersenne, *Harmonie universelle*, "Livre second, des chants," 90.

There was thus some measure kept by counting the syllables of the office, and the brothers would introduce pauses where appropriate but included no melodic developments. Punishments were specified for any brothers who broke these rules. One Parisian edition from 1632 put it as follows: "If someone introduces a chant with notes, which is against the fourth chapter of the Rules of this Order, if he is superior, even if he is general, he will be deposed from his office, made unfit for any kind of office of this Order."[73] Whether these rules were actually enforced is difficult to tell, and there were most likely great regional differences.[74] The form of text recitation practiced by the Minims apparently did not conflict with the large collection of music books in the Minim library or Mersenne's attention to more elaborate forms of music making. The members of the order would also take part in festivities with more complex music, for example the Minims at Trinità dei Monti in Rome, and in the Minim church of Place Royal, where Jacques Mauduit's requiem was performed.[75]

Although the Minims, together with several reformed orders like the Capuchins, went perhaps the furthest in their abandonment of melodic development, the question of music in the liturgy was still far from resolved. Although they would become one of the primary promotors of the evangelical and doctrinal uses of music, the Jesuit order was also hesitant in their use of music in their early years, initially forbidding their members to sing the office.[76] Composed music in the liturgy was still a rarity in many religious institutions, and in many parishes, chapels, and monasteries, the majority of the liturgy was performed by "chant," albeit embellished or improvised upon. In terms of performance, there was no single tradition but rather a fragmented landscape. As Philippe Vendrix and Peter Bennett have shown, the performance of chant was subject to debate in early seventeenth-century France, with several alternatives put forth by, for example, the Oratorians and the Royal Benedictine Abbey under the guidance of Antoine Boësset.[77] These reforms centered around rhythm and text declamation, issues the Minims and Mersenne specifically struggled with.

73 "Si quelqu'un introduit un chant avec la notte, ce qui est contre le quatriéme chapitre de la Regle de cét Ordre, si celuy-là est General, il sera deflors deposé de son office, & deformais inhabile à toute sorte d'office de cét Ordre." De Paule, *Les regles des freres*, 158–59.

74 In 1754 Benedict XIV would sign an edict that would allow the singing of Gregorian chant for the Minims. Pons, *Droit ecclésiastique*, 3:201–2.

75 Boiteux, "Fêtes et identité"; Mersenne, *Harmonie universelle*, "Livre septiesme des instrumens de percussion," 63.

76 Filippi, "'Their sound hath gone forth.'" For the hesitant use of music among the early Jesuits, see Bertoglio, *Reforming Music*, 482–84.

77 Vendrix, "Pour les grands"; Bennett, *Music and Power*, 193–237.

Mersenne himself commented on the straight and monotone nature of the chant of the Minims and the Capuchins, which was "like the line," in his assessment.[78] He did find that the chant of the Capuchins was "more cheerful and more agreeable" ("plus gay & plus agreable"), as the brothers would sing the final verses more slowly and loudly.[79] But little is known about his own musical activities, although De Coste wrote that Mersenne was frequently found singing psalms, especially Psalm 22 ("Dominus regit me") and Psalm 150, with its exhortation to praise God with musical instruments.[80] In his *Harmonie universelle*, Mersenne gives the melody for the "O Seigneur mon Dieu que tu es grand en toute grandeur" as he himself would sing it. His goal here is to indicate that his starting pitch was composed of fifty vibrations of air per second.[81] Throughout his works, he includes psalms or songs, for example at the end of his *Traité*. And in his *Harmonicorum*, he gives twelve examples of the church modes (tones) (fig. 0.2) and would later write that the modes "5, 6 and 12 seem to me to be the most beautiful," but "each one can choose the one that will suit him best for his particular consolation."[82] He always remained clear on the theological importance of music in service, writing that "church songs" [*chants Ecclesiastiques*] "excite devotion when they are well sung."[83] Later in his life, Mersenne supported the changes made by chant reformers like François Bourgoing, writing in his *David françois* (Paris, 1641) that these new chants help "to excite everyone to worship God with heart and mouth."[84] Here and in many other cases, Mersenne went beyond the recitation "without notes" instructed by the order, seemingly without any reservations. Although much is uncertain about his relation to music practice, it is clear that music was not only an intellectual concern for Mersenne but also a deeply moral and emotional activity, as music moved him to tears when he listened to a musical rendition of Psalm 22 or the verse *Est deus pastor mihi* (The Lord is my shepherd).[85]

78 "le chant des Minimes & des Capucins, est semblable à la ligne." Mersenne, *Traité*, 35.

79 Mersenne, *Harmonie universelle*, "Livre second, des chants," 91.

80 De Coste, *La vie*, 102–3.

81 Mersenne, *Harmonie universelle*, "Livre troisiesme des instrumens," 149

82 Mersenne, *Harmonicorum*, "De generibus, & modis," 112; "le 5, le 6, et le 12 me semblent les plus beaux: mais chacun peut choisir celuy qui luy agreera davantage pour sa consolation particuliere." Mersenne, *Harmonie universelle*, "Livre second, des chants," 96.

83 chants Ecclesiastiques qui excitent la devotion lors qu'ils sont bien chantez." Mersenne, *Harmonie universelle*, "Livre second, des chants," 95.

84 "pour exciter tout le monde à loüer Dieu de coeur & de bouche." Mersenne, "L'approbation" in Bourgoing, *David françois*, n.p. Cited in Vendrix, "Pour les grands," 93.

85 Mersenne, *Quaestiones*, col. 1622 and 1684.

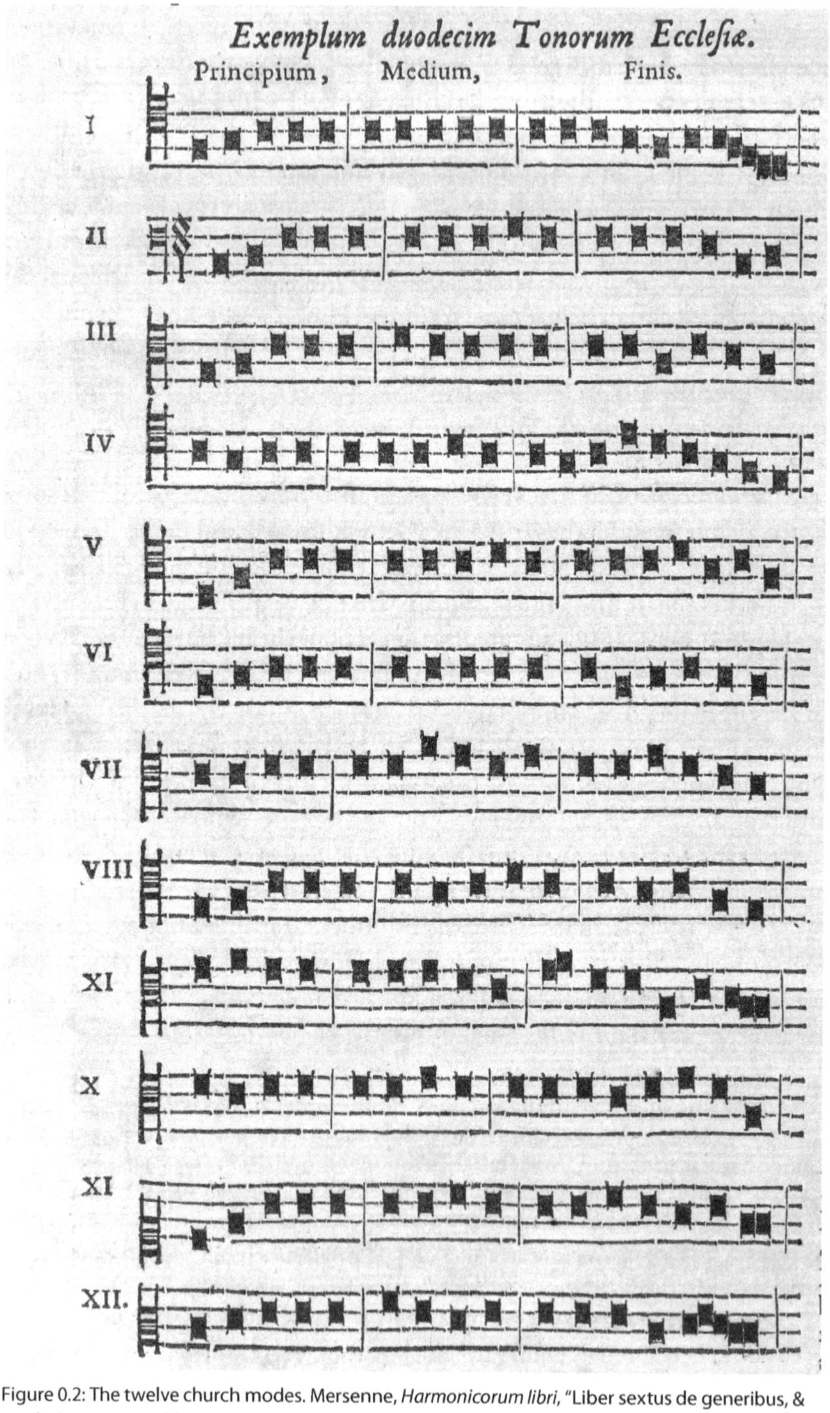

Figure 0.2: The twelve church modes. Mersenne, *Harmonicorum libri*, "Liber sextus de generibus, & modis," 112. Paris, Bibliothèque nationale de France, département Réserve des livres rares, RES-V-580.

Although the Minims abstained from many forms of music in the liturgy, music did feature in the stories surrounding their patron Francis of Paola. In the retelling of his life by the seventeenth-century Minim historian Claude du Vivier, Vivier recounted how during Francis's life, a man called François Carbonello from the town of Paola became so annoyed with the order that he furiously marched to the convent. Ready to confront Francis and possibly harm him, he was about to open the door of his cell when suddenly, in the words of Vivier, "he could hear from within the chamber a sound so melodious, [as if] the angels were accompanying the holy man." According to Vivier, "it [the music] overthrew the sourness of this mind [that] pretended offense." It was music "used by angels, but unusual for men."[86] Another Minim historian from the early seventeenth century commented upon the same anecdote and wrote, "Several times the blessed spirits were heard inside his chamber, making divinely melodious concerts and harmonies, and voices and instruments that rendered a very sweet music and harmony."[87] Some Minims, such as John François Binans, a Minim active in Provence in the early seventeenth century, was described as a great amateur of music who found that music was a "means of elevating his spirit to God" and frequently sang spiritual airs.[88] Mersenne's attention to music as a spiritual practice and exercise was thus no exception for the order, but the issue that still demands investigation is the extent to which his attention to more elaborate forms of music making collided with the norms of the order.

As I discuss in chapter 5, his contemporaries remarked that Mersenne famously had a bad singing voice, but apart from these accounts, there is not much known about his own musical skills. Mersenne would have encountered music theory and music lessons at the Jesuit college in La Fléche, where the students performed plays, danced, and most likely also sang. Music would have also been discussed in class as part of the

86 "Il entend du dedans de la chambre sortir un son si tres-melodieux, les Anges estre de compagnie au sainct Homme: ... qu'elle renversa les aigreurs de cét esprit pretendue offense. ... chant usité aux Anges, inusité aux hommes." Du Vivier, *Vie et miracles*, 142–43. Eric Nelson has studied the role of Vivier and other seventeenth-century Minim historians in Nelson, "Remembering the Martyrdom."

87 "plusieurs foison entendit dedans sa chambrette les bien-heurex esprits, faisant des concerts & accords divinement melodieux, & instrumens de voix qui rendoient une tres-douce musique & harmonie." Dony-d'Attichy, *Histoire general*, bk. 1, 19. There was a popular iconographic tradition depicting Saint Francis of Assisi, after whom Francis of Paola is named, while listening to angels making music; see Dell'Antonio, *Listening as Spiritual Practice*, 25–30. Francis of Paola is not frequently depicted in this way; see Vaast, "L'esthétique musicale," 259.

88 "comme un moyen fort propre à hausser & eslever son esprit en Dieu." Dony-d'Attichy, *Histoire generale*, bk. 1, 394.

mathematical or mixed-mathematical sciences. Early modern textbooks frequently commented upon music as a subordinate science of arithmetic or discussed the physics of sound in the Aristotelian tradition. Books like Jacques Lefèvre d'Étaples's *De musica libris* (Paris, 1496) or Heinrich Glarean's *Isagoge in musicen* (Basel, 1515) were used to teach students the principles of music, relying still heavily on late antique sources such as Boethius.[89] At the Jesuit college of La Flèche, the theory of music was discussed as part of the mathematical sciences, with music being a science subordinate to arithmetic. Jesuit reformers in the late sixteenth century still recommended d'Étaples' *De musica libris* (Paris, 1496), which discusses concepts such as notes, intervals, consonances, proportions, and tetrachord species (*genera*), working from late antique sources such as Boethius.[90] As music was a part of many young students' education in the time of Mersenne, a theoretical or scholarly interest in music was no exception. Men in his circle like René Descartes, Pierre Gassendi, Ismael Boulliau, Girard Desargues, Gilles Personne de Roberval, and François de La Mothe le Vayer all wrote their own musical treatises or contributed parts to Mersenne's work.[91]

Music was already extensively discussed in Mersenne's earliest works, such as *Quaestiones* and *La verité*. In the former, he devoted much space to what is known about ancient Greek music, and in *La verité*, music's mathematical nature is used to demonstrate that under the right circumstances, true knowledge on appearances is possible. In these years, Mersenne already hinted at a larger work on music.[92] A larger outline of this book was included in his 1627 *Traité de l'harmonie universelle*. He envisioned a book on music that would discuss the theory and practice of ancient and modern music, the nature of sound, the human voice, musical instruments, the world as an

89 Mai Groote, "Studying Music."

90 Gozza, "A Renaissance Mathematics"; Rasch, "René Descartes and Isaac Beeckman," 294. On Lefèvre's attention to music among his mathematical reform, see Oosterhoff, *Making Mathematical Culture*, 150–61.

91 This has already been pointed out by Albert Cohen in Cohen, *Music in the French Royal Academy*, 5. Descartes's treatise *Compendium musicae* (Utrecht, 1650) is most well known. Gassendi's treatise *Manuductio ad theorem seu partem speculativam musicae* can be found in Gassendi, *Opera omnia*, 5:633–58. It has been translated into French in Gassendi, *Iniation à la théorie*. Ismael Boulliau's treatise on music has been studied very little and is located in Paris, Observatoire de Paris, MS a, b, 5, 11, 351–64. Roberval wrote a treatise entitled "Elementa musicae" now in Paris, Bibliothèque nationale, MS fr. 9119, fols. 374–470v. Mersenne included a small treatise on the method for learning to sing by Girard Desargues; see Mersenne, *Harmonie universelle*, "Livre sixiesme de l'art de bien chanter," 332–42. La Mothe Le Vayer's "Discours sceptique sur la musique" was first published in Mersenne's *Questions harmoniques*.

92 Mersenne, *Observationes et emendationes*, col. 439–40; Mersenne, *La verité*, 370–71.

instrument of music, and the sounds and voices present in paradise.[93] Around the same time, he worked on a manuscript entitled *Livre de la nature des sons* that remained unpublished but formed the basis for *Harmonie universelle*.[94] It seems that he interrupted the process of writing *Harmonie universelle* in 1634 to publish several shorter books, outlining his epistemological and scholarly principles: *Questions theologiques* (Paris, 1634), *Questions inouyes* (Paris, 1634), *Questions harmoniques* (Paris, 1634), and *Les preludes de l'harmonie universelle* (Paris, 1634). Many of the topics of these books returned in the *Harmonie universelle* and its Latin counterpart, *Harmonicorum libri XII* (Paris, 1635).[95] After struggling with different printers for three years, Mersenne finally added the last book of his *Harmonie universelle* in 1637. The work consists of individual "books" or treatises on the physics of sound, the voice, singing, the consonances, musical composition, mechanics, and the different musical instruments (Appendix A and B). Although the subtitle describes the book as "the theory and practice of music," its content goes well beyond this, as it emphasizes the role of music as an instrument for universal harmony.

In 1627 Mersenne deplored the state of musical science, opening his *Traité* with these words: "I have often amazed myself that the science of music has remained so imperfect up till the present."[96] Looking at the state of the sciences in the early seventeenth century, Mersenne saw improvement everywhere. In fields such as mathematics, optics, or mechanics, he found that the moderns could match the achievements of the ancients, or even eclipse them. But not in the field of music. Music, Mersenne writes, "has been so abandoned that it seems that it is no longer practiced except for profit, or sensual delight, without considering its principal use, which is to lead to virtue and to regulate morals."[97] The situation seemed especially dire

93 Mersenne, *Traité*, n.p. ("Sommaire des seize livres de la Musique").

94 Paris, Bibliothèque national, MS 2884. See Fabbri, "Genesis"; Buccolini, "Mersenne Translator of Bacon?"; Taveau, "Le manuscrit 2884." In a letter to Peiresc, Mersenne writes that he had been working on *Harmonie universelle* for a decade, see Mersenne to Peiresc, 20 March 1634, CM 4, 81. I would like to thank Laurence Taveau for making this valuable study and edition of the manuscript available to me.

95 Mersenne's *Harmonicorum libri* is the Latin version of *Harmonie universelle*, according to Mersenne written for those who cannot read French. *Harmonicorum libri* is the smaller book and cannot be considered a direct translation, since it contains many passages that are not found in the French *Harmonie universelle*, and vice versa. According to Mersenne, he shortened the Latin book because of the costs. See Mersenne to Peiresc 17 November 1635, CM 5, 477.

96 "Je me suis souvent estonné de ce que la science de la Musique est demeurée si imparfaite jusques à present." Mersenne, *Traité*, n.p. ("Preface au lecteur").

97 "La Geometrie, l'Algebre, l'Astronomie, la Perspective, la Catoptrique, la Dioptique, & les Mechaniques ont acquis une grande perfection au siecle où nous sommes, mais la Musique a tellement esté abandonnée, qu'il semble qu'elle ne soit plus maniée que pour le lucre, ou pour la

to Mersenne, since the ancients had already pointed to the importance of music: Pythagoras and Plato privileged music as a tool for universal knowledge and an essential part of any political entity. But observing the state of music in France, Mersenne found nothing of this; his contemporaries seemed only to care about sensual music and only practiced music very imperfectly. Music was a science first and foremost, and even if all music making was banned, its theories should still be studied, as sound could shed more light on philosophy than all the other sensory objects.[98]

Music was of course not only a science. Mersenne emphasized the multiple nature of music in his work. He also described it as an "intellectual virtue."[99] In his *Les preludes*, he reached back to Aristotle for the importance of sound in shaping the morals, as sounds are based on movement and virtuous habits are similarly shaped by repeated actions. Because of this, humans had a special relation to sound: "we observe this love and affection for sounds when we sometimes sing without thinking about what we are doing."[100] Music, by its principles, makes certain virtues clear, which are called intellectual since they guide humans toward a divine truth. But thirdly, music is also an art, since it has rules and precepts about the way to make harmonies, melodies. In this Mersenne compared music to logic, since it its object is order and can be expressed in categories, universalities, syllogism, figures, and demonstrations; ethics, since it concerns will and the life of humans; and mechanics, since it can be "reduced to art" and is in the service of humans.[101] Music is all three fields of knowledge at once, ranging from the *musica speculativa* (reflective music) to the practical music making.

Mersenne wrote at a time when many of these classifications were increasingly questioned and reshuffled. The period around 1600 saw several changes in the way music was investigated and conceptualized. Throughout the Middle Ages, music was classified as a mathematical science, as part of the *quadrivium*, together with geometry, arithmetic, and astronomy. But from the sixteenth century onward, it was increasingly considered part of the language arts (so-called *trivium*). Music was studied alongside other humanistic subjects such as poetics and rhetoric. As Ann Moyer writes, "By the late sixteenth century the scholarly methods of late humanism, rather than Pythagorean

volupté, sans considerer son principal usage qui est d'acheminer à la vertu, & regler les moeurs." Mersenne, *Traité*, n.p. ("Preface au lecteur").

98 Mersenne, *Harmonie universelle*, "Livre premier des consonance," 88.

99 "une vertu intellectuelle." Mersenne, *Traité*, 14.

100 "or l'on remarque cét amour, & cette affection que l'on a pour les sons, lors que l'on chante quelquefois sans penser à ce que l'on fait." Mersenne, *Les preludes*, 214.

101 "reduisent en art." Mersenne, *Traité*, 15.

mathematics, came instead to serve more and more as the ways to define and study music."[102] In her work on early modern England, Linda Phyllis Austern similarly argues that "by the turn of the seventeenth century, more commentators emphasized the human response to music than its sublime origins or numerical proportions."[103] Music was increasingly seen as something culturally dependent; what appears beautiful to one group might offend others.

In France a similar discussion took place and Mersenne's own work can be seen as part of this trend, as he emphasized the passions and individual taste of listeners.[104] Some authors took this even further. Arguing for his use of licenses and compositional freedom against "the censors," the French composer Jacques de Gouy defended himself as follows: "We are no longer in the time of Pythagoras, when it was enough to say the master put it this way; if it is permissible to argue today against Aristotle, I believe it is no less permissible to argue against those who established the rules of music, seeing that they have much less foundation than the principles of philosophy."[105] De Gouy summarized his argument as follows: "Rules were invented for beauty and grace: but if they are not observed, other rules may be invented to achieve the same end."[106] His appeal to Aristotle was timely; multiple mysterious movements sprung up in Paris, associated with anti-Aristotelian thought. In 1624 three men refuted Aristotle's philosophy publicly, but they were removed from their teaching offices and convicted to exile.[107] If one could doubt the principles of Aristotle, what foundation do the rules of music have?

Science Defended

As this book shows, Mersenne struggled with this question regarding the rational principles of music for most of his life. A central feature of

102 Moyer, "Musical Scholarship in Italy," 185. For the permanence of the mathematical model, see Vendrix, "Music and Model in the Renaissance."
103 Austern, *Both from the Ears and Mind*, 165.
104 Mace, "Marin Mersenne"; Duncan, "Persuading the Affections"; Psychoyou, "Trajectories of Musical Acoustics."
105 "Nous ne sommes plus du temps de Pythagoras, où c'estoit assez de dire le Maistre l'a dit ainsi; S'il est permis de disputer aujourd'huy contre Aristote, je croy qu'il n'est pas moins permis de disputer contre ceux qui ont estably les reigles de la Musique, veu qu'elles ont bien moins de fondement, que les principes de la Philosophie." De Gouy, *Airs à quatre parties*, n.p. ("Preface").
106 "Les reigles on esté inventées pour la beauté & pour la grace: mais si en les observant on ne la rencontre pas, il est permis d'en inventer d'autres pour arriver à ce but." De Gouy, *Airs à quatre parties*, n.p. ("Preface").
107 Garber, "On the Frontlines."

Mersenne's work is his defense of the sciences, a position that has been explained in various ways. Mersenne published his books during a time of increasing skepticism regarding the foundations of knowledge. The works of Sextus Empiricus were printed and popularized through the works of Michel de Montaigne and Pierre Charron.[108] One of the central questions in French skepticism was what the role of the senses should be in acquiring knowledge.[109] Can one trust what one sees or hears? Which experiences are considered scientific, and which not? The examples of how the senses fail humans were numerous. The skeptic Francisco Sanches mentioned in his *Quod nihil scitur* (Lyon, 1581) the example of looking at a coin through a glass of water. It was not known how a coin appears bigger or smaller depending on the position or material. He also mentioned fallacies of the ear, such as the perception of an echo as a second original sound.[110] Since the senses are untrustworthy, the entire foundation of knowledge was suspicious for many skeptics. The writer and friend of Mersenne François de La Mothe le Vayer described it as follows:

> Since we cannot know anything except by means of the senses, which are said to be the gates of our soul into which nothing enters except through them, should we not be very suspicious of all our knowledge, given the natural weakness of these very senses, their usual defects and their errors, which are so frequently apparent?[111]

Most famously, Mersenne's close correspondent Descartes turned inward to find any truths that might stand the test of time. In this view, knowledge is only possible if one rids of sensory experiences and focuses on reason. But such skepticism also had its antagonists. More orthodox scholars considered skepticism particularly dangerous. Skeptics were associated with various libertine and anti-Catholic movements. The fear was that any doubt regarding the senses or the possibility of knowledge would quickly lead to doubting the fundamental tenets of society.

As Melinda Latour has shown, "Music proved to be a perfect example of the problem of knowledge in these debates as it was clear that human musical

108 The standard work continues to be Popkin, *The History of Scepticism*. Popkin's work has been criticized for its emphasis on a single form of skepticism. Instead, others have suggested that there were many forms, see Perler, "Was There a Pyrrhonian Crisis."

109 Clarke, *French Philosophy*, 34.

110 Sanches, *That Nothing Is Known*, 252.

111 La Mothe le Vayer, *Opuscule ou petit traité sceptique*, 114–15. Translation from Clarke, *French Philosophy*, 46.

practices modeled a complex interplay of physical skill and mechanics, sense perception and understanding."[112] The ancient philosopher Sextus Empiricus already wrote a book that aimed to refute the dogmatic and expert claims concerning music, something that was taken up again in the sixteenth century.[113] Mersenne's work offers numerous passages that express a skeptical standpoint toward music. In his 1625 *La verité*, Mersenne stages a conversation between an alchemist, a skeptic, and a Christian philosopher. Regarding music, the skeptic says, "Music is nothing more than an appearance, since what I find pleasant, someone else will find discordant according to their mood."[114] There were no arguments as to why the intervals of the fourth or fifth are more harmonious than the seventh or major second, since musical appreciation differs from person to person. The skeptic's position on music is once more brought to the fore in his *Questions harmoniques*. Here it is stated that "many believe that music is not a science, but a mechanical art, whose rules are based on the senses, particularly the ear."[115] The book also includes an entire "Discours sceptique sur la musique," by François de la Mothe le Vayer that Mersenne added without the author's name and without his permission.[116] La Mothe le Vayer attacks music at various levels, claiming, for example, that most music teachers are vile characters. But he goes beyond these accusations to question the notion of world harmony itself: those who love the truth might recognize that it is better to imagine no worldly harmony at all than to suggest there is one. Ultimately, La Mothe le Vayer suggested that the idea of the harmony of spheres started when a drunk musician "imagined that the sound of the jars and glasses was that of heaven."[117] With reference to the first mode of skepticism (variations among animals), he writes that

112 Latour, *The Voice of Virtue*, 163.

113 Sextus Empiricus, *Against the Musicians*; Latour, *The Voice of Virtue*, 166–69. For other examples of attacks on music, see Gerbino, "Skeptics and Believers."

114 "La musique n'est rien qu'en apparence, puis que ceque je treuve agreable, un autre le treuve discordant selon l'esprit." Mersenne, *La verité*, 32.

115 "Plusieurs croyent que la Musique n'est pas une science, & qu'elle n'est qu'un art mechanique, dont les regles sont fondées sur les sens, & particulierement sur celuy de l'oreille." Mersenne, *Questions harmoniques*, 80.

116 In 1640 La Mothe le Vayer issued another version of the text, as he was unhappy to see the earlier version printed imperfectly. He originally included more praise to Mersenne and had not written in such a way that it was ready to be made public. La Mothe le Vayer, *Petit discours de l'immortalité de l'ame*, 179. La Mothe le Vayer would continue to visit Mersenne and move in his circles, see Chapelain to Balzac, 9 October 1639, CM 8, 528.

117 "s'imagina que le son des pots & des verres estoit celuy des Cieux." La Mothe Le Vayer, "Discours sceptique sur la musique," 138.

"nothing can be established with certainty in this pretended science ... what is discordant in our ears is melodious in that of the ox and the snake."[118]

In many ways Mersenne's work reacted against these skeptical currents. According to the Minim, skepticism could only lead to undermining the foundations of the church and state. However, he agreed with the skeptics that the senses are difficult to trust. Richard Popkin has therefore described his position as "mitigated skepticism," arguing that Mersenne partially agreed with skeptics but that he remained convinced that one could come to reliable statements on the world with the help of mathematics, common sense, and instruments.[119] Music received a special role in combatting such ideas, as Mersenne repeatedly emphasized regarding its special status as a science with certainties.

The question of the status and use of music was asked widely. Courtier literature, for example, often described music as effeminate and sensuous, and any professional activity in the realm of music was not advised. Rob Wegman has shown that there was a crisis underway in musical thought, at least since the late fifteenth century, when a generation of critics argued that polyphony was wasteful, immoral, and effeminizing.[120] There were books that promoted lute playing, for example Nicholas Faret's *L'honneste homme* (Paris, 1630), but Faret did not advise music making because of its respectability or spiritual purpose; rather, it was to please "the masters and mistresses" at court.[121] As Jeanice Brooks writes, "Music figured significantly in several strands of French thinking about noble status and its demarcation."[122] This had little to do with imitating the divine harmonies present in the world but was closely related to an understanding of music based on social constructions and conventions.

118 "Peut-estre que ce qui discorde en nostre oreille, est melodie en celle du boeuf & du serpent ... qu'on ne peut rien establir de certain en ceste pretenduë science." La Mothe Le Vayer, "Discours sceptique sur la musique," 125–26. On La Mothe le Vayer's skeptical approach to music, see Luppi, "Musiche immaginarie"; Mackensen, "System und Kritik"; Rentsch, *Die Höfflichkeit musikalischer Form*, 186–96. A similar criticism is already stated in Mersenne's *Traité*; see Fabbri, *De l'utilité de l'harmonie*, 30.

119 Popkin, *The History of Scepticism*, 112–27.

120 Wegman, *The Crisis of Music*. Andrew Dell'Antonio has argued how "active listening" was fostered within Jesuit pedagogy, emphasizing listening as an active process rather than passive, as the Jesuits were concerned with the dangers of music for "the stable subjectivity and identity of the noble male." Dell'Antonio, *Listening as Spiritual Practice*, 9. For the early modern critique that music is effeminizing, see Austern, "'Alluring the auditorie to effeminacie'"; Latour, *The Voice of Virtue*, 170–77.

121 "puis que nos Mâistres & nos Mâistresses s'y plaisent." Faret, *L'honneste-homme*, 27.

122 Brooks, *Courtly Song*, 164.

Some discouraged music making all together. In 1638 the question of whether music does more harm than good was asked at one of Théophraste Renaudot's conferences in Paris. Two speakers both weighed the value of music. One speaker pointed to music's beneficial uses in medicine and honoring God. Another found that music effeminizes, excites to "foul pleasures," and "blinds the eyes for understanding."[123] Mersenne joined this debate in his own publications by asking if it respectable to gain pleasure out of music: should one be ashamed of an interest in music, or is it essential for the learned men and *honnête hommes*?[124] He disagreed with those who compared the pleasure of music to the pleasures "of meats, smells, and other objects that feed the senses."[125] Unlike those things, music was not a "brutal pleasure that makes the use of reason disappear."[126] Even when there is a sensual or "brutal" element present in musical listening, when humans consider the reasons of harmony, music is elevated "to the status of intellectual things."[127] Such a listener would be devoted to God and the eternal pleasures of the afterlife rather than momentary delights of this life.

This book thus shows how Mersenne sought stable ground for musical science in order to elevate its status. A better understanding of the workings and physics of music would advance musical practitioners so that they would be welcomed rather than scorned by all those in high society. The physician Christoph de Villiers described his achievements as follows: "From a vilified and misunderstood science, you have made a science that will forever be known as a model of excellence with all others." In short, Villier wrote, "you have made a servant of servants into a lady of honor who will soon become a princess."[128] Whether he was successful remains to be decided.

123 "mais sur tout, incitant aux sales voluptez, dont les chansons & le jeu d'instrumens sont les plus communes allumettes assoupissans les yeux de nostre entendement." Renaudot, *Troisieme centurie des questions*, 479–80.

124 Mersenne, *Questions harmoniques*, 1.

125 "des viandes, des odeurs, & des autres objets qui repaissent les sens." Mersenne, *Questions harmoniques*, 12.

126 "comme le plaisir brutal qui fait perdre l'usage de la raison." Mersenne, *Questions harmoniques*, 14.

127 "Et bien que ce plaisir soit brutal, neantmoins il devient raisonnables, lors que l'homme en use comme il doit, d'autant qu'il en contemple les raisons, & l'esleve jusque à la dignité des choses intellectuelles." Mersenne, *Questions harmoniques*, 53.

128 "d'une science avile et mesprisee vous en avez fait une qui aura pour tousjours le renom d'entree en parangon avec toutes les autres ... vous avez fait d'une servant des servants une dame d'honneur qui dans peu de terms doit estre une princesse." Villiers to Mersenne, 13 May 1634, CM 4, 119–20.

As this book shows, in defending the status of music, Mersenne arrived at a mitigated position, arguing that no absolute certainties are possible but that controlled observations, experiments, and mathematics could allow one to approach at least some certainty.

Structure

The first chapter of the book delves into the world of listening and musical science in the early modern period, showing how Mersenne turned music into the art of measuring sounds. It departs from a little-explored passage in his *Questions theologiques* where he imagines an independent field of study for sound, called "psophologie." Although this science would call for a high number of experiments and empirical observations, the role of the ear remained highly ambiguous for Mersenne; although it functioned as a conduit for knowledge, hearing needed to be guided by mathematics. The chapter discusses Mersenne's notion of sound and hearing before delving into two case studies that show that his turn toward measurement was always within certain bounds; the ear was not deemed accurate enough to investigate the minute differences of sound necessary for such a science of sound to exist. His turn toward measurement thus remained a promise for future generations and the afterlife, where sounds in their full perfection would be unveiled for the blessed.

The second chapter considers the role of musical instruments in Mersenne's work. This chapter investigates what information Mersenne hoped to find at the maker's workshop and how he went about gathering such knowledge. I situate Mersenne's interest in craft knowledge within the larger attention to artisanal practices among seventeenth-century scholars. Although scholars increasingly turned to workshops, they also drew sharp boundaries between those who worked with their mind and those who worked with their hands. After this investigation of the social world of instruments makers, I consider their use as experimental objects. Looking at Mersenne's efforts to demonstrate the naturalness of the consonances, I show that it was ultimately not the monochord but the trumpet that took on a central role for Mersenne. In contrast to his contemporaries, he found the nature of music not in animals or the human voice but in instruments. I show how Mersenne collected materials on the trumpet, what experiments he and his collaborators performed, and which observations were utilized in his pursuit of harmony in the world around him.

Chapter 3 turns to methods of collecting and corresponding. Although Mersenne's role as "postmaster" or "secretary" is often noted, there has been little engagement with how such methods of circulation and collecting were utilized in his musical writings and how the collection of observations fits Mersenne's broader epistemology. Offering one way to read his desire for ever more observations and examples, I emphasize here the role of "curiosity," the term by which many of Mersenne's contemporaries described his state of mind and scope of interest. I show how Mersenne interpreted curiosity as a state of searching, moving through many different subjects. Whereas some of his contemporaries dismissed the cabinet of curiosity, it became a central place of inquiry for Mersenne. From this metaphorical collecting, I make a link to the material collecting of observations, descriptions, and images, showing how Mersenne gathered observations and images in his correspondence. Together with his correspondents, he investigated images, marble plates, and plaques, taking part in a wider community of antiquarians and collectors. Both the metaphorical collecting of sounds and the material collecting of objects were interpreted by Mersenne as an imitation of the divine, capable of revealing something about nature.

The fourth chapter uncovers new routes of inquiry in the relation between musical scholarship and transcultural contacts by investigating the moment when scientific debates on harmony and rationality in music gained new urgency in the context of increased mercantile contacts and colonialism. I show how for several years in the early 1630s, the antiquarian Nicolas-Claude Fabri de Peiresc circulated requests to the eastern Mediterranean about various aspects of music on behalf of Mersenne. The requests asked about the ways of singing and music in a wide variety of cultures across the Mediterranean and were addressed to missionaries, travelers, and merchants in Cairo, Tunis, and Aleppo. Although this endeavor is often overlooked, I argue that it contains key insights to understanding how musical scholarship in early modern Europe turned toward music outside Europe. Mersenne's interest in musical materials from the Ottoman Empire were closely tied to his defense of the European diatonic system as the most natural of all musical systems and the revival of theoretical principles from ancient Greece.

Chapter 5 brings the study full circle by looking at the capacity of reason to explain aesthetic experiences. The alignment of what is rational with what is perceived as beautiful was a central conundrum of the early modern period. The period paid increasing attention to the capabilities of the subjective listener and was deeply conflicted on whether art should follow a prescribed set of rules. In the field of music, some writers continued to relate beauty to rational principles, while others highlighted aesthetic sensibilities, often described by a "je ne sais quoi." Whereas some compared the composer to an architect following the rules of perspective and geometry, others emphasized the role of inspiration and artistic licenses. This chapter investigates how these problems were articulated in the work of Mersenne, pointing to an increasing tension between rules, taste, mind, and ear. My case studies include the role of combinatorics, definitions of consonance, and the application of compositional rules. I argue that Mersenne allowed "taste," "chance," and "inspiration" to become heuristic categories while he also emphasized the existence of certainties. Countering those who argued for the primacy of taste, Mersenne argued that there were certain musico-mathematical rules, but on the level of the individual response, it was impossible to establish axiomatic principles. Even as individual taste gained a larger role, sound's close association to the divine and the intellectual realm authorized musical harmony as evidence for the harmonies beyond the sensorial realm.

Taken together, these five chapters aim to give insight into Mersenne's conceptualization of what it meant to study music in the early seventeenth century. Close examination of his work reveals cracks and fissures in his universalistic approach but simultaneously reveals a monumental attempt to mobilize different traditions, practices, and ways of music making to show definitively that music is a science, and possibly one of the most important ones.

Bibliography

Manuscript sources

Paris, Observatoire de Paris, MS a, b, 5, 11, 351–64. Ismael Boulliau's manuscript on music.

Paris, Bibliothèque nationale de France, MS fr. 9119, fols. 374–470v. Gilles Personne de Roberval's "Elementa musicae."

Printed primary sources

Augustine. *De musica liber VI. A Critical Edition with a Translation and an Introduction by Martin Jacobsson*. Stockholm: Almqvist & Wiksell, 2002.

Bodin, Jean. *The Six Bookes of a Common-Weale. Written by J. Bodin a Famous Lawyer, and a Man of Great Experience in Matters of State*. London: Bishop, 1606.

Bourgoing, Francis. *Le David françois, ou traité de la saincte psalmodie, contenant plusieurs advis ou enseignements méthodiques, pour ceux qui sont appliquez par estat à chanter les loüanges de Dieu. Avec un petit traité touchant la saincte messe*. Paris: Huré, 1641.

Cicero, Marcus Tullius. *On the Republic and The Laws*. Translated with an introduction, notes, and indexes by David Fott. Ithaca/London: Cornell University Press, 2014.

De Coste, Hilarion. *La vie du R.P. Marin Mersenne*. Paris: Cramoisy, 1649.

De Gouy, Jacques. *Airs à quatre parties, sur la paraphrase des pseaumes de Messire Antoine Godeau, évesque de Grasse*. Paris: Ballard, 1650.

De Paule, François. *Les regles des freres, des soeurs, et des fideles de l'un & de l'autre sexe de l'Ordre des Minimes*. Paris: Martin, 1632.

De Paule, François. *Regulae fratrum et sororum ac fidelium utriusque sexus Ordinis Minimorum*. Paris: Dehors, 1631.

Descartes, René. *Compendium musicae*. Utrecht: Van Zijl, 1650.

Dony-d'Attichy, Louis. *Histoire generale de l'ordre sacre des Minimes. Divisee en huict livres*. Paris: Cramoisy, 1624.

Du Caurroy, Eustache. *Fantasies a III. IIII. V. et VI. parties*. Paris: Ballard, 1610.

Du Caurroy, Eustache. *Preces ecclesticae ad numeros musices redactae*. Paris: Ballard, 1609.

Du Vivier, Claude. *Vie et miracles de Sainct Francois de Paule Instituteur de Lordre des Freres Minimes*. Paris: Cramoisy, 1609.

Faret. Nicolas. *L'honneste-homme ou, l'art de plaire a la court*. Paris: Du Bray, 1630.

Fludd, Robert. *The Temple of Music*. Translated by Peter Hauge. London/New York: Routledge, 2017.

Gassendi, Pierre. *Epistolica exercitatio, in qua principia philosophiae Roberti Fluddi*. Paris: Cramoisy, 1630.

Gassendi, Pierre. *Iniation à la théorie ou partie speculative de la musique*. Introduction, translation, and notes by Patrice Bailhache. Turnhout: Brepols, 2005.

Gassendi, Pierre. *Opera omnia*. 6 Vols. Lyon: Anisson, 1658.

Greengrass, Mark, Leslie, Michael, and Hannon, Mark. The Hartlib Papers. Published by The Digital Humanities Institute, University of Sheffield (2013). Available at https://www.dhi.ac.uk/hartlib.

Hobbes, Thomas. *Angli Malmesburiensis philosophi vita*. London: Cooke, 1682.

Kepler, Johannes. *The Harmony of the World*. Translated by E. J. Aiton, A. M. Duncan, and J. V. Field. Philadelphia: American Philosophical Society, 1997.

La Mothe le Vayer, François, de. "Discours sceptique sur la musique." In Marin Mersenne, *Questions harmoniques. Dans lesquelles sont contenuës plusieurs choses remarquables pour la physique, pour la morale, et pour les autres sciences*, 84–165. Paris: Villery, 1634.

La Mothe le Vayer, François de. *Opuscule ou petit traité sceptique sur cette commune façon de parler: N'avoir pas le sens commun*. Paris: Sommaville, 1646.

La Mothe le Vayer, François de. *Petit discours de l'immortalité de l'ame. Avec le corollaire, & un discours sceptique sur la musique. Seconde edition.* Paris: Camusat, 1640.

La Primaudaye, Pierre de. *The French Academie Fully Discoursed and Finished in Foure Bookes.* London: Legat for Adams, 1618.

Le Jeune, Claude. *Le Printemps.* Paris: Ballard, 1603.

Mersenne, Marin. *Correspondance du P. M. Mersenne.* Edited by Paul Tannery, Cornelis de Waard, Bernard Rochot, Armand Beaulieu, and René Pintard. Paris: Centre National de la Recherche Scientifique et al., 1932–88.

Mersenne, Marin. *Harmonie universelle contenant la théorie et la pratique de la musique (Paris, 1636). Édition fascimilé de l'exemplaire conserve à la Bibliothèque des Arts et Métiers et annoté par l'auteur.* Introduction by François Lesure. Paris: Éditions du Centre National de la Recherche Scientifique, 1963.

Mersenne, Marin. *L'impiété des deistes, et des plus subtils Libertins découverte, & refute par raisons de Theologie, & de Philosophie.* Paris: Billaine, 1624.

Mersenne, Marin. *Observationes et emendationes ad Francisci Giorgii Veneti problemata.* Paris: Cramoisy, 1623.

Mersenne, Marin. *Les preludes de l'harmonie universelle ou questions curieuses. Utiles aux predicateurs, aux theologiens, aux astrologues, aux medicins & aux philosophes.* Paris: Guenon, 1634.

Mersenne, Marin. *Questions harmoniques: Dans lesquelles sont contenuës plusieurs choses remarquables pour la physique, pour la morale, et pour les autres sciences.* Paris: Villery, 1634.

Mersenne, Marin. *Traité de l'harmonie universelle.* Paris: Baudry, 1627.

Marin Mersenne. *La verité des sciences, contre les sceptiques ou pyrrhoniens.* Paris: Du Bray, 1625.

Niceron, Jean-François. *La perspective curieuse ou magie artificielle des effets merveilleux.* Paris: Billaine, 1638.

Plato. *Republic.* Translated by C. D. C. Reeve. Indianapolis/Cambridge: Hackett, 2004.

Renaudot, Théophraste. *Troisieme centurie des questions traictees aux conferences du Bureau d'Adresse, depuis le 18. fevrier 1636. jusques au 17. janvier 1639.* Paris: Bureau D'Adresse, 1639.

Sanches, Francisco. *That Nothing Is Known (Qvod Nihil Scitur).* Translated by Douglas F. S. Thomson. Cambridge: Cambridge University Press, 1988.

Sextus Empiricus. *Against the Musicians (Adversus musicos). A New Critical Text and Translation on Facing Pages, with an Introduction, Annotations, and Indices Verborum and Nominum et Rerum by Denise Davidson Greaves.* Lincoln/London: University of Nebraska Press, 1986.

Thuillier, René. *Traduction nouvelle des regles, du correctoire, et du ceremonial de l'Ordre des Minimes.* Paris: Giffard, 1703.

Tyard, Pontus de. *Les discours philosophiques*. Paris: L'Angelier, 1587.

Secondary literature

Ammann, Peter J. "The Musical Theory and Philosophy of Robert Fludd." *Journal of the Warburg and Courtauld Institutes* 31, no. 1 (1967): 198–227.

Austern, Linda Phyllis. "'Alluring the auditorie to effeminacie': Music and the Idea of the Feminine in Early Modern England." *Music & Letters* 74, no. 3 (1993): 343–354.

Austern, Linda Phyllis. *Both from the Ears and Mind: Thinking about Music in Early Modern England*. Chicago: University of Chicago University Press, 2020.

Baker, Geoffrey. *Imposing Harmony: Music and Society in Colonial Cuzco*. Durham, NC: Duke University Press, 2008.

Bank, Katie. *Knowledge Building in Early Modern English Music*. New York: Routledge, 2021.

Beltrán, José "Nature *au natural* in Late Seventeenth-Century France." In *Ad vivum? Visual Materials and the Vocabulary of Life-Likeness in Europe before 1800*. Edited by Thomas Balfe, Joanna Wooddall, and Claus Zittel, 272–93. Leiden: Brill, 2019.

Bennett, Peter. *Music and Power at the Court of Louis XIII: Sounding the Liturgy in Early Modern France*. Cambridge: Cambridge University Press, 2021.

Blasius, Leslie. "Mapping the Terrain." In *The Cambridge History of Western Music Theory*, edited by Thomas Christensen, 25–45. Cambridge: Cambridge University Press, 2002.

Boiteux, Martine. "Fêtes et identité nationale à la Trinité-des-Monts." In *La chiesa e il convento della Trinità dei Monti: ricerche, nuove letture, restauri*, edited by Colette Di Matteo and Sebastiano Roberto, 68–75. Rome: De Luca Editori d'Arte, 2016.

Brooks, Jeanice. *Courtly Song in Late Sixteenth-Century France*. Chicago: University of Chicago Press, 2000.

Buccolini, Claudio. L'exégèse du dernier Mersenne et le cartésianisme." In *The Philosophers and the Bible: The Debate on Sacred Scripture in Early Modern Thought*, 233–47. Leiden: Brill, 2021.

Buccolini, Claudio. "Mersenne Translator of Bacon?" *Journal of Early Modern Studies* 27, no. 1 (2013): 33–59.

Christensen, Thomas. "Mersenne and the Mechanics of Musical Proportion." In *Proportions: science, musique, peinture & architecture*, edited by Sabine Rommevaux, Philippe Vendrix and Vasco Zara, 247–60. Turnhout: Brepols, 2011.

Christensen, Thomas. "Mersenne's Sound World." In *Structures of Feeling in Seventeenth-Century Cultural Expression*, edited by Susan McClary, 60–89. Toronto: University of Toronto Press, 2013.

Clarke, Desmond. *French Philosophy, 1572–1675*. Oxford: Oxford University Press, 2016.

Coelho, Victor, ed. *Music and Science in the Age of Galileo*. Dordrecht: Kluwer, 1992.

Coelho, Victor. "Preface." In *Music and Science in the Age of Galileo*, edited by Victor Coelho, ix–xii. Dordrecht: Kluwer, 1992.

Cohen, Albert. *Music in the French Royal Academy of Sciences: A Study in the Evolution of Musical Thought*. Princeton: Princeton University Press, 1981.

Cohen, H. Floris. "Music as Science and as Art: The Sixteenth/Seventeenth-Century Destruction of World Harmony." In *The Making of the Humanities, Volume 1: Early Modern Europe*, edited by Rens Bod, Jaap Maat, and Thijs Weststeijn, 59–71. Amsterdam: Amsterdam University Press, 2010.

Cohen, H. Floris. *Quantifying Music: The Science of Music at the First Stage of the Scientific Revolution, 1580–1650*. Dordrecht: Reidel, 1984.

Copenhaver, Brian. "The Occultist Tradition and Its Critics." In *The Cambridge History of Seventeenth-Century Philosophy*, vol. 1 of 2 vols., edited by Daniel Garber and Michael Ayers, 454–512. Cambridge: Cambridge University Press, 1998.

De Buzon, Frédéric. "Harmonie et métaphysique: Mersenne face a Kepler." *Les Études philosophiques* 1, no. 2 (1994): 119–28.

De Rosa, Agostino and Alessio Bortot. "Anamorphosis: Between Perspective and Catoptrics." *Handbook of the Mathematics of the Arts and Sciences*, edited by Bharath Sriraman, 243–289. Cham: Springer, 2021.

De Rosa, Agostino, Alessio Bortot, and Luca Ruggiero D'Elia. "The Digital Reconstruction of the Minim's Convent in Place Royal and its Optical-Perspective Games." *Disegnarecon* 14, no. 27 (2021): 1–14.

Dear, Peter. *Mersenne and the Learning of the Schools*. Ithaca: Cornell University Press, 1988.

Dell'Antonio, Andrew. *Listening as Spiritual Practice in Early Modern Italy*. Berkeley: University of California Press, 2011.

Dostrovsky, Sigalia. "Early Vibration Theory: Physics and Music in the Seventeenth Century." *Archive for the History of Exact Sciences* 14, no. 3 (1975): 169–218.

Drake, Stillman. "Renaissance Music and Experimental Science." *Journal of the History of Ideas* 31, no. 4 (1970): 483–500.

Duncan, David Allen. "Persuading the Affections: Rhetorical Theory and Mersenne's Advice to Harmonics Orators." In *French Musical Thought, 1600–1800*, edited by Georgia Cowart, 149–75. Ann Arbor: UMI Research Press, 1989.

Fabbri, Natacha. "Genesis of Mersenne's *Harmonie universelle*: The Manuscript *Livre de la Nature des Sons*." *Nuncius: Journal of the History of Science* 22, no. 2 (2007): 287–308.

Fabbri, Natacha. "Hérésie et orthodoxie dans la pensée mersennienne: vers la 'concordia' de la République chrétienne." *Historia philosophica* 2 (2004): 13–34.

Fabbri, Natacha. *De l'utilité de l'harmonie: filosofia, scienza e musica in Mersenne, Descartes e Galileo*. Pisa: Edizioni della Normale, 2008.

Fabbri, Natacha. "Questioning Fludd, Kepler and Galileo: Mersenne's Harmonious Universe." In *Space, Imagination, and the Cosmos from Antiquity to the Early Modern Period*, edited by Frederik A. Bakker, Delphine Bellis, and Carla Rita Palmerino, 201–32. Cham: Springer, 2018.

Filippi, Daniel V. "'Their sound hath gone forth into all the earth': Music and Sound in the Ministries of Early Modern Jesuits: Introduction." *Journal of Jesuit Studies* 3 (2016): 357–64.

Garber, Daniel. "On the Frontlines of the Scientific Revolution: How Mersenne Learned to Love Galileo." *Perspectives on Science* 12, no. 2 (2004): 135–63.

Garber, Daniel. "Philosophia, Historia, Mathematica: Shifting Sands in the Disciplinary Geography of the Seventeenth Century." In *Scientia in Early Modern Philosophy Seventeenth-Century Thinkers on Demonstrative Knowledge from First Principles*, edited by Tom Sorell, G. A. Rogers, and Jill Kraye, 1–17. Dordrecht: Springer, 2010.

Gaukroger, Stephen. *Descartes: An Intellectual Biography*. Oxford: Clarendon Press, 1995.

Gerbino, Giuseppe. "Skeptics and Believers: Music, Warfare, and the Political of Renaissance Italy according to Francesco Bocchi." *The Musical Quarterly* 90, no. 3–4 (2007): 578–603.

Gouk, Penelope. *Music, Science, and Natural Magic in Seventeenth-Century England*. New Haven: Yale University Press, 1999.

Gozza, Paolo, ed. *Number to Sound: The Musical Way to the Scientific Revolution*. Dordrecht: Kluwer, 2000.

Gozza, Paolo. "A Renaissance Mathematics: The Music of Descartes." In *Number to Sound: The Musical Way to the Scientific Revolution*, edited by Paola Gozza, 155–72. Dordrecht: Kluwer, 2002.

Guillo, Laurent. "Sous la main du père Mersenne: la bibliothèque de musique des Minimes de la place Royal (Paris, XVIe–XVII siècles)." In *Collectionner la musique*. Vol. 3 of 3 vols.: *Erudits collectionneurs*, edited by Denis Herlin, Catherine Massip, and Valérie De Wispelaere, 29–80. Turnhout: Brepols, 2015.

Hollander, John. *Untuning of the Sky: Ideas of Music in English Poetry, 1500–1700*. Hamden: Archon Books, 1993.

Honisch, Erika Supria. "Encounters with Music in Rudolf II's Prague." *Austrian History Yearbook* 52 (2021): 64–80.

Jackson, Myles. "Music and Science during the Scientific Revolution." *Perspectives on Science* 9, no. 1 (2001): 106–15.

Kassler, Jamie C. *The Beginnings of the Modern Philosophy of Music in England: Francis North's A Philosophical Essay of Musick (1677) with Comments of Isaac Newton, Roger North and in the Philosophical Transactions*. London/New York: Routledge, 2004.

Kassler, Jamie C. *Inner Music: Hobbes, Hooke, and North on Internal Character.* Madison/London: Fairleigh Dickinson University Press, 1995.

Kellman, Jordan. "Mendicants, Minimalism, and Method: Franciscan Scientific Travel in the Early Modern French Atlantic." *Journal of Early Modern History* 26 (2022): 10–37.

Kent, Emily. "'Minimizing' Marin Mersenne: Tracing the Intellectual Culture of a Seventeenth-Century Parisian Convent." *History of Universities* 36, no. 2 (2023): 58–79.

Krakovitch, Odile. "L'architecture des trois couvents minimes de Paris." In *Saint François de Paule et les Minimes en France de la fin du XVe au XVIIIe siècle,* edited by André Vauchez and Pierre Benoist, 229–48. Tours: Presses universitaires François-Rabelais, 2011.

Krakovitch, Odile. "La vie intellectuelle dans les trois couvents minimes de la Place Royale, de Nigeon et de Vincennes." *Bulletin de la Société de l'Histoire de Paris et de L'Ile-de-France* 109 (1983): 23–175.

Latour, Melinda. *The Voice of Virtue: Moral Song and the Practice of French Stoicism, 1574–1652.* Oxford: Oxford University Press, 2023.

Lenoble, Robert. *Mersenne ou la naissance du méchanisme.* Paris, 1943.

Luppi, Andrea. "Musiche immaginarie e critica scettica nel Discours di François de la Mothe." *Studi Musicali* 20, no. 1 (1991): 111–40.

Mace, Dean T. "Marin Mersenne on Language and Music." *Journal of Music Theory* 14, no. 1 (1970): 2–34.

Mackensen, Karsten. "System und Kritik. Zur Krise des Musikwissens bei Johann Mattheson." *International Review of the Aesthetics and Sociology of Music* 40, no. 2 (2009): 179–205.

Mai Groote, Inga. "Studying Music and Arithmetic with Glarean: Contextualizing the Epitomes and Annotationes among the Sources for Glarean's Teaching." In *Heinrich Glarean's Books: The Intellectual World of a Sixteenth-Century Musical Humanist,* edited by Iain Fenlon and Inga Mai Groote, 195–222. Cambridge: Cambridge University Press, 2013.

Mehl, Éduouard. "Gassendi in the Philosophical Debate: Stakes of the Essay Concerning the Principles of Robert Fludd's Philosophy (1630)." In *Pierre Gassendi: Humanism, Science, and the Birth of Modern Philosophy,* edited by Delphine Bellis, Daniel Garber, Carla Rita Palmerino, 169–92. New York: Routledge, 2023.

Moyer, Ann E. "Musical Scholarship in Italy at the End of the Renaissance, 1500–1650: From Veritas to Verisimilitude." In *The Reclassification of Knowledge in Early Modern Europe,* edited by Donald R. Kelley, 185–202. Rochester: University of Rochester Press, 1997.

Nelson, Eric. "Remembering the Martyrdom of Saint Francis of Paola: History, Memory and Minim Identity in Seventeenth-Century France." *History and Memory* 26, no. 2 (2014): 76–105.

Oosterhoff, Richard J. *Making Mathematical Culture: University and Print in the Circle of Lefèvre d'Étaples*. Oxford: Oxford University Press, 2018.

Palisca, Claude V. *Music and Ideas in the Sixteenth and Seventeenth Centuries*. Urbana: University of Illinois Press, 2017.

Palisca, Claude V. "Scientific Empiricism in Musical Thought." In *Seventeenth-Century Science and the Arts*, edited by Hedley H. Reys, 91–137. Princeton: Princeton University Press, 1961.

Peil, Dietmar. "Concordia discors: Anmerkungen zu einem politischen Harmoniemodell von der Antike bis in die Neuzeit." In *Geistliche Denkformen in der Literatur des Mittelalter*, edited by Klaus Grübmüller, Ruth Schmidt-Wiegand, and Klaus Speckenbach, 401–34. Munich: Wilhelm Fink, 1984.

Perler, Dominik. "Was There a Pyrrhonian Crisis in Early Modern Philosophy? A Critical Notice of Richard H. Popkin." *Archiv für Geschichte der Philosophie* 86, no. 2 (2006): 209–20.

Phillips, Henry. *Church and Culture in Seventeenth-Century France*. Cambridge: Cambridge University Press, 1997.

Pierre, Benoist, and Andre Vauchez, eds. *Saint François de Paule et les Minimes en France de la fin du XVe au XVIIIe siècle*. Tours: Presse Universitaire François Rabelais, 2011.

Popkin, Richard. *The History of Scepticism: From Savonarola to Bayle*. Oxford: Oxford University Press, 2003.

Pons, André. *Droit ecclésiastique et musique sacrée. Tome III: Décadence et réforme du Chant liturgique*. Vol. 3 of 5 vols. St-Maurice: Editions St-Augustine, 1958.

Psychoyou, Théodora. "Trajectories of Musical Acoustics in Seventeenth-Century French Theory and Experiment: Sound, Noise, and the Authorities of the Past." In *Music and Science from Leonardo to Galileo*, edited by Rudolf Rasch, 361–375. Turnhout: Brepols, 2022.

Rasch, Rudolf. "René Descartes and Isaac Beeckman: The Philosopher and the Schoolmaster." In *Music and Science from Leonardo to Galileo*, edited by Rudolf Rasch, 277–304. Turnhout: Brepols, 2022.

Reiss, Timothy. *Knowledge, Discovery and Imagination in Early Modern Europe: The Rise of Aesthetic Rationalism*. Cambridge: Cambridge University Press, 1997.

Rentsch, Ivana. *Die Höflichkeit musikalischer Form: Tänzerische und anthropologische Grundlagen der frühen Instrumentalmusik*. Kassel: Bärenreiter, 2012.

Rochemonteix, Camille de. *Un college de jésuites aux XVIIe & XVIIIe siècles: le Collége Henri IV de La Flèche*. Le Mans: Leguicheux, 1889.

Rothman, Aviva. *The Pursuit of Harmony: Kepler on Cosmos, Confession, and Community*. Chicago: University of Chicago Press, 2017.

Shapin, Steven. *The Scientific Revolution*. Chicago: University of Chicago Press, 1996.

Shaw, Robert L. J. "The Minims, Their Rule, and French Religious Society at the Turn of the Sixteenth Century." *Viator* 51, no. 2 (2020): 335–87.

Sherr, Richard. "Plainsong in the Age of Polyphony." In *The Cambridge History of Fifteenth-Century Music*, edited by Anna Maria Busse Berger and Jesse Rodin, 771–84. Cambridge: Cambridge University Press, 2015.

Taussig, Sylvie. *L'examen de la philosophie de Fludd de Pierre Gassendi par ses hors-texte*. Pisa/Roma: Fabrizio Serra, 2009.

Taveau, Laurence. "Le manuscrit 2884 du P. Mersenne à la bibliothèque de l'Arsenal: Étude et édition critique." PhD diss., Université de Paris-Sorbonne, Laboratoire IREMUS, 2018.

Vaast, Corrine. "L'esthétique musicale du père Mersenne et l'austérité musicale des Minimes." In *Saint François de Paule et les Minimes en France de la fin du XVe au XVIIIe siècle*, edited by Pierre Benoist and André Vauchez, 249–71. Tours: Presses universitaires François-Rabelais, 2011.

Van Orden, Kate. *Music, Discipline, and Arms in Early Modern France*. Chicago: University of Chicago Press, 2005.

Vendrix, Philippe. "Music and Model in the Renaissance." In *Music and Mathematics in Late Medieval and Early Modern Europe*, edited by Philippe Vendrix, 9–21. Turnhout: Brepols, 2008.

Vendrix, Philippe. "Pour les grands et les autres: la réforme oratorienne du plain-chant." In *Plain-chant et liturgie en France au xviie siècle*, edited by Jean Duron, 87–96. Versailles: Éditions du Centre de musique baroque de Versailles, Klincksieck, Fondation Royaumont, 1997.

Wald-Fuhrmann, Melanie. "'Discors Concordia': Diskursverflechtungen in Athanasius Kirchers musikalischem Staatsmodell." *Zeitschrift für Ästhetik und allgemeine Kunstwissenschaft* 53 (2008): 285–321.

Wald-Fuhrmann, Melanie. "Musik und Universalwissenschaft: Musik als Universalwissenschaft." In *Musik und kulturelle Identität: Bericht über den XIII. Internationalen Kongress der Gesellschaft für Musikforschung*, vol. 2 of 3 vols., edited by Detlef Altenburg and Rainer Bayreuther, 315–24. Kassel: Bärenreiter, 2012.

Walker, Daniel Pickering. *Studies in Musical Science in the Late Renaissance*. London/Leiden: Warburg Institute/Brill, 1978.

Wegman, Rob. *The Crisis of Music in Early Modern Europe, 1470–1530*. New York/London: Routledge, 2005.

Whitmore, P. J. S. *The Order of Minims in Seventeenth-Century France*. The Hague: Nijhoff, 1967.

Yates, Frances. *The French Academies of the Sixteenth Century*. London: Warburg Institute, 1947.

1. Music, the Measure of Sound

Abstract: The first chapter of the book delves into the world of listening and musical science in the early modern period, showing how Marin Mersenne turned music into the art of measuring sounds. It departs from an underexplored passage in his *Questions theologiques* from 1634 in which he imagines an independent field of study for sound, called *psophologie*. Although this science would call for countless experiments and empirical observations, the role of the ear remained highly ambiguous for Mersenne; although it functioned as a conduit for knowledge, hearing could only give an approximation and demanded guidance.

Keywords: listening, harmonics, ear, empiricism, acoustics, musical science

In his *Questions theologiques* from 1634, Mersenne proposed establishing a new science of sound. This science would investigate the sounds of all the natural bodies and would be called *psophologie*, after the Greek word for sound *psophos*.[1] By using the word *psophos*, Mersenne indicated that he was interested in all sounds rather than solely musical or articulate sounds (*phthongos*).[2] Sound was particularly suitable for learning about the properties of things, he argued, and through sound one could study objects, represent them, and communicate about them to others. Since all humans hear sounds and all objects make sound, it is the closest thing to a natural or universal language. Mersenne even thought that once a better

1 "A sçavoir si l'on peut établir une nouvelles science des sons, qui soit nommée Psophologie, ou de tel autre nom que l'on voudra." Mersenne, *Questions inouyes, Questions harmoniques, Questions théologiques*, 403. Not all the survived versions of *Questions theologiques* include this question, since Mersenne added these pages only when he was forced to remove his discussion of Galileo Galilei. As only few copies include this part, I here cite the 1985 edition of Mersenne's 1634 works.

2 Aristotle used "psophos" and "psophein" for sound and things sounding. Aristotle, *On the Soul*, 36–39 (II.8).

Van der Miesen, Leendert. *Marin Mersenne and the Study of Harmony: From Sound to Music*. Taylor & Francis Group, 2025.
DOI 10.5117/9789048564132_CH01

understanding of sound was achieved, it would be possible to distinguish a good wine by listening to the sound of the cask or to know how many people are in a room by knocking on the walls. It was this belief in the power of sound that led Mersenne to investigate the acoustic properties of bells, strings, cannons, and planetary spheres. But for now, he had to conclude that the human senses were limited and the ear was not capable of the hearing powers demanded by this idealized science. Thorough knowledge on sound was still lacking; audible phenomena awaited further observation, measurement, and comparison.

Mersenne included his description of psophologie only after the Parisian censors had rejected his comments about the condemnation of Galileo Galilei by the Catholic Church, forcing him to find less volatile subject matter. He wanted to send the work to Rome and thus had to remove some of his comments on Galilei.[3] As such, it might be considered an afterthought, but the project of measuring sound was in many ways continued in *Harmonie universelle* and *Harmonicorum libri XII*.[4] Throughout these works, Mersenne included his acoustic measurements of metals, gut strings, different types of wood, and echoes. In his measurements and experiments, Mersenne seem to be motivated by precision more than certainty.[5] He went to great lengths to describe measuring the distances between syllabic echoes or to count the vibrations of a string without expecting to find any definitive answer, instead aiming to demonstrate the rationality and commensurability of the harmonies present in the world.

Mersenne did not pursue the idea of a separate science of sound further. Rather, the study of sound was firmly engrained within his musical investigations and in his design for *Harmonie universelle*. Although the subtitle for the book is "the theory and practice of music," it starts with a book on the nature and physics of sound. The study of sound in its diverse manifestations was central to his endeavor to study harmony and music. In his unpublished manuscript *Livre de la nature des sons*, he formulates it as follows:

> Because sound is to music what unity is to arithmetic, the point to geometry, the instant to chronology, the pole to astronomy, the perspectival point to optics, the center of gravity to all things that weigh. The center

3 Mersenne to Peiresc, 28 July 1634, CM 4, 267–68. I abbreviate the *Correspondance du P. Marin Mersenne* with CM and the number of the volume.

4 For different emphasis on Mersenne's psophologie, see Gominet-Brun, "Une curieuse invention."

5 I take this formulation from Crombie, *Styles of Scientific Thinking*, 815.

> of the earth to the universe, thought to logic, the natural body to physics, and being to metaphysics; which is why sound is like the pivot on which all music builds its axioms, its propositions, and its theorems, and without which it cannot exist; which has made me decide to begin with sounds, and to deal with them amply in this work.[6]

Sound is thus the founding principle, the pivot of music. All music's axioms, propositions, and theorems would not exist without sound. Comparing it to the different sciences of arithmetic, geometry and astronomy, sound is the principle of music, its most fundamental element that should be the starting point for all configurations. In contrast to the emancipation of the scientific study of sound from music advocated by later generations, Mersenne's physical and mechanical investigations of sound were firmly ingrained in his effort to understand harmony in its widest sense. As he put in his *Questions harmoniques*, "God made the matter and causes of sounds and music, which he helps to compose harmony."[7] The discipline of music necessitates an understanding of its most basic element, sound. As Bettina Varwig has argued, in such a view "any sounds took place within or were anchored by that everlasting harmonic hum of matter and spirit."[8]

Mersenne's words echo the opening of a very different work on music, namely René Descartes's *Compendium musicae* (Utrecht, 1650). Descartes started his treatise in a much more succinct way with the opening words "Music, whose object is sound."[9] He wrote this short treatise around 1618 when he was in the Netherlands, around ten years earlier than Mersenne's *Livre de la nature du son*. Descartes's words have often been interpreted as

6 "car le son est à la Musique ce qu'est l'unité à l'Arithmetique, le point à la Geometrie, l'instant à la Chronologie, le pole à l'Astronomie le point de rencontre à l'optique, le centre de pesanteur aus fardeaus. Le centre de la terre à l'univers, la pensée à la Logique, le corps naturel à la Physique, et l'estre à la Metaphysique, c'est pourquoi le son est comme le pivot sur lequel toute la Musique bâtit ses axiomes ses propositions, et ses theoremes, et sans lequel elle ne peut subsister; ce qui m'a fait resoudre à commencer par les sons, et à en traiter amplement dans cet oeuvre." Paris, Bibliothèque nationale de France, MS 2884, fol. 18r. Transcription from Taveau, "Le manuscrit 2884," 175.

7 "Dieu ait fait la matiere & les causes des sons, & de la Musique, & qu'il aide a composer l'harmonie." Mersenne, *Questions harmoniques*, 56.

8 Varwig, *Music in the Flesh*, 62.

9 "Hujus objectum est sonus." Descartes, *Compendium musicae*, 2. Descartes wrote the text in 1618 during his stay in the Low Countries, where he shared it with Isaac Beeckman. Descartes referred to his "little treatise" ("petit traité") in a letter to Mersenne in 1629: Descartes to Mersenne, 18 December 1629, CM 2, 350. See Rasch, "René Descartes and Isaac Beeckman."

a fundamental shift in musical thought. Suzannah Clark and Alexander Rehding see Descartes's words as signaling "the change from music as a divine force to music as a material phenomenon."[10] Rather than lofty words on the excellence of music or the harmony of the spheres, Descartes's treatise is succinct and exhibits the same emphasis on order, clarity, and rules that his later writings would display. Compared to the lofty, metaphorical, and theologically motivated words of Mersenne, Descartes's treatise seems to look forward, operating on an epistemic rupture. Music's turn toward the material qualities of sound is often treated as a departure from music's status as the science of universal harmony, but in the context of Mersenne's work, these words take on a different meaning. Mersenne's words show that music as material phenomenon can not only coexist with music as divine force; in fact, the former is motivated by the later. Both Mersenne and Descartes saw sound as the object of investigation in music.[11] Locating any clear move away from harmony as a universalizing force and toward an attention to the material and physical qualities of sound is thus incredibly complex.

This chapter explores these tensions in the work of Mersenne as he tried to establish music as a role of measuring sound but simultaneously struggled to incorporate empirical methods into his larger framework. Mersenne's work shows that the materiality of music and the materiality of sound were not opposed but that they fundamentally reinforced one another. As becomes evident in Mersenne's theological orientation of the science of sound in his *Questions theologiques*, his empiricism cannot be separated from a conviction that the world is a harmonious whole created by God. The science of psophologie was not only empirical but also deeply eschatological and theological. As he wrote, when bodies and objects make sound, they seem to say, "my creator gave me language in order to teach you my properties and

10 Clark and Rehding, "Introduction," 6. Others have similarly identified an epistemic shift in the work of Descartes. See Wymeersch, *Descartes*; Moreno, *Musical Representations*; Erlmann, *Reason and Resonance*.

11 Descartes and Mersenne were not singular in their attention to sound in defining music. Robert Fludd's *Templum musicae* (Oppenheim, 1618) arrives at a very similar definition, writing, "The matter or subject of music is sound." Fludd, *The Temple of Music*, 168. Although his work on universal harmony is often treated as looking backward, Fludd's words come much closer to a proposed epistemic rupture than some other contemporary works. Johannes Alstedt or Pierre Gassendi, for example, continued to define music with "the subject of music is harmonic song" or "music is the art of singing or modulating": "Subjectum musicae est cantilena harmonica" from Alsted, *Encyclopaedia*, 1195; "Musica est canendi, seu modulandi ars," from Gassendi, *Opera omnia*, 5:633. Augustine famously gave the definition of music as "the science of modulating properly" ("scientia bene modulandi").

powers, which you should use to thank him."[12] His writings nimbly alternate between observations on the mechanics of a string and contemplations of the divine or "perpetual harmony" that surrounds humans:

> The last returns of the chord are so small that if everything in the world, for example the earth, walls, and everything we see and touch, were to move by similar turns and returns, we could not perceive it in any way; so that all the bodies of the world can make a perpetual harmony, even though no one can hear it, and that we have reason to humiliate ourselves in our ignorance, which we cannot remedy until it pleases God to release us from the obligation we have to the stupidity of the senses.[13]

Mersenne's worldview was essentially harmonious; a cosmic order was divided proportionally by God. Like Augustine, Mersenne saw music as the best entry point for understanding cosmic harmony.[14] It was sound, more than light, that best helps to understand the surrounding world, God, and salvation. Consequently, listening was an activity that connected humans to the intellectual realm. The science of sound, the friar argued, "is more advantageous for making the author of the universe known to us than the other sciences, in that it gives us greater knowledge of the properties of what is sensible."[15] The sounds of objects should be studied and in turn used to thank God as creator of the world. Sound, Mersenne argued, "made us conceive divine revelation and made us embrace true religion."[16] This theological conviction was central to his scholarly projects. Sound as an object of scientific investigation and as proof of divine harmony were two

12 "mon Createur m'a donné le langage pour t'enseigner mes proprietez et ma puissance, dont tu dois user pour l'en remercier." Mersenne, *Questions inouyes, Questions harmoniques, Questions théologiques*, 403.

13 "Les derniers retours de la chorde sont si petits que si tout ce qui est au monde, par exemple la terre, les murailles, & tout ce que nous voyons, & ce que nous touchons se mouvoir par de semblables tours & retours, nous ne pourrions l'appercevoir en aucune façon; de sorte que tous les corps du monde peuvent faire une perpetuelle harmonie, quoy que nul ne l'entende, & que nous avons sujet de nous humilier dans nostre ignorance, à laquelle nous ne pouvons remedier jusques à ce qu'il plaise à Dieu de nous delivrer de l'obligation que nous avons à la stupidité des sens." Mersenne, *Harmonie universelle*, "Livre troisiesme du mouvement," 168.

14 Weber, "Pascal and Music"; Dear, *Mersenne*; Vendrix, "L'augustinisme musical."

15 "la science des sons approche advantage de cette racine, qu'elle est plus propre pour nous faire cognoistre l'autheur de l'univers, que les autres sciences, si elle nous donne une plus grande cognoissance des proprietez de ce qui est sensible." Mersenne, *Questions inouyes, Questions harmoniques, Questions théologiques*, 404.

16 "que le son, qui nous a fait concevoir la revelation Divine, & nous a fait embrasser la vraye Religion." Mersenne, *Questions theologiques*, 23.

forces that continually exerted influence on Mersenne and those around him.

This chapter investigates how Mersenne's empiricism was shaped by these two forces as he attempted to measure, quantify, and measure sound in order to understand harmony. Here I am especially interested in the empirical qualities of Mersenne's work. I will do so by first discussing the idea of measurement in music, followed by the question of music and motion, and then I consider two case studies: how Mersenne approached measuring materials and measuring musical proportions. With the help of precise observation and experimentation he hoped that musical science would have more certainties that it thus fare had. At the same time, he was always quick to point out that humans could not know this structure of the universe, since "we did not exist, when its [nature's] foundations were laid, and its effects do not lead us clearly enough to their source to convince us, and compel us to stop there by the force of a perfect demonstration."[17] Unquestionable truths, especially in the Aristotelian sense as eternal and unchanging, were not attainable for the human intellect, according to Mersenne. Since music relies on the senses and physics, mathematical causal demonstrations do not exist, he admitted in the opening to his book on the consonances.[18] Here he was influenced by the large volume of skeptical literature of his day, which emphasized that humans could never make any claim of finding the truth. Mersenne argued that nature consisted of order but that its precise configuration was beyond the human horizon. He held that it was beyond the power of the human intellect to attain certain knowledge concerning the hidden, true nature of things, but he believed that intellect could at least arrive at knowledge concerning some regularities in natural phenomena. This was sufficient and scientific.[19]

"Music in all things"

In his earliest publication that dealt with music, Mersenne writes that "music is in all things."[20] It is clear that music in this mention is much more than sounding music. Mersenne acknowledges this when he positions it as the

17 "dautant que nous n'estions pas, quand l'on a posé ses fondemens, & que ses effets ne nous meinent pas assez evidemment à la source pour nous convaincre, & pour nous contraindre de nous y arrester par la force d'une parfaite demonstration." Mersenne *Questions theologiques*, 117.

18 Mersenne, *Harmonie universelle*, "Livre premier des consonances," n.p. ("Preface"). For the search for certainty in the (mixed) mathematical sciences, see Dear, *Discipline and Experience*, 40–44.

19 On this position, see also Lenoble, *Mersenne*, 336–82; Fabbri, "Questioning Fludd."

20 "Musica est in omnibus rebus." Mersenne, *Quaestiones*, col. 1570.

first of the mixed mathematical sciences, as music considers quantity but is "conjoined to the perceptible things." Music considers "the number of sounds" (*numerum sonorum*) when it considers singing and can be called "human music" when it is considered the music of the body, "worldly" (*mundana*) when it considers the harmonies of the world, "elemental" when it considers the elements, "heavenly" when it considers the heavens, "angelic music" when it considers the angels, "divine music" when it considers the divine operations and trinity, and "archetypal music" (*archeptica Musica*) when it considers the divine mind to which all music corresponds.[21]

From this range of topics, it becomes clear that music is understood as much more than sounding vocal or instrumental music. Mersenne followed the Platonic or Augustinian tradition of music, in which music is described as a science dealing with its quantifiable properties, closely related to moral questions.[22] Codifying and transmitting antique sources, the fifth-century philosopher Boethius described music a science of acquiring speculative (beyond the practical or pedagogical) knowledge on the world, with harmony playing a central role.[23] The goal of all the mathematical disciplines of the quadrivium (which, together with the trivium, formed the liberal arts), Boethius wrote, was the search for truth.[24] Music was traditionally divided into three parts, comprising the music of the universe (*musica mundana*), the music of the human body and soul (*musica humana*), and singing and instrumental music (*musica instrumentalis*). The latter, actual sounding music, was understood as a first step in understanding world harmony rather than the true end of music. The cosmic harmonies were seen as embodying perfection, whereas instrumental music could only approximate this. The pleasure of knowing "the harmonic proportions that are kept in the fabric of the world and all its parts," Mersenne argued, would be a thousand times more delightful "than the concerts of the Hebrews, the Greeks, the Italians, the French and all the nations of the world."[25]

21 "rei sensibili conjunctam quarum prima numerum sonorum, considerans Musica vocatur"; "Humana"; "Mundana"; "Coelestis"; "Angelica"; "Divina"; "archeptica Musica." Mersenne, *Quaestiones*, col. 1204.

22 Mackensen, *Musik und die Ordnung der Dinge*, 17. Joscelyn Godwin has emphasized the longevity of these ideas: Godwin, *The Harmony of the Spheres*.

23 On music as a universal science in the Boethian tradition, see Isherwood, *Music in Service of the King*, 1–54; Leinkauf, "Kirchers Musikverständnis"; Wald-Fuhrmann, *Welterkenntnis aus Musik*; Heilmann, *Boethius' Musiktheorie*; Mackensen, *Musik und die Ordnung der Dinge*.

24 Boethius, *Fundamentals of Music*, 2 (I.1).

25 "les raisons harmoniques qu'il a gardées en la fabrique du monde, & de toutes ses parties, cette connoissance nous raviroit mille fois davantage que les concerts des Hebrieux, de la Grece, des Italiens, des François, & de toutes les nations du monde." Mersenne, *Traité*, 19.

Mersenne was not alone in identifying music as a universal science. This thought was widely present in the late sixteenth and early seventeenth centuries, especially in France. Of all the sciences, Pontus de Tyard wrote, music was an "image of the entire encyclopedia," encompassing all disciplines and virtues. Those who strove for divine knowledge would necessarily have to master the science of music.[26] This ideal was echoed in the early seventeenth century by David de Flurance Rivault, who wrote that "music is the science that understands all, fills all, and forms all."[27] And around the same time as Mersenne, Nicolas Bergier wrote in his *Musique speculative* (ca. 1608), "This is where the ancients got the name of music for the understanding of all things, in which they learned these numbers and measures, not confused, and undifferentiated, but united by judgment and reason."[28] Music thus presents a model for uniting opposites and heterogeneous elements, for uniting sense and reason, as it can be mathematized while also being experientially perceived. The Jesuit author Étienne Binet wrote in 1623, "To lessen the bitterness of our poor life, God has given us the sweetness of music, which is the refrain and echo of the harmonious songs of Heaven, and an ingenious array of all the proportions and pleasures that nature has sown throughout this universe, which lives only by the cadence and sway [*branle*] of heaven."[29] Bergier told a similar story with his *La musique speculative*. According to him, "speculative music" is nothing other than "the consideration of numbers and measures."[30] Music was thus a privileged science, as the proportions and numbers studied in music are applicable to the wider world.

Music was projected as a divine force present in all things, becoming a model or meta-discipline for all other fields. This idea had a long history as

26 "image de toute l'Encyclopedie." Tyard, *Solitaire second*, 70–71.

27 "Musique, science qui comprend tout, remplit tout, forme tout." Rivault, *L'art d'embellir*, 84v.

28 "c'est d'ou vient que les Anciens ont estendu le nom de Musique sur la cognoissance de toutes les choses, esquelles Ilz ont apperceu ces nombres et ces mesures, non pas confuses, et indifferentes, mais unies ensemble par jugement et raison." Bergier, *La musique speculative*, 16.

29 "Pour desaigrir les amertumes de nostre pauvre vie, Dieu nous a donné les douceurs de la Musique: qui est le refrain & l'écho des chansons harmonieuses du Ciel, & un ingenieux amas de toutes les proportions, & plaisirs que la nature, a semez par l'estenduë de cét Univers qui ne vit qu'à la cadence, & au branle des Cieux." Binet, *Essay*, 470. Mersenne came to a similar description in a letter to Constantijn Huygens, writing, "Music, and consequently airs, are made especially and principally to charm the mind and the ear, and to make us pass life with a little sweetness amidst the bitterness that is encountered" ("la musique, et par consequent les airs, sont faictz particulierement et principalement pour charmer l'esprit et l'oreille et pour nous faire passer la vie avec un peu de douceur parmy les amertumes qui s'y rencontrent"). Mersenne to Constantijn Huygens, 14 November 1640, CM 10, 237.

30 "la Musique speculative, … la consideration des nombres et des mesures." Bergier, *La musique speculative*, 16. On this tradition, see especially Jeserich, *Musica Naturalis*.

well. According to Roman author Cassiodorus, “The discipline of music … extends through all acts of our life.” It is the “discipline of proper harmony” and the discipline “that deals with numbers.”[31] The Bishop Isidore of Seville claimed that “without music, no discipline can be perfected, for nothing is without music.”[32] Such ideas continued to be repeated in the early modern period. A sixteenth-century music treatise formulated it as follows: “there is nothing beneath heaven or even above heaven that is not the subject of music.”[33] The seventeenth-century engineer Salomon de Caus went as far as to declare music the only science that is natural. In contrast to the sciences such as arithmetic, mechanics, and geometry, De Caus wrote, “this science of music does not come from human invention.”[34] In this view, musical proportions give privileged access to the soul or harmony of the world, which is present in humans but also in the wider world itself. These harmonies are already there; music just makes them audible and measurable.

Mersenne’s adoption of these categories of music in his *Quaestiones* can be seen as a mere tribute to this tradition. He never seemed particularly interested in locating musical harmony in the proportions of the planets or orbits, although the topic is discussed in his *Traité*. In this work, he describes the “worldly, celestial, and elemental” music as the order of all the bodies in the universe.[35] But in the same book, he criticizes contemporaries like Robert Fludd, who were not precise enough in their notion of a harmonious world soul, who had “no other foundation than the imagination.”[36] Distancing himself from Fludd and others, Mersenne argued that this approach was quantitative, and that the natural order of the world and universe at large should be measurable in some way, imitating God himself, projected as a divine mathematician:

> The divine wisdom has arranged all things in weight, number, and measure, so that there is nothing that does not relate to all the other things of the world, and [nothing that] does not bear unmistakable evidence to

31 Cassiodorus, *Institutions of Divine and Secular Learning*, 217 (II.2.).

32 Seville, *The Etymologies*, 95 (III.17).

33 “nihilque item sub coelo vel supra coelum quod Musicae non subsit.” Vanneo, *Recanetum de musica aurea*, 1r.

34 “ceste science de Musique ne vient d’aucune invention humaine.” De Caus, *Institution harmonique*, n.p. (“A la tres-illustre”).

35 “Mondaine, Celeste & Elementaire.” Mersenne, *Traité*, 68. See also Buzon, “Harmonie et métaphysique,” 122. As Natacha Fabbri has shown, Mersenne shifted “the attention from the level of perception to the level of production,” comparing the planets to other sounding bodies such as cannonballs. Fabbri, “Questioning Fludd,” 233.

36 “n’ont presque autre fondement que l’imagination.” Mersenne, *Traité*, 421.

> the industry of the sovereign mathematician, who has so measured all the circles, and all the other figures of heaven, earth, and all that is in between, so that there is nothing imperfect.[37]

Since the world was arranged in weight, number, and measure, as stated in the Wisdom of Solomon 11:20, all things in the world bear testimony to God as mathematician, inscribing proportions in the heavens and earth. These were not just analogies. As Peter Dear has shown, for Mersenne "the object of mathematics ... had a prototypical existence in the intellect, preeminently that of God."[38] Although numbers can be measured in the world, their origin is much more divine, as Mersenne writes in *Les preludes*: "number is so pure and simple that its principle is found even in God, who is one, and who is accompanied by the sacred trinity."[39] In his *Questions harmoniques*, he wrote that the theory of music (and the theory of the sciences in general) "responds in some way to the inner operations of God and to his divine ideas."[40] Since the sciences are based on reason and reason is divine, the sciences correspond in some way to the most inner workings of God. This correspondence was not imaginative but could be measured and observed.

Mathematics continued to hold this special position for Mersenne. For example, he writes in his *La verité*, "For there are no sciences, after theology, which offer to us, and makes us see, so many wonders as mathematics does, which raises the mind above itself and forces it to recognize a divinity."[41] Comparing the certitude of the different sciences in his *Questions theologiques*, he argues that the principles of physics "have almost no certainty compared to those of pure mathematics."[42] Physics is understood here as

37 "Aussi la sagesse Divine a tellement disposé toutes choses en poids, en nombre & en mesure, qu'il n'y a rien qui n'ait rapport avec toutes les autres choses du monde, & qui ne rende un evident témoignage de l'industrie du souverain Mathematicien, qui a tellement compassé tous les cercles, & toutes les autres figures du ciel, de la terre, & de toute ce qui y est compris, qu'il n'y a rien d'imparfait." Mersenne, *Traité*, 23.

38 Dear, *Mersenne*, 71.

39 "le nombre est si pur et si simple que son principe se trouve mesme en Dieu, lequel est un, & qui est accompagné du sacré Ternaire." Mersenne, *Les preludes*, n.p. ("A Monsieur").

40 "qui respond en quelque maniere aux operations interieures de Dieu, & à ses divines idées." Mersenne, *Questions harmoniques*, 251.

41 "Car il n'y a point de sciences, apres la Theologie, qui nous proposent, & nous fassent voir tant de merveilles comme font les Mathematiques, lesquelles élevent l'esprit par dessus soy-mesme, & le forcent de reconnoistre une divinité." Mersenne, *La verité*, n.p. ("Preface"). Later in the same book, he calls mathematics "the mistress and queen of the other sciences" ("la maistresse, & la reyne des autres sciences"). Mersenne, *La verité*, 233.

42 "la Physique, qui n'a quasi nulle certitude en comparaison de celle des pures Mathematiques." Mersenne, *Questions theologiques*, 178–79.

the Aristotelian science dealing with the principles of change in all natural bodies. Music and optics, understood as subalternate sciences of arithmetic and geometry, had a higher degree of certainty, but not as much as pure mathematics, since "they presuppose the senses and their operations."[43] Mathematics was essential to understanding many other sciences, such as mechanics, music, optics, medicine, chemistry, and law.[44] Mersenne propagated his own particular blend of mathematical practice, which Peter Dear has called an "alternative mathematical natural philosophy to replace essentialist physics."[45] Other sciences could increase their certainty and epistemological foundation by relying on the science of mathematics. In his *Harmonie universelle*, he writes, "It [music] would like to share in the certitude of geometry and arithmetic, if possible, so that her principles cannot be contested by the Pyrrhonian [skeptics] and doubters."[46]

The tension between certainty and uncertainty is a constant presence in Mersenne's writings on music. Although music would like to be as uncontested as mathematics, he believed that music was too far removed from pure mathematics to have complete certainty. In his *Harmonie universelle*, he writes, "I do not want anyone to take the words 'demonstrate' and 'determine,' which I use at the beginning of the propositions in the same sense and in the same meaning as in geometry, but only as another word for to know, or examine."[47] It is too difficult to truly demonstrate something in physics as one can in mathematics, Mersenne acknowledges, and no perfect demonstrations exist in the realm of music. Commentating upon music's status as a science in the early modern period, Eric Bianchi writes, "A musical composition might well manifest or exemplify the effects of a distant mathematical cause. Considered

43 "parce qu'elles presupposent les sens." Mersenne, *Questions theologiques*, 179.

44 Mersenne, *La verité*, 235–47.

45 Dear, *Mersenne*, 72.

46 "Elle desireroit d'estre participante de la certitude de la Geometrie, & de l'Arithmetique, s'il estoit possible, afin que ses Principes ne luy peussent estre contestez par les Pyrrhoniens, & les Doutans." Mersenne, *Harmonie universelle*, "Livre sixiesme des orgues," n.p. ("A Monsieur Pascal").

47 "je ne desire pas que l'on prenne les dictions *demonstrer* & *determiner* dont j'use souvent au commencement des Propositions, au mesme sens, & en la mesme signification qu'en Geometrie, mais seulement comme l'autre diction à sçavoir, ou examiner &c. dont je me sers pour mesme sujet, car je sçay qu'il est trop difficile de pouvoir demonstrer aucune chose dans la Physique, si l'on prend la demonstration à la rigueur." Mersenne, *Harmonie universelle*, "Livre premier des consonances," n.p. ("Preface & advertissement"). Although music was not solely dependent on the science of arithmetic, it was most intimately connected with it, in the same way in which optics relied on geometry. Comparing the sciences of optics and music in his *Questions theologiques*, he considered the study of light more certain, as it comes "closer to demonstration" ("plus proche de la demonstration"). Mersenne, *Questions theologiques*, 182–83. The notion of demonstration in the realm of music is also discussed in chapter 5.

ontologically, that same composition was too far removed from the cause to be scientifically demonstrative; it was too physical, too particular, too embodied."[48] This holds true for Mersenne, but it was precisely music's embodied nature that made it into an especially important science, as music could function as a channel for what was beyond the sensual realm. In his dedication of the *Traité*, he writes on the goal of a contemplative music that would be based on "a harmonic proportion of these celestial and divine movements with those that are natural to human reason."[49] Music was not solely a science of numbers but rather something that had to be lived and experienced, as it aligned human reason to the wider harmonies in the world, providing access to divine wisdom by means of numerical ratios. Music provided another type of access to divine knowledge, one that could not be fully rationalized: "it [music] explains with a wonderfully effective eloquence the mysteries of religion by singing the praises of God."[50] This is why, Mersenne argued, the ancients instructed the youth in music, so that music's ability to moderate remained in their mind, and they would be able to resist "the troubles and disturbances that occur during the course of life."[51] Mersenne's conceptualization of music as an encyclopedic field of knowledge thus belonged to a broader movement that viewed music as the study of numbers as they appear in the sensory world, but for Mersenne this was not fully rationalized; music was a performative and moral practice that provided a special insight into divine knowledge.

Music and Motion

Throughout his lifetime, Mersenne pondered the fallibility of the human senses. Although the ear and eye were more connected to reason than the

48 Bianchi, "Scholars, Friends, Plagiarists," 95. The Jesuit Giuseppe Biancani, for example, distinguished pure mathematics (geometry and arithmetic) from applied mathematics, such as mechanics, astronomy, optics, and music. According to Biancani, "Music has its own demonstrations." Cited and translated in Mancosu, *Philosophy of Mathematics*, 208. In France the music theorist Antoine de Cousu wrote that "music is noble, excellent and very certain, since it is a part of mixed mathematics, and has very truthful demonstrations, which it bases on its own principles" ("la Musique est noble, excellente, & tres-certaine; puisqu'elle est une partie mixte des Mathematiques, & qu'elle a des demonstrations tres-veritables, qu'elle fonde sur ses propres principes"). De Cousu, *La musique universelle*, 4.

49 "une proportion harmonique de ses celestes & divins mouvemens, avec ceux qui sont naturels à la raison humaine." Mersenne, *Traité*, n.p. ("A Monsieur de Refuge").

50 "qu'elle explique par une eloquence merveilleusement efficace les mysteres de la Religion en chantant les loüanges de Dieu." Mersenne, *Traité*, n.p. ("A Monsieur Coutel").

51 "la premiere teinture de sa moderation demeurast dans l'esprit, & resistast aux troubles & perturbations qui surviennent durant le cours de la vie." Mersenne, *Traité*, n.p. ("A Monsieur Coutel").

senses of smell and touch, they were easily mistaken as well. He listed numerous uncertainties in the perception of sound, such as distance, strength, and location (for example, acoustic effects like the echo). One of his central examples was the existence of inner voices: "we find that it cannot discern whether a sound is exterior, or if it is made within ourselves."[52] Echoing Augustine on the realistic quality that inner voices can have while also potentially explaining the visitations of angels, Mersenne maintained that humans can never be certain that perceived objects are really present—whether the sounds we hear are really sounds existing outside in the world, or just "hums and noises" (*bourdonnemens, et aux bruits*) made inside the ear. This unreliability of the senses was central to Mersenne's epistemology and endeavor to establish a certain science of music.

Although aware of the increasingly sophisticated descriptions of the anatomy of the ear around 1600, Mersenne only took up newer theories of audition to a limited degree; anatomical knowledge played only a very small role in how he described the process of listening itself. The ear itself was even described as secondary. This must not be understood as a lack of interest into the workings of the ear. Mersenne devoted much attention to the physiology of the ear, performed tests with musicians on the perception of consonances, and investigated the limits of audibility. The lack of attention to the physical characteristics of the ear aligned with his emphasis on the central role of the mind in auditory perception. According to Mersenne, the ear itself does not know sound but is an instrument or transmitter of sound: it passes information on to the mind.[53]

Following Augustine, Mersenne saw reason as the distinguishing feature in sensory perception between humans and animals. Animals can hear sound as well, sometimes even better than humans, Mersenne admitted, but animals mechanically act on this input, like the automatic wheels of a clock. When humans listen, Mersenne argued, they consider the nature of sounds, their properties, and their differences.[54] This is why Mersenne opposed philosophical movements such as skepticism (both Pyrrhonian

52 "l'on trouvera qu'il ne peut discerner si le son est exterieur, ou s'il se fait au dedans de nous mesmes." Mersenne, *Harmonie universelle*, "Traitez de la voix," 80. Augustine included several examples of inner voices, for example in Augustine, *Confessions*, 184 (X.13).

53 Mersenne, *Harmonie universelle*, "Livre premier de la nature," 79. As Peter Dear has shown, Mersenne relied heavily on Augustine in his description of the how the ear perceives sound. Dear, *Mersenne*, 101–2. For the developments in the understanding of the ear, see Wardhaugh, *Music, Science, Experiment in England*, 59–96.

54 Mersenne, *Harmonie universelle*, "Traitez de la voix," 79. See also Crombie, "Music, Mathematics and Medical Science."

and Academic skepticism) with such vehemence. According to Mersenne, reducing man's knowledge to the senses alone would position animals and humans on the same level. This was an untenable position for him.[55] Like Augustine, he singled out humans' privileged position in the creation and the role of order in auditory perception.

What exactly sound is and what causes or transmits it was hotly debated in the early modern period. One of the most common descriptions of what constitutes sound was still derived from Aristotle and his followers. In this view, sound is a sensible quality, similar to color and flavor, which means that its existence ties the perceiving subject and the sounding object together. In this view, a perceiver needs to exist for a sound to exist. Sounds can be potential or actual, meaning that they are stored in objects and can be produced under circumstances when they encounter a different sounding object. Objects that do not bring forth any sounds, such as wool or sponges, have no potential or actual sounds. A bronze object has a potential sound and can produce a sound between the object and organ of hearing. Sound could never be identified solely with the object, or the hearer, but meandered somewhere in between. Many early modern authors took up the Aristotelian conception of sound, albeit with adaptation, such as the Jesuit philosopher Francisco Suárez in his commentary on Aristotle's *De anima* (Lyon, 1635). Suárez here reaffirms the idea that sound is a sensible quality between the impacted sound body and the listener, and as such, it cannot not be identified with the movement of air. Sound and motion exist apart from one another, so they cannot not be essences of each other. Rather, sound is a quality that falls upon the motion of air or another medium.[56] In a letter to Mersenne, the physician Jan-Baptista van Helmont also put this idea forward, describing sound as a quality generated by air but spread out through motionless air. Although he did not see it as an immaterial quality like light or color, sound could not be identified with air itself.[57]

Mersenne rejected this Aristotelian notion of sound and argued that sound is already sound before it is heard. Throughout his works, he propagated a more mechanistic understanding of sound: sound must be understood as motion. He defined it as "a movement of external or internal air capable of being heard" and suggested that all motions emit sound, regardless of

55 Mersenne, *La verité*, n.p. ("A Monsieur Frere du Roy").

56 "Sonus est qualitias sensibilis." Suárez, *Complectens tractatum*, 150 (III.xix). On sound as a sensible quality in scholastic philosophies, see Pasnau, "Sensible Qualities"; Heider, "Suárez on Sound."

57 Van Helmont to Mersenne, 15 January 1631, CM 3, 35.

whether they are heard.[58] In his *Harmonicorum*, he formulated it as follows: "The proper subject, or object, of music is motion, as it moves hearing."[59] This definition did not use any reference to audible species or sensible qualities and instead made sound an object much more like light, allowing it to be measured and quantified. According to Mersenne, all moving objects emit tiny tremors that set air in motion, and when these motions fall in the realm of audibility, they can be heard. The speed of these motions determines their pitch: "there is no other cause of the pitch of sound, than the scarcity of beatings, that is to say, the small number of tremors and beatings of air."[60] He found many arguments to describe sound as something physical, such as the feeling one gets standing in the presence of a loud church organ, or the fact that some deaf persons could feel the movement of air during thunder or standing close to a loud cannon. That sound relates to motion had been repeated throughout antiquity and the Middle Ages, but these statements put them on equal footing at an unprecedented level.[61] With the emphasis on sound as a composite of vibrations of air, hearing was turned into a process of counting: "we can say that the action of hearing is nothing other than the number of the beats of the air, whether the soul tells them without us apperceiving it, or that it feels the number that touches it."[62]

Mersenne argued that the ear counts the number of air vibrations but that it is not precise enough to hear these individual motions.[63] As he puts it in his description of psophologie in his *Questions theologiques*: "all impressions that objects make on our senses are nothing other than a kind of sound."[64]

58 "de maniere que le Son se peut definir *un mouvement de l'air exterieur ou interieur capable d'estre ouy*." Mersenne, *Harmonie universelle*, "Livre premier de la nature," 2. On this shift, see also Valleriani, "Galileo's Abandoned Project."

59 "Proprium musicae subjectum, vel objectum est motus, prout auditum movet." Mersenne, *Harmonicorum*, "De natura & proprietatibus sonorum," 1.

60 "il n'y a point d'autre cause de la gravité des Sons, que la rareté des battemens, c'est à dire que le petit nombre des secousses & tremblemens de l'air." Mersenne, *Harmonie universelle*, "Livre premier de la nature," 12.

61 The Euclidian *Sectio canonis* (composed around 300 BCE) already identified sound with motion: if nothing moved, nothing would be heard; see Barker, *Greek Musical Writings*, 191.

62 "qu'on peut dire que l'action de l'ouye n'est autre chose que le desnombrement des battemens de l'air, soit que l'ame les conte sans que nous l'appercevions, ou qu'elle sente le nombre qui la touche." Mersenne, *Harmonie universelle*, "Livre premier de la nature," 23. As Théodora Psychoyou has shown, Mersenne broadened his musical investigations to include sounds more broadly but did not study "noise" specifically in the way later generations of scholars would; see Psychoyou, "Trajectories of Musical Acoustics," 364–65.

63 Mersenne, *Harmonie universelle*, "Livre premier des consonances," 5–6.

64 "toutes les impressions que les objects font sur nos sens, ne sont autre chose qu'une espece de sons." Mersenne, *Questions inouyes, Questions harmoniques, Questions théologiques*, 405.

All motions, and therefore all sensations, were turned into a subspecies of sound. Or as he summarized it in the Neoplatonic maxim: "all things are in all things" (*omnia in omnibus*).[65] An understanding of sound could contribute to all other disciplines, since it showed how sensations are created by motions that in turn can be quantified. The flapping of a bird's wings, the imperceptible movement of walls and floors, and the flickering of a candle were for Mersenne all motions that could be described as sound. The scholar was perhaps thinking of his own childhood in Maine as the son of farm surveyors when he wrote on the proportions that workers make when rhythmically beating wheat. Proportions were everywhere, also in the movements of farmers beating wheat (threshing), or the whips of carriage drivers, or craftsman "who make several kinds of noises and clanging while working, whose combinations and unions give some kind of satisfaction."[66] All these motions could also be described with musical terminology; the senses just did not perceive them as such. If one could perceive these sounds emitted by all motions in the world, he believed that they would bring forth a beautiful harmony. By identifying sound with music, Mersenne shifted the study of sound from the domain of Aristotelian physics to the realms of mechanics and mathematics, fields that were less tied up with scholastic philosophy and dogmas of the church. Although he would write that his own notion of sound could coexist with Aristotle's, it was alone already sufficient to explain different acoustic effects.[67] As Peter Dear and Thomas Christensen have articulated, rather than subsuming music into the study of mechanics, mechanical principles were enclosed into the realm of universal harmony.[68]

His experiments with strings served as a central proof that sound could be quantified. Mersenne took it as a special point of pride that he was able to quantify this motion, as contemporaries like Galileo Galilei found that it was impossible. He argued that the vibration of strings could be easily counted with longer strings. In *Harmonie universelle*, he explained

65 "toutes choses sont en toutes choses, *omnia in omnibus*." Mersenne, *Questions inouyes, Questions harmoniques, Questions théologiques*, 405.

66 "plusieurs autres artisans, qui font plusieurs sortes de bruits & de cliquetis en travaillant, dont les rencontres & les unions donnent quelque sorte de satisfaction." Mersenne, *Harmonie universelle*, n.p. ("Seconde observation").

67 Mersenne did not follow Aristotle's categories of essences and causes. He wrote in his *L'impiéte*, "We can only see quantity, shape, light, or the color of things" ("D'abondant nous ne pouvons rien voir que la quantité, la figure, la lumiere, ou la couleur des choses, car nous n'appercevons point les accidens interne"). Mersenne, *L'impiéte des deistes*, 372. For more on Aristotle's position in early modern France, see Martin, *Subverting Aristotle*.

68 Dear, *Mersenne*, 117–69; Christensen, "Mersenne."

his procedure as follows: The first step was to find a reference point, for which he used an organ pipe tuned at chapel pitch and a string tuned to the lowest note his voice could go. One needed a string, which was, according to Mersenne, large enough to complete the experiment, "since it does not vibrate too quickly, since it gives the opportunity to count its returns."[69] From here he altered the tension and length of the string until he arrived to a point where he could observe the individual vibrations. According to his account, this was sufficient to count the vibrations and approximate the number for every possible pitch. The results of this experiment are shared throughout *Harmonie universelle* but often contradict each other. For example, Mersenne wrote at one point that the lowest voices in France make forty-eight beatings per second, and elsewhere he wrote that this number is eighty-four. But he himself was not worried about such differences. The goal was not to find an absolute number but to demonstrate that some approximations of air movements are possible and to urge others to perform the experiment as well.[70]

Believing that he found a privileged entry point in the study of sound, Mersenne saw uses for these numbers everywhere, from medicine to music theory. In his discussion of the different consonances, for example, vibration took on a central role. The question of why humans find the octave a beautiful interval was reformulated as the problem of "forty-eight vibrations" in one ear and "ninety-six" in the other.[71] The octave and all other consonances were described in terms of air vibration, since these were considered to be a more precise descriptions of the physical world.[72] He argued, "We cannot represent sound better than by the number of beatings [*battemens*], since they are in no way different from sound, which we call sonorous or sounding number."[73] To demonstrate how such numbers could be used in the realm of practice, he printed the *air de cour* "Divine Amarillis" by the Parisian composer Antoine Boësset. But instead of staff notation, Mersenne represented the composition with numbers that stood for the amount of air vibrations for each individual pitch (fig. 1.1

69 "d'autant qu'elle ne tremble pas trop viste, & qu'elle donne loisir de conter ses retours." Mersenne, *Harmonie universelle*, "Livre troisiemes des mouvements," 169.

70 Mersenne, *Harmonie universelle*, "Livre troisiemes des mouvements," 171.

71 "quarante-huit battemens d'air d'un costé, & par nonante & six de l'autre." Mersenne, *Harmonie universelle*, "Livre premier des consonances," 1.

72 Mersenne, *Harmonie universelle*, "Livre premier des consonances," 3.

73 "l'on ne peut mieux representer le son que par le nombre desdits battemens, puis qu'ils ne sont nullement differens du son, que l'on appelle nombre sonant ou sonore." Mersenne, *Harmonie universelle*, "Livre troisiesme des instrumens a chordes," 147.

Tablature des retours ou mouuemens que font les chordes, ou les voix qui chantent l'air d'Anthoine Boëſſet Intendant de la Muſique de la chambre du Roy, & de la Reyne.

Baſſe.			Taille.			Haute-contre.			Deſſus.		
lettres	meſures	retours	lettres	meſures	retours	lettres	meſures	retours	lettres	meſures	retours
G	1/3	32	F	1/3	64	A	1 2/3	320	D	1/3	106 2/3
F	4 1/3	390	E	1/6	30	G	2/3	144	C	3 2/3	940
E	1	90	D	2	320	F	7	1344	♯	1/2	133 1/6
✕e	1/3	28 4/5	C	5 1/6	764	E	3 1/3	600	B	4 9/11	1232
D	3 1/2	280	B	1 2/3	165 1/3	✕e	3 1/3	67 9/15	A	5 5/12	1280
C	3 1/2	252	A	4 1/6	500	D	6 5/6	994	G	3 1/2	108
B	3	192	G	5 1/3	606	C	2 1/6	312	✕f	2/3	66 2/3
A	1	60	✕f	2/3	33 1/3				F	2 1/2	216
G	2 1/2	136	F	2 1/2	225				E	2/3	120
F	2 1/2	120									
ſomme 1579.			ſomme 3007.			ſomme 3781.			ſomme 4202 1/2.		

N iiij

Figure 1.1: Table of air vibrations. Mersenne, *Harmonie universelle*, "Livre troisiesme des instrumens a chordes," 143. Paris, Bibliothèque nationale de France, département Arsenal, FOL-S-1449.

and fig. 1.2). With this astounding table, Mersenne transferred the notes of the four-part vocal piece by Boësset to the number of vibrations that a string would need to produce that pitch.[74] Although he suggested that it could be used in the same way as a musical score, the goal was more demonstrative than practical. By adding the numbers of the different pitches together, Mersenne aimed to demonstrate that even the physical basis of sound (imperceptible beatings of air) followed harmonious numbers. Since Mersenne believed that air vibrations operated proportionally, he thought that it was possible to count and subtract these numbers and come to (an almost) harmonious sum at the bottom of the table. He argued that "we can now consider how many returns each part of music more or less must

74 Boesset's composition was first published by Pierre Ballard in 1630 in four voices and later in a version for voice and lute in 1632.

Figure 1.2: Anthoine Boësset, "Divine Amarillis" in lute tablature. Mersenne, *Harmonie universelle*, "Livre second, des instrumens a chordes," 90. Paris, Bibliothèque nationale de France, département Arsenal, FOL-S-1449.

make, compared to the other, to make a perfect concert."[75] Although he did not arrive at perfect numbers, the approximations of 1579 or 3007 were close enough to remain convinced that he was approaching harmony by adding numbers of vibrations.

Not all his contemporaries saw use in these numbers or trusted Mersenne's methods. Correspondents like Isaac Beeckman were skeptical about its practical use.[76] Scholars such as Athanasius Kircher or the Dutch mathematician Dirck Rembrandtsz van Nierop struggled to reproduce Mersenne's intricate counting. Kircher wrote that he attempted to count a music string but found that when the movement of the string was visible to the eye, no sound was perceivable. And when he tightened the string, he heard a sound, but his eye

75 "l'on peut considerer de combien chaque partie de Musique doit faire plus ou moins de retours l'une que l'autre pour rendre un concert parfait." Mersenne, *Harmonie universelle*, "Livre troisiesme des instrumens a chordes," 144.

76 Beeckman to Mersenne, around March 1629, CM 2, 219.

could not observe the movements. According to the Jesuit, this was nature's way of saying that the audible sensation of a tone could never collide with the visual perception of the string's vibrations by the eye.[77] But Mersenne's obsession with the vibration of air as the smallest denominator of sound was present in many aspects of his work. The movement of air became a central key to uncovering the secrets of the harmony. The study of sound was turned into the study of vibration. Quantifying the movement of air with the help of his string experiments, Mersenne was now able to describe all types of motions in musical terms. For example, two birds—one moving its wings at twice the speed of the other bird—were described by Mersenne as an octave without anyone being able to hear it. Although Mersenne defined sound as the motion of air, he always pointed to the divine origin of these motions. Commenting upon the harmonious numbers emitted by musical strings, Mersenne argued that it was the musician's duty to consider the movements of the string, which obediently followed their fingers, and follow God to partake in "the great concert of the entire world."[78]

Sounding Out Materials

Like his predecessor Vincenzo Galilei, Mersenne thought that there were few things "that cannot be weighed, numbered, and measured."[79] The most obvious tools of demonstration in the realm of music were strings and pipes. Although one might be able to demonstrate the pitches with "rods, vases, copper and silver coins, and other things," these were harder to express numerically.[80] To measure something like sound was, however, a difficult feat. Around the same time, the German music theorist Johannes Lippius went in a different direction, describing sound as a quantity, with measurable qualities of duration, volume, and pitch.[81] Sound is the audible quality of air, and every quality takes on the properties of its body, so sound must also have quantity.[82] The volume of sound would also be quantified in the measurable quantity of breath needed to produce it, but he, too, found it difficult to express quantities numerically. One can give note values in

77 Kircher, *Musurgia universalis*, IV, 428–30; Van Nierop, *Wis-konstige musyka*, 19–20.

78 "le grand concert de tout le monde." Mersenne, *Harmonie universelle*, "Livre troisiesme des instrumens a chordes," 146.

79 Galilei, "A Special Discourse," 181.

80 Galilei, "A Special Discourse," 185.

81 Lippius, *Synopsis musicae*, B4v; Rivera, *German Music Theory*, 50.

82 On Lippius's disolvement of quantity and quality, see Rivera, *German Music Theory*, 55.

terms of lengths and approximations in terms of loudness, but how can one measure the volume of sound?

Like Galilei and Lippius, Mersenne struggled with how one could measure sound effectively and in a standardized manner. In his *Traité*, he already wrote that "we can know each body by the sound it makes."[83] One problem, though, was how to measure this sound, as there was no pitch standardization. For this he expressed a desire for an international academy of music that would concern itself with these matters and find a method to standardize pitch so that the same pitch could be heard in Rome, Paris, and anywhere else in the world. He advocated for an academy of music in every part of the world and a standardized way of determining string width and tension, as "we would find the same unison in Rome as in Paris, or anywhere else in the world."[84] He also looked for methods to measure sound in terms of time but found the current options lacking, concluding, "It is therefore very difficult to find a proper measure for sound, as we have no known measure in nature that is fixed and reliable."[85]

Around the same time as his work on *Traité*, Mersenne devoted himself to a different project that survived in manuscript form as *Livre de la nature des sons*, a work that is considered a preparatory work for the first book on sound in *Harmonie universelle*. The simultaneity of the two projects, one more concerned with the physics of sound and one more concerned with a more abstract notion of harmony, makes clear that they belonged to the same impetus. The topic of measuring returns as well. The tenth proposition, for example, declares that "sound often represents the size, and other qualities, of the bodies by which it is produced."[86] Mersenne tied together sound and the body that produces it, calling sound the image of the body that forms it. And with changes in the material, shape, and size of the body, the potential sound changes. This is clear from the simple fact that larger bodies produce lower sounds, but Mersenne thought that one could go much further. In reference to the study of metoposcopy—that is, the art of reading the lines of foreheads—and physiognomy, he argued that one should be able to know bodies through

83 "comme on peut connoistre chaque corps par le son qu'il fait." Mersenne, *Traité*, 9.

84 "on trouvera le mesme unisson à Rome qu'à Paris, ou en quelqu'autre lieu du monde." Mersenne, *Traité*, 50.

85 "Il est donc fort difficile de trouver une mesure propre pour le son, car nous n'en avons point de connu en la nature qui soit fixe & asseuré." Mersenne, *Traité*, 48.

86 "Le son represente souvent la grandeur, et les autres qualitez des corps par lesquels il a esté produit." Paris, Bibliothèque nationale de France, MS 2884, fol. 4r. Transcription from Taveau, "Le manuscrit 2884," 59.

their sounds, and vice versa.[87] Since objects can be known through their sounds, Mersenne envisioned future scholars being able to know the temperament of individuals and persons in the very same way. Musicians, in turn, could use this knowledge to know what music their listeners would prefer by looking at their faces. Although perhaps a future ideal, the topic does return in *Harmonie universelle*, where he wrote that one gets to know a person better through sound than through touch.[88] But Mersenne had to acknowledge the difficulties here. "We are often mistaken in this matter," he noted in his unpublished manuscript, "for different sounds are sometimes so similar that the ear and human industry cannot recognize the difference."[89] The human ear was simply not precise enough to observe all sounds and their differences.

In his manuscript, Mersenne connected the fallibility of the senses to the recently perfected microscope, which he most likely encountered as early as 1623. Just as many small animals cannot be seen with the naked eye, there must be other sensations and movements that are not sensible. When he explored the idea of knowing objects through their sounds in 1628, he also wrote, "This is why I have no doubt that the mite feels many things within us that we do not know, but it does not perceive the great movements of our bodies, such as the movements of the arms and legs."[90] The microscope had made clear once more how many things hide beneath what is perceptible to the human senses. In his publication *L'usage de la raison* (Paris, 1623), Mersenne wrote regarding the microscope (or magnifying glasses in general) that "it only takes the smallest animal in the world to make us ashamed of our ignorance."[91] The smallest of animals, the mite, had managed to undermine many scientific claims. The microscope, however, showed that there was

87 "Cet axiome pourroit encore donner occasion aus Metoposcopes, ou Physionomes d'establir une nouvelle science pour cognoistre par la vois des hommes." Paris, Bibliothèque nationale de France, MS 2884, fol. 4v. Transcription from: Taveau, "Le manuscrit 2884," 61. See also Dandrey, "La phoniscopie."

88 Mersenne, *Harmonie universelle*, "Traitez de la voix," 8.

89 "qu'on est souvent deceu en cette matiere car les sons divers sont quelquefois si semblables, que l'oreille, et l'industrie de l'homme ne peut en recognoistre la difference." Paris, Bibliothèque nationale de France MS 2884, fol. 4v. Transcription from Taveau, "Le manuscrit 2884," 63.

90 "C'est pourquoi je ne doute nullement que le ciron ne sente beaucoup de choses dans nous, que nous ignorons, mais il n'apperçoit non plus les grands mouvemens de nos corps." Paris, Bibliothèque nationale de France MS 2884, fol. 4v. Transcription from Taveau, "Le manuscrit 2884," 63. Accounts of the microscope actively travelled at the time, for example by Nicolas-Claude Fabri de Peiresc; see Brunet, "Peiresc et le microscope"; Ruellet, "'Voir cette merveille.'"

91 "il ne faut que le plus petit animal du monde pour nous faire rougir de honte de nostre ignorance." Mersenne, *L'usage de la raison*, 10. In his *Quaestiones*, Mersenne specifically singles out how mites move their tails like peacocks. Mersenne, *Quaestiones*, col. 1835.

more to these creatures than previously thought, namely that they have legs and perhaps even veins. Mersenne easily transferred this the realm of sound: there must be sounds imperceptible to the human ears, just as some things are invisible to the naked eye. When dealing with the physical world, humans cannot claim certainty on whether they can perceive all that is there.

In his correspondence, he shared numerous experiments with different materials, but several of his correspondents were skeptical. Villiers wrote to Mersenne that it still must be established whether two blocks from the same species of wood have the same pitch.[92] Later on he suggested that some drier woods sound higher due to "the air inside, which is less constricted and freer and more subtilized."[93] Ten years earlier, the chemist Henry de Stanihurst had already pointed out such difficulties, writing on the different qualities of brass and copper mixtures. These differences are so small that the ear is "not delicate enough in discerning it." He continued,

> Just as it is certain that the major semitone differs from the minor in the diatonic music by a comma [a very small musical interval], but the voice of the man nor the distribution of the keys on the instrument do not distinguish it, and the ear does not perceive it so clearly that they cannot ordinarily pass one for the other either on the voice or on the instruments. The differences, too, that mathematicians find in their demonstrations are so small that human eyes cannot discern them.[94]

Stanihurst found that humans were unfit to notice such small distances. Humans could not even produce such small tone distances, let alone hear them. The chemist suggested that uniting such observations was doomed to fail, as there were many minor differences in the sound of different metals that the human ear could not pick up.

Regardless of these hesitations, Mersenne listened to and measured many different materials such as wood, metal, gut, and stone in line with

92 Villiers to Mersenne, 25 February 1635, CM 5, 62–63.

93 "de l'air interieur qui est moins resseré et plus libre, et subtilisé." Villiers to Mersenne, 25 March 1635, CM 5, 119.

94 "si l'oreille n'est grandement delicate à la discerner. Tout ainsy qu'il est certain que le demy ton majeur differe du mineur en la musique diatonique d'un comma, mais la voix de l'homme, ny la distribution des touches sur l'instrument ne le distinguent pas, et l'oreille ne l'apperçoit pas si nettement qu'ilz ne puissent passer d'ordinaire l'un pour l'autre tant au voys qu'aux instrumentz. Les differences aussy que les mathematiciens trouvent par leurs demonstrations sont si petites que les yeux des hommes ne les peuvent discerner." Stanihurst to Mersenne, September 1625, CM 1, 282. Stanihurst was part of an intellectual circle in Rouen that also included Jehan Titelouze and Robert Cornier; see Cohen, "Jehan Titelouze."

the research program of the science of psophologie set out in the *Questions theologiques*. He found that the even the slightest change in a body would already create a different sound. No language could describe the qualities of sounds, as he wrote that "one cannot establish any science of them [sound qualities] because they almost go to infinity."[95] In his *Harmonie universelle*, Mersenne recounted some of his experiments, for example with different types of wood and small metal bells. The latter experiment involved the construction of different bells of the same size with different metal compositions, such as gold, silver, copper, brass, tin, lead, and antimony. The bells were measured, and the differences in pitches were determined. For this task, Mersenne relied on a local musician whose "clear ears" he trusted.[96] But even for experienced listeners such as Mersenne's anonymous musician, it was difficult to determine the pitches of such small metal objects. The musician eventually gave two contradicting accounts, both printed by Mersenne in his *Harmonicorum*. The two results differed significantly, with some discrepancies of the interval of a third between the two observations. In his characteristic manner, Mersenne did not try to explain these discrepancies but presented both conflicting findings side by side. He performed a very similar experiment with wood, which also did not lead to any conclusions, testing the sound of twelve blocks of different woods of the same size. From these experiments, he could observe that a block of fir is a diminished fourth higher and a block of cherry wood a minor third higher than a block of sycamore. But apart from these observations, he could not obtain any useable results, concluding, "The ear is not capable of noticing this sufficiently to judge the differences of wood by their sounds."[97]

Mersenne had to admit that even expert listeners could determine these minor differences, halting his idea of a science that would measure all materials and their sounds. Throughout his books there are references to such expert listeners who validated his own listening experiences, acting as corroboration of his ideas and demonstrating that even when Mersenne did not name his collaborators, his project was very much a collective one. On the topic of an experiment with the sound of metal, he wrote, "The very exact experiences we have made multiple times in the presence of

95 "l'on ne peut en establir une science, à raison qu'ils vont presque à l'infiny." Mersenne, *Harmonie universelle*, "Livre premier de la nature," 29.

96 "notis Musicorum Practicorum .... auribus purgatissimis." Mersenne, *Harmonicorum*, "De campanis," 154; Mersenne, *Harmonie universelle*, "Livre septiesme des instrumens de percussion," 24–25.

97 "neantmoins l'oreille n'est pas capable de le remarquer suffisamment pour juger de la difference des bois par leurs sons." Mersenne, *Harmonie universelle*, "Livre troisiesme du mouvement," 181.

excellent geometers and musicians, whose ears are very knowledgeable and delicate."[98] Several years earlier, he wrote that he had experimented "with those who are considered to have the best ears of France" but did not disclose who these expert listeners were.[99]

One particular reader of Mersenne was especially critical of the experiment. Several years after publication of *Harmonie universelle*, the English Edmund Chilmead commented upon the experiments in his manuscript on sound. As Chilmead pointed out, the gold and brass bells emitted the same pitch in Mersenne's results, which would be unlikely. The fault could lie in the quality of the bells and their size. Mersenne only experimented with smaller bells, and according to Chilmead, the experiment would be better with larger bells.[100] Larger bells would emit a clearer sound and would make the experiment more certain. Chilmead not only criticized the size of the bells but also the hearing capabilities of Mersenne and his earwitnesses. According to Chilmead, "the Ears he used" were not reliable. Mersenne made the mistake of trusting the ears of the musicians, which were not so "purged & accurate, as he pretends."[101] In Rome Kircher reproduced Mersenne's bell experiment as well, aided by local Roman musicians, but the Jesuit could not obtain the same results as Mersenne: "whatever I did, I could not discern the same things that Mersenne had presented."[102] In a world with few standards for the study of sound, the measurement of the sounds of different materials Mersenne envisioned in his *Questions theologiques* had to wait.

The Science of Harmonics

Wooden blocks and small metal bells were difficult objects to study. There was already another discipline that investigated strings, namely harmonics. Harmonics was understood as the study of mathematical relations that

98 "les experiences tres-exactes que nous en avons faites plusieurs fois en la presence d'excellens Geometres & Musiciens, dont l'oreille est tres-sçavante & delicate." Mersenne, *Harmonie universelle*, "Livre troisiesme du mouvement," 175.

99 "Il suffit que je die ce que j'ay experimenté, et ce que j'ay veu pratiquer à ceux lesquels on estime avoir la meilleure oreille de la France." Paris, Bibliothèque nationale de France, MS 2884, fol. 37r. Transcription from Taveau, "Le manuscrit 2884," 335.

100 Chilmead, "An Examination." Oxford, Bodleian Library, MS Tanner 204, 52. For Chilmead's work on sound in general, see Feingold and Gouk, "An Early Critique."

101 Chilmead, "An Examination." Oxford, Bodleian Library, MS Tanner 204, 51.

102 "at quicquid agerem easdem quas Mersennus posuit, deprehendere non potui." Kircher, *Musurgia universalis*, VI, 528.

provide the foundation of audible music. As will become clear, Mersenne followed Ptolemy in his definitions and conceptualizations of the science of music and the role of the musical scholar. According to Ptolemy, "Harmonic knowledge is the power that grasps the distinctions related to high and low pitch in sounds."[103] The principal task of the harmonicist is constructing the harmonic canon (*kanōn*). The canon is the division of the monochord, aimed at uniting what is reasonable and agreeable to the senses. Its criteria were hearing and reason. The two did not stand on equal footing; hearing could only approximate and provided a rough outline. Because sensory knowledge dealt with matter and the fallibility of the human observer, it would always have its uncertainties. Reason must be imposed to establish order and certainty, for example with the help of instruments. The senses needed the "crutch" of reason. This was a process of verifying and eliminating those sense perceptions that did not conform to reason: "straightening [*kanonizein*] those things in sense perception that are inadequate to reveal the truth."[104] As Boethius would later write, hearing only approximates: "the sense discovers something confused, yet close to the truth."[105] Only reason could fully judge on the perfection of objects.

In this treatise, Ptolemy included a description of the different schools in harmonics in ancient Greece, a division that would be adopted by subsequent authors. First, there were the followers of Pythagoras, who famously gave no credence to the ear. They emphasized the centrality of perfect ratios, especially of the first four whole numbers (1, 2, 3, 4, or the numbers of the *tetractys*). The second group was the school of Aristoxenus (himself a follower of Aristoteles), which gave the senses a central role. This school abandoned the emphasis on whole numbers and ratios, deviating from the Pythagorean way of thinking of harmony in terms of numbers. The judgement of the ear was central. As Boethius would write, Aristoxenus "yielded all things to the judgement of the ears."[106] Accordingly, in Aristoxenus's *Elementa harmonica*, the distance of a fourth is described as two and a half whole tones and an octave as six whole tones, something that Pythagorean authors argued against.[107] Aristoxenus divided the whole tone into two, three, and four equal parts, arriving at irrational numbers. Summarizing these different

103 Ptolemy, *Harmonics*, I.1, in Barker, *Greek Musical Writings*, 276.

104 Ptolemy, *Harmonics*, I.2, in Barker, *Greek Musical Writings*, 278. For Ptolemy's understanding of reason and sense, see Barker, *Scientific Method*, 14–32. On how these passages were interpreted in the Middle Ages, see Hicks, *Composing the World*.

105 Boethius, *Fundamentals of Music*, 163.

106 Boethius, *Fundamentals of Music*, 88.

107 Aristoxenus, "Elementa harmonica," II.46, in Barker, *Greek Musical Writings*, 142.

schools, Ptolemy found faults with both and argued that a middle group must be established. Sense and reason must be united.[108] Boethius would codify this several centuries later, writing that Ptolemy's goal of harmonics as "that there is not irreconcilable between the ears and reason."[109] Ptolemy already connected the practical elements of harmonics to harmonies of the universe. In the third book of his treatise, he explained how tuning is present in all things, including human souls and the movement of the heavens. The goal of the harmonicist, according to Ptolemy, was to observe regularities in nature in the zodiac, stars, planets, spheres, and humans themselves. Harmony was something extremely rational—something that could be mathematically demonstrated—and at the same time an overarching cosmic concept connecting all creatures with the universe.[110]

This classical tradition was central to Mersenne's conceptualization of the task of the musical scholar. Like the ancient scholar Ptolemy, he aimed to establish the authority of reason over the world of materials and human perception. His empirical methods were closely modeled on these classical examples, although he argued that the ancients did not go far enough. And even Ptolemy's own sense of hearing seemed to have been flawed, as he once wrote.[111] Many passages from Ptolemy were adopted and incorporated in Mersenne's writings. Like Ptolemy, Mersenne singled out sight and hearing as the two senses that could demonstrate this order the best. With the help of a monochord or drawing compass, the sensations perceived by the ears or eyes could be controlled and disciplined. Mersenne paraphrased Ptolemy's words, for example in his *Les preludes*:

> For it belongs to learned men, who devote their life and their study to contemplation, to demonstrate that the works of nature are well ordered, and that there is nothing confused and nothing done by chance. Particularly when it concerns seeing and hearing, which approach reason more than the other senses and which help us to understand the sciences and to praise, contemplate, and admire the works of God, the excellence, and grandeur of the maker.[112]

108 Ptolemy, *Harmonics*, I.2, in Barker, *Greek Musical Writings*, 279.

109 Boethius, *Fundamentals of Music*, V.3, 165.

110 Ptolemy, *Harmonics*, III.3, in Barker, *Greek Musical Writings*, 371.

111 Mersenne, *Harmonie universelle*, "Livre second des instrumens a chordes," 56.

112 "Car il appartient aux hommes sçavans qui employent leur vie, & leur estude à la contemplation, de monstrer que les oeuvres de la nature sont bien ordonnées, & qu'il n'y a rien qui soit confus, ou qui se fasse par hazard, particulierement dans ce qui concerne la veuë & l'oüie, qui approchent plus de la raison, que les autres sens, & qui nous servent pour apprendre les

The task of the scholar was to demonstrate the well-regulated order within nature. Instead of reveling in particulars, Mersenne emphasized the role of order. The order of nature was for Mersenne a deeply theological, moral, and epistemological question. He does make an addition to the Greek author, namely that the works of "learned men" should be used for praising, contemplating, and admiring God. As Ptolemy writes, it is the task of the harmonicist "to preserve in all respects the rational postulates of the *kanōn* as never in any way conflicting with the perceptions that correspond to most people's estimation."[113] Or in the words of Mersenne: "it is the musician's task to preserve or to find the ratios of this rule [the harmonic canon] that agrees with hearing, following the sentiment of the majority of men."[114] In his efforts to study sound and its harmonies, Mersenne followed the classical formulation of the science of harmonics.

The principal task of the harmonicist is constructing the harmonic canon (*kanōn*), the division of the monochord (fig. 1.3) to get all the right intervals. Traditionally, the monochord had been the classical instrument to demonstrate the consonances, with little practical use in music. In his own work Mersenne continued to uphold the monochord as the ultimate object to demonstrate the rationality of music. As he wrote, it is "the most suitable instrument and the most exact for regulating the pitches and harmony."[115] He saw the instrument reflected in all other stringed instruments, for example in the fingerboard of a lute or the neck of a violin. Because of its qualities in demonstrating the proportions of harmony, Mersenne envisioned the monochord as a measuring instrument, comparing it to the ruler or compass of a geometer.[116] The monochord should unite what is reasonable (according to mathematical proportions) and what is agreeable to the ear, but the sense of hearing was at times completely eliminated. The topic of tuning an instrument without relying on the ear, but instead on "some science and art," returns at several moments in his writings.[117] Already in

sciences, & pour loüer, contempler & admirer les oeuvres de Dieu, & l'excellence, & la grandeur de l'ouvrier." Mersenne, *Les preludes*, 163.

113 Ptolemy, *Harmonics*, I.2, in Barker, *Greek Musical Writings*, 278. On the science of harmonics in the early modern period, see Gouk, "The Role of Harmonics."

114 "car cest l'office du Musicien de conserver ou de trouver les raisons de ladite regle, qui s'accordent avec l'oüie, suivant le sentiment de la plus grande partie des hommes." Mersenne, *Les preludes*, 162.

115 "l'Instrument le plus propre & le plus exact pour regler les sons & l'harmonie." Mersenne, *Harmonie universelle*, "Traité des instrumens a chordes," 15.

116 Mersenne, *Harmonie universelle*, "Traité des instrumens a chordes," 16.

117 "par quelque science & artifice." Mersenne, *Harmonie universelle*, "Livre sixiesme des orgues," 366.

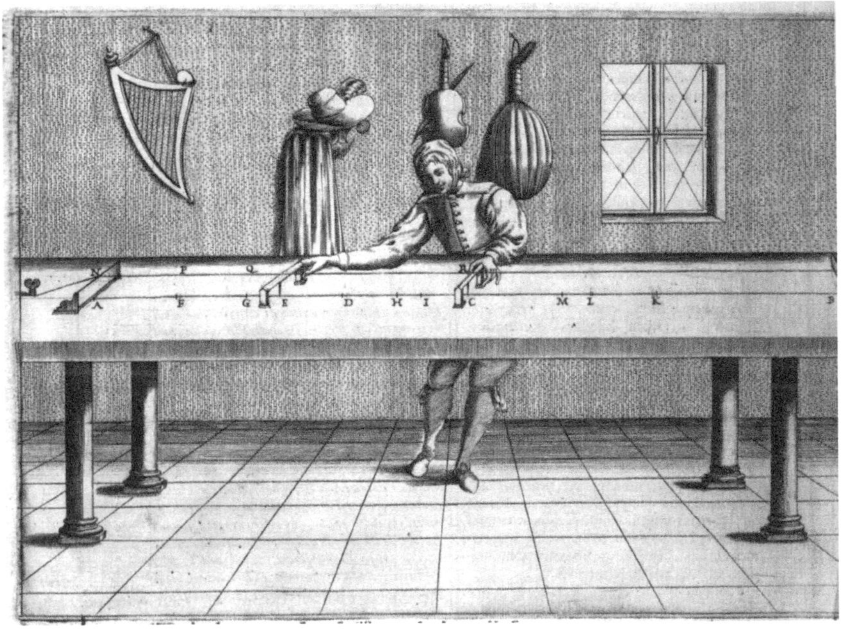

Figure 1.3: The monochord. Bettini, *Apiaria universae*, 8 ("Musica"). Augsburg, Staats- und Stadtbibliothek – 2 Math 11 -2 / Münchener DigitalisierungsZentrum.

his manuscript of 1628, he explained that it was possible to tune different strings to unison by solely relying on touch or vision. With the help of pieces of straw, one could tune two strings in unison by means of its sympathetic vibration. Such a method was more certain than relying on the ear, Mersenne argued, which he considered fallible. He wrote that he himself had performed this experiment multiple times, each time "without the ear, as I have done several times."[118]

This emphasis on eliminating the ear stemmed from a longer tradition what wrestled with the relationship between sense and reason. Mersenne's predecessor Pontus de Tyard wrote that the senses are easily mistaken, such as taste "in the choice of foods" or vision in small differences of colors.[119] So one should not be surprised, he argued, that "hearing remains dumb" or at least unsuccessful in reporting the true distances of intervals.[120] True knowledge lies outside the realm of the senses. Like Mersenne, the Jesuit Mario Bettini tried to achieve such knowledge by eliminating the ear. Bettini

118 "sans l'oreille, comme j'ay fait plusieurs fois." Paris, Bibliothèque nationale de France, MS 2884, fol. 26r. Transcription from Taveau, "Le manuscrit 2884," 239.

119 "au chois des viandes." Tyard, *Solitaire second*, 170.

120 "l'ouye demeure stupide." Tyard, *Solitaire second*, 170.

argued in 1642 that a deaf person could tune the monochord exactly to the right intervals by simply following his mathematical rules (fig. 1.3). Without any previous knowledge and any hearing capabilities, this person would be able to arrive at "the most exact consonances," Bettini wrote.[121] In his depiction of the experiment, a young man is depicted with a scarf wrapped around his ears, possibly signaling that he is excluding the judgment of his ears. Mersenne perhaps did something similar when he wanted to calculate the consonances by sight and measurement only. In his own text, Bettini described how one should divide string AB into two parts to get the octave, to get the fifth in three parts, to get the fourth in four parts, etc. If followed exactly, one could divide the monochord without making it sound, and according to Bettini, one should arrive at the same proportions when one when searching for them with the ear. Mersenne was similarly interested in eliminating the sense of hearing, relying only on mathematics completely. Writing about his own instructions for deaf persons (discussed in a different context in chapter 2), he noted, "Here are the general rules for making all kinds of agreements, which will serve here as proof and demonstration, as we have shown elsewhere that they are true and infallible."[122] These rules did not depend on the senses but were rather based on irrefutable mathematical principles.

The need to verify the senses with reason became most clear in the perception of the consonances with the help of the monochord. The demonstration

121 "Etiam à surdo, & pratica musices ignaro monochordum, & musica alia instrumenta posse ad exactisimas consonantias attemperari." Bettini, *Apiaria universae*, 8 ("Musica").

122 "Or voicy les regles generales, dont il faut user pour faire toutes sortes d'accords, lesquels serviront icy de preuve, & de Demonstration, d'autant que nous avons fait voir ailleurs, qu'elles sont veritables & infaillibles." Mersenne, *Harmonie universelle*, "Livre troisiesme des instrumens a chordes," 123. Like many other early modern (natural) philosophers, Mersenne had an interest in deafness and hearing disabilities. In August 1634, he wrote to Peiresc, "If you have made some observations to teach the deaf and mute to learn to read, write, and speak, I would be very happy to be informed of them, for I deal with them in my book" ("Si vous aviez aussi fait quelques observations pour apprendre les sourds et les muets à apprendre à lire, escrire et parler, je serois bien aise d'en estre averti, car j'en traite dans mon livre"). Mersenne to Peiresc, 2 August 1634, CM 4, 280. Peiresc seems to have supplied Mersenne with a written description of one of his relatives, Monsieur de Roumoulles de Linceaux, who had an account book with images. See Peiresc to Gassendi, 4 February 1633, CM 3, 375–76; Peiresc to Gassendi, 22 February 1633, CM 3, 378; Tamizey de Larroque, *Notice inédite*. Mersenne also read books like Jacques Roland de Belebat's *Aglossostomographie* (Paris, 1630), on a young boy without a tongue, and cites the book in his *Harmonie universelle*: "Traitez de la voix," 77–78. Focusing on sixteenth-century Geneva and early modern England, Anna Kvíčalová and Rosamund Oates have shown that there was a wider interest among preachers in deafness and hearing loss; see Kvíčalová, "Hearing Difference"; Oates, "Speaking in Hands." For the interest of physicians in hearing disabilities, see Verwaal, "Fluid Deafness."

of the proportions of the consonances (2:1 an octave, 3:2 a fifth, etc.) was cited throughout the medieval and early modern periods as evidence that the harmonies of music follow mathematical proportions. But as Mersenne pointed out, one could shorten or lengthen a string by only a small amount and still recognize the pitch. The perfection of Pythagorean tuning was hardly necessary for the ear to be satisfied. Already early on, the Minim investigated precisely to what extent the musical intervals must follow their mathematical proportions in order to sound "right." In Mersenne's hands, the monochord was no longer a demonstrative device but rather a sound source to investigate human hearing by experimental means. He argued the following in the early manuscript *Livre de la nature des sons*:

> Experience shows that we can add or diminish the octave without the ear being able to perceive it. Because if we touch two strings, one which has 1000 parts, and the other 2001 parts, it is impossible for the ear to recognize the difference of the true octave .... But it is difficult to determine how big the difference must be to perceive it ... depending on the least or greatest delicacy of the ear.[123]

Here Mersenne pointed to one of the greatest difficulties of musical science during his lifetime: how to incorporate the ear into measurements of the consonances. As he noted, a perfect octave might be created by a 2:1 proportion, but if one moved the bridge of a monochord ever so slightly, to a 2,001:1 proportion, one still heard the octave. Later in his *Harmonie universelle*, he expressed a similar idea: because the human ear is not delicate enough, it does not matter whether an octave is tuned according to the ratio of 1000/499 instead of 1000/500.[124] But when both proportions sound agreeable to the human ear, why would one prefer one ratio over the other? Although most early modern theorists repeated the idea that the ear only provides a rough outline, Mersenne was especially interest in how rough this outline exactly was, paying more attention to the physiological workings of the ear than his predecessors. The monochord served not just as a device to demonstrate proportions but also as an instrument to probe the limits of hearing.

123 "L'experience fait voir qu'on peut ajouter, ou diminuer à l'octave sans que l'oreille le puisse appercevoir, car si on touche deux chordes l'une desquelle ait 1000 parties, et l'autre 2001 il est impossible que l'oreille reconoisse la difference de cet intervalle d'avec la vraye octave .... Mais il est plus difficile de determiner combien il faut que la difference soit grande pour l'appercevoir, ... selon la moindre, ou la plus grande delicatesse de l'oreille." Paris, Bibliothèque nationale de France, MS 2884, fol. 37v. Transcription from Taveau, "Le manuscrit 2884," 333–35.
124 Mersenne, *Harmonie universelle*, "Livre premier des consonances," 47.

Mersenne shared his difficulties with aligning the mathematical world of measuring strings with the sensory world of human hearing in 1636 with the mathematician Pierre de Fermat. The first surviving letter between the two men is worth quoting extensively. Here Fermat wrote,

> I have always believed that it is very difficult to shake and break down the principles of science. Because being founded on the laborious experience of those who have investigated them, it seems very difficult to make them more precise, and it is even more useless to call reason to the rescue of the senses, since in its operations, it always presupposes those of the exact and truthful senses. So, in my opinion and for these reasons, it would be very difficult to find a different proportion apart from the double [2:1] that would make the octave more exactly than this [proportion]. I must admit that there are infinite numbers of them, which will indeed make different harmonies, and yet the difference will not be understood by the most delicate hearing; and from this one might conclude that perhaps the true octave does not consist precisely in double proportions. But since we have been unable to discover any significant error in this principle which the ancients have bestowed to us, let us pay them the respect of believing it to be true, until the opposite is apparent to us.[125]

Since sound always presupposes the senses, Fermat argued, it is pointless to call to reason. Just as there are "glasses that make visible things that were not visible before," there might be instruments that will fundamentally change the ways in which sounds are understood and differentiated. Mersenne had also suggested that new sounds would be discovered, similar to the

125 "J'ai toujours cru qu'il étoit bien malaisé de secouer et détruire les principes des sciences. Car, étant fondés sur l'expérience laborieuse de ceux qui les ont recherchés, il semble qu'il est bien malaisé d'en faire de plus précises, et il est encore plus inutile d'appeler la raison au secours des sens, puisque dans ses operations, elle présuppose toujours celles des sens exactes et veritables. De sorte que, par mon sentiment et par ces raisons, j'estime qu'il seroit bien malaisé de trouver une proportion different de la double qui fît l'octave plus exactement que celle-là. Je vous avoue bien qu'il y en a infinies qui effectivement feront des accords différents et desquels néanmoins la différence ne sera pas comprise par l'ouie la plus délicate qui puisse être; et de là on pourroit conclure que peut-être la vraie octave ne consiste pas précisément en la proportions double. Mais puisque, en ce principe que les Anciens nous ont baillé, nous n'avons jusques à présent su découvrir d'erreur sensible, rendons-leur ce respect de le croire véritablement, jusques à ce que le contraire nous ait apparu." Fermat to Mersenne, 26 April 1636, CM 6, 51–52. According to Mahoney, Fermat "saw no *a priori* reason to assume that natural phenomena obeyed exact mathematical laws of behavior." Mahoney, *The Mathematical Career*, 370. Although the two men would have agreed on the fallibility of the senses, Mersenne had argued a decade earlier that reason should come to aid to the senses precisely on these grounds.

discovery of new planets by means of the telescope.[126] Although there was no instrument that magnified small sounds or made faraway sounds audible, it is likely that the presence of optical instruments compelled scholars to consider the existence of sounds beyond the limits of audibility. Fermat seemed to have little interest in finding an exact mathematical principle to describe natural phenomena. He also objected to mixing categories between sensory perception and geometrical demonstration:

> Now to seek by reason why the octave is in double proportion is, it seems to me, to treat heterogeneous things: the sound of the octave is the accident and the quality of the double proportion which consists in quantity. Proportion is understood by sight; the agreement it makes, by hearing; and so it seems that we cannot assign a necessary reason why the one suits the other. For, as you know, demonstrative reasons always stops between homogeneous subjects. So, it is better to leave all the questions in your letter to the senses to decide than to alter received maxims that cannot be convinced of being false.[127]

Fermat was possibly referring to Aristotle's *Analytica posteriora*, a work that aims to set out the principles of science and argues that it is incorrect to draw conclusions from one field to another, for example arithmetic and geometry.[128] He seemed to suggest that a truly demonstrative reason for why the double proportion of a string makes an octave cannot be found. According to the mathematician, it is better to stick to what the ancients had taught, as this has not been proven false, and let the senses decide on this matter.

The question of how forgiving the ear could be was not just a theoretical question but also a practical one—for example, in choosing tuning or temperaments. Temperament is the tuning of a scale so that some or all

126 "des lunettes qui rendent visibles les choses qui ne l'estoient pas auparavant." Fermat to Mersenne, 26 April 1636, CM 6, 52; Mersenne, *Harmonie universelle*, "Livre premier des consonances," 84.

127 "Or de chercher par raison pourquoi l'octave est en proportion double c'est, ce me semble, traiter les choses hétérogènes: le son de l'octave est l'accident et la qualité de la proportion double qui consiste en quantité. La proportion se comprend par le vue; l'accord qu'elle fait, par l'ouie; et ainsi il semble qu'on ne sauroit assigner une raison nécessaire pourquoi est-ce que l'un convient à l'autre. Car, comme vous savez, les raisons démonstratives s'arrêtent toujours entre des sujets homogènes. De sorte qu'il vaut mieux laisser decider aux sens toutes les questions de votre lettre que d'altérer des maximes reçues et qu'on ne sauroit convaincre de faux." Fermat to Mersenne, 26 April 1636, CM 6, 52.

128 Aristotle, *Posterior Analytics*, 12 (I.7).

compound notes are slightly impure. Some theorists accepted a form of what we now call equal temperament for certain instruments. Throughout the sixteenth century, fretted instruments like the lute or viol were thought to be suitable for equal temperament, but its use for keyboard instruments or singing was more controversial.[129] For Mersenne, the question of whether one should allow a form of equal temperament was closely related to his response to skepticism. In his *Questions harmoniques*, for example, he responds to the skeptic interlocutor on the uncertainty regarding the mathematical ratios in music. Mersenne writes that he encountered "excellent geometricians" who denied the existence of perfect ratios in music. Instead, they proposed a form of equal temperament, arguing that musical intervals are inexplicable or irrational. Mersenne describes these mathematicians as follows:

> One meets excellent geometers ... who believe that all the reasons for the degrees and intervals are inexplicable, nonsensical [*sourdes*], and irrational; for they maintain that all the whole tone and semitones are equal, that three ditones [major thirds] make a pure octave, that the augmented fifth is no different from a minor sixth, that the diminished fifth and the tritone [augmented fourth] are the same: that the practice and composition of music is much better, or easier, by following equal whole tones and semitones than by using the theory that maintains their inequality, and finally, that the consonances and degrees produced by lutes, viols, and other instruments, as well as the ear, testify to this equality.[130]

According to the advocators of equal temperament, an augmented fifth is no different than a minor sixth. And these advocators argued that many problems in tuning could be avoided by dividing the octave into twelve equal notes. Although this was already done on instruments like the lute, adopting equal temperament for other instruments or voices was more controversial. But as becomes clear from Mersenne's quotation, the advocates

129 On Mersenne within the history of tuning, see especially Lindley, "Mersenne on Keyboard Tuning."

130 "l'on rencontre d'excellens Geometres ... qui croyent que les raisons de tous les degrez & intervalles sont inexplicables, ou sourdes, & irrationnelles; car ils maintiennent que tous les tons, & les demy-tons sont égaux: que trois ditons font l'octave juste: que la quinte superfluë n'est point differente de la sexte mineure: que la fausse quinte & le triton sont une mesme chose: que la pratique & la composition de la Musique est beaucoup meilleure, ou plus aisée en suivant l'egalité des tons & des demy tons, qu'en usant de la theorie qui met leur inegalité: & finalement que les consonances, & les degrez qui se font sur les luths, les violes, & les autres instrumens, & quant & quant que les oreilles, tesmoignent ceste egalité." Mersenne, *Questions harmoniques*, 81–82.

of temperament argued that not only instruments attest to the equality of intervals but also the ear. One of these theorists was the Dutch engineer Simon Stevin, who in his *Van de spiegheling der singconst* (ca. 1600–1620) proposed a form of what is today called equal temperament. Interestingly, Stevin connected his acceptance of this tuning system to the human ear, since human hearing was already imperfect. There was no use in trying to approach perfect proportions in music, Stevin argued, since "no human hearing, however keen it may be, is able to fit two tones quite surely in their perfection."[131]

Stevin's suggestion was one of the more radical solutions of the difficulty of uniting sense and reason around 1600. Most theorists continued to uphold the idea that one should strive toward perfect proportions in music. Gioseffo Zarlino, for example, described the intellect as the seat of judgement in hearing. According to Zarlino, reason deals with unchallenged truths, whereas the senses deal with things that change. The composer argued that two proportions are at work when humans listen to music, an "imagined" proportion of our ears and a "true" proportion rectified by our intellect.[132] Although perfect proportions might not exist in nature, one should still try to approach these proportions, since the mind is able to recognize them. Others argued that ears must get used to new tunings. Vincenzo Galilei, for example, wrote that ears are used to equal tempered intervals on the lute, but not on keyboard instruments. Hearing a keyboard tuned in equal temperament, Galilei wrote, "we cannot escape having the sense be offended in certain particular places."[133] But according to Galilei, this arose from the fact that human ears are not yet used to these tunings, disagreeing with Zarlino that the mind necessarily seeks out perfect ratios. Mersenne's reference to the "excellent geometers" perhaps reveals that with the help of

131 Stevin, "Vande Spiegheling der Singconst," 459. Stevin had already written a rough draft of this treatise in the 1580s. Mersenne's correspondent Isaac Beeckman knew of Stevin's temperament as early as 1613, and he most likely discussed it with Descartes. See Cohen, *Quantifying Music*, 184. It is not known whether Mersenne discussed Stevin's temperament during his visit to Isaac Beeckman in Dordrecht in 1630, but as Rudolf Rasch has pointed out, his description from his *Questions harmoniques* suggests that he might have. Descartes would in turn write to Mersenne that this argument was absurd: "As for the reasons your musicians give for denying the proportions of the consonances, I find them so absurd that I do not know how to respond" ("Pour les raisons qui disent vos musiciens, qui nient les proportions des consonances, je les trouve si absurdes, que je ne sçaurois quasi plus y repondre"). Descartes to Mersenne, end of February 1634, CM 4, 51; Rasch, "Simon Stevin," 293. See also the letter Descartes to Mersenne, 15 May 1634, CM 4, 143. Rudolf Rasch discusses these and other sources in Rasch, "Simon Stevin."

132 Zarlino, *Sopplimenti musicali*, 37–38 (I.13); Fend, "The Changing Functions."

133 Galilei, *Dialogue on Ancient and Modern Music*, 114.

geometrical methods, equal temperament was easily obtained. He himself included several such methods, for example in the book on the organ.[134] As Mark Lindley has observed, the Minim formulated this as a geometrical problem rather than a musical one. In Lindley's words, "Mersenne encouraged organists or organ builders to try out equal temperament, but inhibited them from anticipating a favorable verdict by reminding them of the beauty of other tunings."[135]

In principle, Mersenne was not opposed to the idea of using irrational fractions for expressing the intervals, but at different moments he found that a position such as Galilei's and Stevin's sometimes went too far. Denying an arithmetical rationality to musical intervals could lead to dangerous territories, possibly even denying any reason to the field of music. At other times, his position seemed more relaxed. In his book on the consonances, he wrote that there were no true demonstrations (in a mathematical sense) in the realm of fictions, and states, "That is why everyone is free to follow any opinion they like, according to the truest of reasons: for example, those who prefer to hold that all tones and semitones must be equal ... will not lack reason," singling out the Dutch Stevin and "the Aristoxenians of Italy."[136] It would be difficult to show supporters of equal temperament that the fifth commonly used in music is exactly a proportion of 3:2 when precisely measured. There might be reasonable arguments in favor of equal temperament, but Mersenne continued to be deny the existence of perfect proportions. In many ways, these were unattainable in practice. For theoretical purposes, however, these ratios must be maintained. Mersenne was, however, not clear on the precise purpose. In the *Les preludes,* for example, he was convinced that listeners do not "derive as much satisfaction as if all the concords were perfect."[137] Two years later, he wrote that it was enough to approach perfect proportions in practice, as the ear cannot perceive the smallest movements of air.[138] Reason might strive for perfect ratios, but practice could give a rough outline.

134 Mersenne, *Harmonie universelle*, "Livre sixiesme des orgues," 408–12. Richard Oosterhoff has observed how the sixteenth-century mathematician Lefèvre d'Étaples "moves from arithmetic to geometry, led more firmly by the ear." Oosterhoff, *Making Mathematical Culture*, 156.

135 Lindley, "Mersenne on Keyboard Tuning," 184–85.

136 "C'est pourquoy chacun est libre de suivre telle opinion qu'il voudra, selon les raisons les plus vrayes semblables: par exemple, ceux qui aymeront mieux tenir que tous les tons & les demitons doivent estre esgaux ... comme fait Stevin ... & les Aristoxeniens d'Italie ... ne manqueront pas de raison." Mersenne, *Harmonie universelle*, "Livre premier des consonances," n.p. ("Preface").

137 "l'on n'en reçoit pourtant pas tant de contentement que si tous les accords estoient parfaits." Mersenne, *Les preludes*, 169.

138 Mersenne, *Harmonie universelle* "Livre premier des consonances," 47.

Although irrational fractions were permitted and rough outlines were enough for musical practice, Mersenne did not relinquish his quest for order. Throughout his work, he emphasized the need to find order. With his endorsement of the special status of perfect mathematical proportions, Mersenne positioned himself in the tradition of harmonics as described by Ptolemy and Boethius. The Ptolemaic ideal of uniting sense and reason was central to Mersenne's endeavor and his conceptualization of listening. Hearing was envisioned as an approximation, an outline, that needs to be "straightened" by the crutch of reason. But here, as in other positions, he was remarkable flexible in his ideas and positions. In several positions he followed the idea that the mind takes pleasure in the simpler (arithmetic) proportions. In his early manuscript *Livre de la nature des sons*, Mersenne wrote, "The imagination or the mind understands more easily the divisions of the consonances," concluding, "Nature takes more pleasure in what is simple and always chooses the most easy."[139] Later in his *Harmonie universelle*, he would suggest that "the mind and ear of listeners have a certain relationship to the tones and accents."[140] Following Augustine, Mersenne suggested these numbers could be innate, engraved in the mind and memory. These numbers of the soul are immutable and eternal, not belonging to any material realm but the angelic realm of "pure mathematics."[141]

At other moments, however, he seemed to have rejected this standpoint. For example, Mersenne dismissed the methods of Johannes Kepler in his *Harmonice mundi*, who connected the consonances to the sides of polygons. The proportion 5:3 found on a heptagon, Kepler argued, could explain the major sixth. The astronomer argued that the human mind imitates the divine by showing delight in the same proportions as the celestial bodies.[142] But according to Mersenne, this argument could not demonstrate why certain proportions are preferred over others. Regarding Plato's argument in his *Timaeus* that the ratios of harmony are engraved in the souls of humans, Mersenne found that this cannot explain why certain numbers are more

139 "l'imagination, ou l'esprit comprend, et trouve plus facilement les divisions esquelles les consonances ... la nature prend plus de plaisir à ce qui est plus simple, et choisit tousjours le plus facile." Paris, Bibliothèque nationale de France, MS 2884, fol. 56v. Transcription from Taveau, "Le manuscrit 2884," 481.

140 "Car l'esprit & l'oreille des auditeurs ont un certain rapport à des tons, & à des accents qui sont capables de les esbranler." Mersenne, *Harmonie universelle*, "Livre sixiesme de l'art de bien chanter," 373.

141 "comme les Anges les voyent; ce qui appartient à la pure Mathematique." Mersenne, *Harmonie universelle*, "Livre sixiesme de l'art de bien chanter," 429. See also Christensen, "Mersenne"; Psychoyou, "Plaisirs de l'esprit."

142 Kepler, *The Harmony of the World*, 129.

harmonious or beautiful than others. The mind could just as easily understand the proportions 7:6 or 8:7, which in music are considered "enemies of harmony."[143] Emphasizing the role of custom, he wrote that "long exercise has the habit of making soft and easy what previously seemed harsh and tedious."[144] Later in his book he would argue that "it is difficult to distinguish and recognize long habit from nature."[145] Eventually Mersenne came to the position that the consonances exist as they do because "practice conforms to it."[146] Mersenne seemed to have accepted the argument of those who suggested there is no absolute or rational reason for these eight consonances, but one should still maintain them. Like Descartes, Mersenne included two ways of dividing the consonances, one based what is pleasing to the ear, and one according to the greatest union of their sound.[147] Finally, Mersenne's minutiose breakdown of music into air vibrations complicated the idea of proportionality and commensurability itself: "there is no number of motions or beatings of air that is not commensurable to all other numbers of motions."[148]

The pleasure of music was thus ultimately something that did not necessarily presuppose the mind, as there were many mindless and ignorant people who could still enjoy music, Mersenne argued. Precisely why certain consonances sound better was still unknown, but Mersenne argued that the science of music was more capable in determining what is pleasant and what is not than the other sciences. He found that scholars in optics still disagreed with each other on how many colors there are, or nobody could agree on what is most pleasing to the tongue or sense of taste. This, Mersenne argued, sets music apart from the other sciences:

143 "ennemies de l'Harmonie." Mersenne, *Harmonie universelle*, "Livre premier des consonances," 87.

144 "Puisque le long exercice a coustume de rendre doux & facile ce qui sembloit auparavant rude & fascheux, je ne doute nullement que les intervalles dissonans, dont j'ay parlé dans cette proposition, à sçavoir les raisons de 7 à 6, & de 8 à 7, qui divisent la Quarte, ne puissent devenir agreables, si l'on s'accoustume à les oüir & à les endurer, & que l'on en use comme il faut dans les recits & dans les concerts, afin d'emouvoir les passions, & pour plusieurs effets, dont la Musique ordinaire est privée." Mersenne, *Harmonie universelle*, "Livre premier des consonances," 89.

145 "qu'il est difficile de distinguer & de reconnoistre la longue coustume d'avec la nature." Mersenne, *Harmonie universelle*, "Livre troisiesme des genres," 150.

146 "puisque la pratique y est conforme." Mersenne, *Harmonie universelle*, "Livre premier des consonances," 87.

147 Mersenne, *Harmonie universelle* "Livre premier des consonances," 82. On this, see Cohen, *Quantifying Music*, 105; Fix, "The Acoustical Paradox," 401.

148 "il n'y a nul nombre de mouvements ou de battements d'air, qui ne soient commensurables à tous autres nombres de mouvemens." Mersenne, *Harmonie universelle* "Livre premier des consonances," 86.

> Even if we have not found a reason that perfectly satisfies, we at least have a greater knowledge of sounds than of the objects of the other senses .... To which we can add that many have a dislike for the better meats and detest the better odors, but there are no humans who do not find the octave and the fifth agreeable. This shows that sounds approach closer to the mind and to that which is intellectual than the objects of the other senses.[149]

Ultimately, hearing was not like the senses of taste and seeing; there are limits to the role of individual differences within music. And since these proportions can be measured, "sound can bring more light to philosophy than any other quality."[150] According to the friar, no listener could truly enjoy a dissonant like a diminished fifth: "no human understanding can cause a diminished fifth ... to give pleasure," just as "nature has set limits to the sea that even the stormiest waves cannot surpass."[151] Mersenne acknowledged that listeners might prefer particular intervals or compositions, but certain axioms can still be established. In a similar way, Augustine had emphasized that the harmony between "single and the double" had been naturally implanted in humans so "that not even the illiterate can remain unaware it."[152] Any offense against this proportion is not only against the rules of music but "against our very sense of hearing," Augustine argued.[153] The human ear itself attests to the preference for harmony.

Sound Arguments

Although he allowed for contingencies and uncertainties, Mersenne looked for regularities and rules to observe regularities in nature. Order remained

149 "Encore que nous n'ayons pas rencontré une raison qui contente parfaitement, l'on a du moins une plus grande connoissance des sons que de l'objet des autres sens .... A quoy l'on peut ajoûter que plusieurs trouvent du degoust aux meilleures viandes, & haissent les meilleures odeurs, & qu'il n'y a point d'hommes qui ne trouvent l'Octave & la Quinte agreables: ce qui monstre que les sons approchent plus pres de l'esprit & de l'intellectuel, que l'objet des autres sens." Mersenne, *Harmonie universelle*, "Livre premier des consonances," 88.

150 "les sons peuvent apporter plus de lumiere à la Philosophie que nulle autre qualité." Mersenne, *Harmonie universelle*, "Livre premier des consonances," 88.

151 "comme la nature a mis des bornes à la mer que tous les flots les plus orageux ne peuvent outrepasser, aussi nul entendement humain ne peut faire qu'une fausse Quinte, c'est à dire moindre qu'elle ne doit estre, ou qu'une fausse Octave puisse donner du plaisir." Mersenne, *Harmonie universelle*, "Livre premier des consonances," 41.

152 Augustine, *The Trinity*, 134 (4.2).

153 Augustine, *The Trinity*, 134 (4.2).

"the source of science and disorder the origin of confusion, which no one enjoys."[154] Combined with sensory observations, mathematics could provide reliable statements on appearances. Mirroring the central role of geometry in the study of light, Mersenne introduced geometry to understand reflected sound. Mersenne followed the Italian Jesuit Giuseppe Biancani in comparing the movement of sound to that of light. "Sounds are reflected like light" when they encounter another body, he proclaimed.[155] And thus, Mersenne stated that the study of music should not only rely on arithmetic, geometry, and physics but also optics, catoptrics, and dioptrics to explain the phenomenon of the echo. At the same time, Mersenne's experimental nature made him concurrently note the difficulties with this idea. The movement of sound was subject to several other factors that may cause it to move differently, making the study of echoes more uncertain than the study of light: "it is not possible to regulate the echo as infallibly as the reflections of light."[156] As the friar concluded in his *Harmonie universelle*, more observations were necessary, and he hoped that others would properly establish a new science of echoes.[157] Whereas for some skeptics like Francisco Sanches, the echo was a sign that the senses are easily deceived, for Mersenne it was proof that regularities in nature do exist, even when the messy world of human experience does not follow mathematics precisely.

Although his description of psophologie in *Questions theologiques* suggests a greater role of observation and the ear, Mersenne's position on the ear was ambivalent. He continuously emphasized that he was interested in an "intellectual music" that would be preferred over any bodily experience. Throughout his career, he emphasized that empiricism could never be employed as the only method; it had to be accompanied by reason. At the same time, the ear was a necessary instrument and object of observation itself. Even when he directed the reader's attention to a cosmic listening, Mersenne investigated contemporary listeners and their capabilities. In this respect, he followed the classical tradition of harmonics, which similarly emphasized the need for reason and mathematics in the study of music. Sensory experience needed to be straightened by reason.

154 "l'ordre est la source de la science, & le desordre est l'origine de la confusion, à la quelle nul ne prend plaisir." Mersenne, *Harmonie universelle*, "Livre second, des chants," 177.

155 "les Sons se reflechissent comme la lumiere." Mersenne, *Harmonie universelle*, "Livre premier de la nature," 18. For Biancani's work and the interest in echoes more generally, see Van der Miesen, "Studying the Echo." On echo as an emblem of uncertainty, see Latour, *The Voice of Virtue*, 184–86.

156 "il n'est pas possible de regler les Echo aussi infailliblement que les reflexions de la lumiere." Mersenne, *Harmonie universelle*, "Livre premier de la nature," 48.

157 Mersenne, *Harmonie universelle*, "Livre premier de la nature," 50.

Mersenne continuously argued that listening was not just something related to the senses but something involving the mind. Since everything consists of motions and vibrations, the whole world could be explained by means of sound. Mersenne remained convinced that God is the author of this harmony and that all human efforts should ultimately direct their attention to the divine. This conviction was not separate from his interest in scientific experiments. For example, in his investigation of overtones, Mersenne continued to adhere to the idea that overtones should be consonants, rejecting observations suggesting otherwise. In his work on temperament, Mersenne remained committed to the arithmetical ratios of proportions, even when his own work questioned the validity of these proportions. And in his demonstrations of echoes, he grounded his work in geometry, since it offered an explanatory model for the regularities in nature, not because it was always true to experience.

Without an accepted acoustic standard—and Mersenne's efforts to count the motions of a string were not accepted by a wider community—experiments could not be compared. His science of psophologie had to wait until the afterlife, where all God's secrets would be revealed to humans. When Joseph Sauveur defined "acoustics" as the science of all sounds in 1700, many of these cosmic resonances had disappeared.[158] Influenced by Cartesian mechanics and Newtonian mathematics, Sauveur aimed to establish acoustics as the study of quantifiable sound, as distinct from the study of music. Whereas Mersenne thought that the secrets of harmony would only be fully revealed to humans in the afterlife, the next generation was less modest.

Bibliography

Manuscript sources

Oxford, Bodleian Library, MS Tanner 204. Edmund Chilmead. "An Examination."

Paris, Bibliothèque nationale de France, MS 2884. Marin Mersenne. "Livre de la nature des sons."

Printed primary sources

Alsted, Johannes Heinrich. *Encyclopaedia septem tomis distincta*. Herborn: n.p., 1630.

Aristotle. *On the Soul and Other Psychological Works*. Translated by Fred D. Miller, Jr. Oxford: Oxford University Press, 2018.

158 See the collection Sauveur, *Collected Writings on Musical Acoustics*.

Aristotle. *Posterior Analytics*. Translated by Jonathan Barnes. Oxford: Oxford University Press, 1993.

Augustine. *Confessions*. Translated by Henry Chadwick. Oxford: Oxford University Press, 2008.

Augustine. *The Trinity*. Translated by Stephen McKenna. Washington, DC: Catholic University of America Press, 1963.

Barker, Andrew. *Greek Musical Writings II: Harmonic and Acoustic Theory*. Vol. 2 of 2 vols. Cambridge: Cambridge University Press, 1989.

Bergier, Nicolas. *La musique speculative*. Edited and translated by Ekkehard Jost. Cologne: Arno Volk Verlag, 1970.

Bettini, Mario. *Apiaria universæ philosophiæ mathematicæ. In quibus paradoxa et nova pleraque machinamenta ad usus eximios traducta, & facillimis demonstrationibus confirmata. Tomus Secundus*. Vol. 2 of 2 vols. Bologna: Ferroni, 1642.

Binet, Étienne [Rene Francois]. *Essay des merveilles de nature, et des plus nobles artifices*. Rouen: Osmon, 1623.

Boethius, Anicius Manlius Severinus. *Fundamentals of Music*. Translated by Calvin M. Bower and edited by Claude V. Palisca. New Haven: Yale University Press, 1989.

Cassiodorus. *Institutions of Divine and Secular Learning and on the Soul*. Translated by James W. Halporn and Mark Vessey. Liverpool: Liverpool University Press, 2004.

De Caus, Salomon. *Institution harmonique*. Frankfurt: Jan Norton, 1615. Reprint, Geneva: Minkoff, 1980.

De Cousu, Antoine. *La musique universelle, contenant toute la pratique, et toute la theorie*. Paris: Ballard, 1658. Reprint, Geneva: Minkoff, 1972.

Descartes, René. *Musicae compendium/Leitfaden der Musik*. Edited by Johannes Brockt. Darmstadt: Wissenschaftliche Buchgesellschaft, 1992.

Digby, Kenelm. *Two Treatises in the One of Which the Nature of Bodies, in the Other, the Nature of Mans Soule Is Looked into in Way of Discovery of the Immortality of Reasonable Soules*. Paris: Blaizot, 1644.

Fludd, Robert. *The Temple of Music*. Translated by Peter Hauge. London/New York: Routledge, 2017.

Galilei, Vincenzo. *Dialogue on Ancient and Modern Music*, translated by Claude V. Palisca. New Haven/London: Yale University Press, 2003.

Galilei, Vincenzo. "A Special Discourse Concerning the Diversity of the Ratios of the Diapason." In *The Florentine Camerata: Documentary Studies and Translations*, edited and translated by Claude V. Palisca, 180–97. New Haven: Yale University Press, 1989.

Gassendi, Pierre. *Opera omnia*. 6 Vols. Lyon: Anisson, 1658.

Kepler, Johannes. *The Harmony of the World*. Translated by E. J. Aiton, A. M. Duncan, and J. V. Field. Philadelphia: American Philosophical Society, 1997.

Kircher, Athanasius. *Musurgia universalis sive ars magna consoni et dissoni in X libros digesta*. Rome: Corbelletti/Grignani, 1650.

Lippius, Johannes. *Synopsis musicae novae omnino verae atque methodociae universae*. Strasbourg: Ledertz, 1612. Reprint, Olms: Hildesheim, 2004.

Mersenne, Marin. *Correspondance du P. M. Mersenne*. Edited by Paul Tannery, Cornelis de Waard, Bernard Rochot, Armand Beaulieu, and René Pintard. Paris: Centre National de la Recherche Scientifique, et al., 1932–88.

Mersenne, Marin. *Harmonicorum libri XII*. Paris: Baudry, 1648. Reprint, Geneva: Minkoff, 1972.

Mersenne, Marin. *Harmonie universelle contenant la théorie et la pratique de la musique (Paris, 1636). Édition fascimilé de l'exemplaire conserve à la Bibliothèque des Arts et Métiers et annoté par l'auteur.* Introduction by François Lesure. Paris: Éditions du Centre National de la Recherche Scientifique, 1963.

Mersenne, Marin. *L'impiété des deistes, et des plus subtils libertins découverte, & refute par raisons de theologie, & de philosophie*. Paris: Billaine, 1624.

Mersenne, Marin. *Les preludes de l'harmonie universelle ou questions curieuses. Utiles aux predicateurs, aux theologiens, aux astrologues, aux medicins & aux philosophes*. Paris: Guenon, 1634.

Mersenne, Marin. *Quaestiones celeberrimae in Genesim, cum accurata textus explicatione: in hoc volumine athei, et deistae impugnantur & expugnantur, & vulgata edition ab hereticorum calumnijs vindicator. Graecorum, & Hebraeorum musica instauratur.* Paris: Cramoisy, 1623.

Mersenne. Marin. *Questions harmoniques. Dans lesquelles sont contenuës plusieurs choses remarquables pour la physique, pour la morale, et pour les autres sciences.* Paris: Villery, 1634.

Mersenne, Marin. *Questions inouyes, Questions harmoniques, Questions théologiques, Les méchaniques de Galilée, Les preludes de l'Harmonie universelle*. Paris: Fayard, 1985.

Mersenne, Marin. *Les questions theologiques, physiques, morales, et mathematiques. Où chacun trouvera du contentement ou de l'exercice*. Paris: Guenon, 1634.

Mersenne, Marin. *Traité de l'harmonie universelle*. Paris: Baudry, 1627.

Mersenne, Marin. *L'usage de la raison, texte revu par Claudio Buccolini*. Paris: Fayard, 2002.

Mersenne, Marin. *La verité des sciences, contre les sceptiques ou pyrrhoniens*. Paris: Du Bray, 1625.

Rivault, David de Flurance. *L'art d'embellir tiré du sens de ce sacré Pradoxe. Estendu en toute sorte de beauté, et és moyens de faire que le corps retire en effect son embellissement des belles qualitez de l'ame*. Paris: Bertault, 1608.

Sauveur, Joseph. *Collected Writings on Musical Acoustics (1700–1713)*. Edited by Rudolf Rasch. Utrecht: The Diapason Press, 1984.

Seville, Isidore of. *The Etymologies*. Translated and edited by Stephen A. Barney, W. J. Lewis, J. A. Beach, and Oliver Berghof. Cambridge: Cambridge University Press, 2006.

Stevin, Simon. "Vande Spiegheling der Singconst." In *Principal Works of Simon Stevin*. Vol. 5 of 5 vols., *Engineering, Music, Civic Life*, edited by Adriaan Daniel Fokker, 413–64. Amsterdam: Swets & Zeitlinger, 1966.

Suárez, Francesco. *Complectens tractatum secundum de opera sex dierum, ac tertium De anima*. Lyon: Boissat, 1635.

Tamizey de Larroque, Philippe. *Notice inédite sur le livre de raison du muet de Laincel d'après les manuscrits de Peiresc*. Digne: Chaspoul et Ve Barbaroux, 1895.

Taveau, Laurence. "Le manuscrit 2884 du P. Mersenne à la bibliothèque de l'Arsenal: Étude et édition critique." PhD diss., Université de Paris-Sorbonne, Laboratoire IREMUS, 2018.

Tyard, Pontus de. *Solitaire second*. Edited by Cathy M. Yandell. Geneva: Droz, 1980.

Van Nierop, Dirck Rembrandtsz. *Wis-konstige musyka: vertoonende de oorsaecke van 't geluyt*. Amsterdam: Goedesbergen, 1659.

Vanneo, Stephano. *Recanetum de musica aurea*. Rome: Apud Valerium Doricum Brixiensem, 1533.

Zarlino, Gioseffo. *Sopplimenti musicali del Rev. M. Gioseffo Zarlino da Chioggia*. Venice: Sanese, 1588.

Secondary literature

Barker, Andrew. *Scientific Method in Ptolemy's Harmonics*. Cambridge: Cambridge University Press, 2001.

Bertoloni Meli, Domenico. *Thinking with Objects: The Transformation of Mechanics in the Seventeenth Century*. Baltimore: The Johns Hopkins University Press, 2006.

Bianchi, Eric. "Scholars, Friends, Plagiarists: The Musician as Author in the Seventeenth Century." *Journal of the American Musicological Society* 70, no. 1 (2017): 61–128.

Brunet, Pierre. "Peiresc et le microscope." *Revue d'histoire des Sciences* 4, no. 2 (1951): 154–58.

Buzon, Frédéric de. "Harmonie et métaphysique: Mersenne face a Kepler." *Les Études philosophiques* 1, no. 2 (1994): 119–28.

Christensen, Thomas. "Mersenne and the Mechanics of Musical Proportion." In *Proportions: science, musique, peinture & architecture*, edited by Sabine Rommevaux, Philippe Vendrix, and Vasco Zara, 247–60. Turnhout: Brepols, 2011.

Clark, Suzannah, and Alexander Rehding. "Introduction." In *Music Theory and Natural Order from the Renaissance to the Early Twentieth Century*, edited by Suzannah Clark and Alexander Rehding, 1–13. Cambridge: Cambridge University Press, 2001.

Cohen, Albert. "Jehan Titelouze as Music Theorist." In *Festa Musicologica: Essays in Honor of George J. Buelow*, edited by Thomas J. Mathiesen and Benito V. Rivera, 391–406. Stuyvesant, NY: Pendragon Press, 1995.

Cohen, H. Floris. *Quantifying Music: The Science of Music at the First Stage of the Scientific Revolution, 1580–1650*. Dordrecht: Reidel, 1984.

Crombie, Alistair. "Music, Mathematics and Medical Science." In *Science, Optics and Music in Medieval and Early Modern Thought*, 363–77. London: Hambledon Press, 1990.

Crombie, Alistair. *Styles of Scientific Thinking in the European Tradition: The History of Argument and Explanation Especially in the Mathematical and Biomedical Sciences and Arts*. Vol. 2 of 2 vols. London: Duckwort, 1994.

Dandrey, Patrick. "La phoniscopie, c'est-à-dire la science de la voix." *Littératures classiques* 12 (1990): 13–76.

Dear, Peter. *Discipline and Experience: The Mathematical Way in the Scientific Revolution*. Chicago: University of Chicago Press, 1995.

Dear, Peter. *Mersenne and the Learning of the Schools*. Ithaca: Cornell University Press, 1988.

Erlmann, Veit. *Reason and Resonance: A History of Modern Aurality*. New York: Zone Books, 2010.

Fabbri, Natacha. "Questioning Fludd, Kepler and Galileo: Mersenne's Harmonious Universe." In *Space, Imagination, and the Cosmos from Antiquity to the Early Modern Period*, edited by Frederik A. Bakker, Delphine Bellis, and Carla Rita Palmerino, 201–32. Cham: Springer, 2018.

Feingold, Mordechai, and Penelope Gouk. "An Early Critique of Bacon's *Sylva Sylvarum*: Edmund Chilmead's Treatise on Sound." *Annals of Science* 40, no. 2 (1983), 139–57.

Fend, Michael. "The Changing Functions of 'Senso' and 'Ragione' in Italian Music Theory of the Late Sixteenth Century." In *The Second Sense: Studies in Hearing and Musical Judgement from Antiquity to the Seventeenth Century*, edited by Charles Burnett, Penelope Gouk, and Michael Fend, 199–221. London: Warburg Institute/University of London, 1991.

Fix, Adam. "The Acoustical Paradox: How Music Was Unmoored and Set Adrift from the Sciences in the Seventeenth Century." In *Music and Science from Leonardo to Galileo*, edited by Rudolf Rasch, 377–420. Turnhout: Brepols, 2022.

Godwin, Joscelyn. *The Harmony of the Spheres: The Pythagorean Tradition in Music*. Rochester: Inners Traditions, 1993.

Gominet-Brun, Julien. "Une curieuse invention du Père Mersenne. La psophologie et la musique du coeur." *Revue Bossuet* 7 (2016): 87–102.

Gouk, Penelope. "The Role of Harmonics in the Scientific Revolution." In *The Cambridge History of Western Music Theory*, edited by Thomas Christensen, 223–45. Cambridge: Cambridge University Press, 2002.

Heider, Daniel. "Suárez on Sound and Hearing." *Anuario Filosófico* 50, no. 2 (2017): 323–44.

Heilman, Anja. *Boethius' Musiktheorie und das Quadrivium: Eine Einführung in den neuplatonischen Hintergrund von "De institutione musica."* Göttingen: Vandenhoeck & Ruprecht, 2007.

Hicks, Andrew. *Composing the World: Harmony in the Medieval Platonic Cosmos.* Oxford: Oxford University Press, 2017.

Isherwood, Robert M. *Music in Service of the King: France in the Seventeenth Century.* Ithaca: Cornell University Press, 1973.

Jeserich, Philipp. *Musica Naturalis: Speculative Music Theory and Poetics, from Saint Augustine to the Late Middle Ages in France.* Baltimore: Johns Hopkins University Press, 2013.

Kvíčalová, Anna. "Hearing Difference in Calvin's Geneva: From Margins to Center." *Sixteenth Century Journal* 49, no. 1 (2018): 25–47.

Latour, Melinda. *The Voice of Virtue: Moral Song and the Practice of French Stoicism, 1574–1652.* Oxford: Oxford University Press, 2023.

Leinkauf, Thomas. "Kirchers Musikverständnis im Kontext der barocken Universalwissenschaft." In *Ars magna musices: Athanasius Kircher und die Universalität der Musik*, edited by Markus Engelhardt and Michael Heinemann, 1–21. Laaber: Laaber Verlag, 2007.

Lindley, Mark. "Mersenne on Keyboard Tuning." *Journal of Music Theory* 24, no. 2 (1980): 166–203.

Mackensen, Karsten. *Musik und die Ordnung der Dinge im ausgehende Mittelalter und in der Frühen Neuzeit.* Frankfurt: Peter Lang, 2017.

Mahoney, Michael Sean. *The Mathematical Career of Pierre de Fermat, 1601–1665.* 2nd ed. Princeton: Princeton University Press, 1994.

Mancosu, Paolo. *Philosophy of Mathematics and Mathematical Practice in the Seventeenth Century.* Oxford: Oxford University Press, 1999.

Martin, Craig. *Subverting Aristotle: Religion, History, and Philosophy in Early Modern Science.* Baltimore: Johns Hopkins University Press, 2014.

Moreno, Jairo. *Musical Representations, Subjects, and Objects: The Construction of Musical Thought in Zarlino, Descartes, Rameau and Weber.* Bloomington: Indiana University Press, 2004.

Oates, Rosamund. "Speaking in Hands: Early Modern Preaching and Signed Languages for the Deaf." *Past & Present* 256, no. 1 (2022): 49–85.

Oosterhoff, Richard J. *Making Mathematical Culture: University and Print in the Circle of Lefèvre d'Étaples.* Oxford: Oxford University Press, 2018.

Pasnau, Robert. "Sensible Qualities: The Case of Sounds." *Journal of the History of Philosophy* 38, no. 1 (2000): 27–40.

Psychoyou, Théodora. "Plaisirs de l'esprit, plaisirs de l'oreilles: anatomie et paradoxes d'un nouveau critère théorique." In *Le plaisir musical en France au XVIIe siècle*, edited by Thierry Fabrier and Manuel Couvreur, 61–78. Liège: Mardaga, 2006.

Psychoyou, Théodora. "Trajectories of Musical Acoustics in Seventeenth-Century French Theory and Experiment: Sound, Noise, and the Authorities of the Past." In *Music and Science from Leonardo to Galileo*, edited by Rudolf Rasch, 361–75. Turnhout: Brepols, 2022.

Rasch, Rudolf. "René Descartes and Isaac Beeckman: The Philosopher and the Schoolmaster." In *Music and Science from Leonardo to Galileo*, edited by Rudolf Rasch, 277–304. Turnhout: Brepols, 2022.

Rasch, Rudolf. "Simon Stevin and the Calculation of Equal Temperament." In *Music and Mathematics in Late Medieval and Early Modern Europe*, edited by Philippe Vendrix, 253–320. Turnhout: Brepols, 2008.

Rivera, Benito V. *German Music Theory in the Early 17th Century: The Treatises of Johannes Lippius*. Ann Arbor: UMI Research Press, 1980.

Ruellet, Aurélien. "'Voir cette merveille et la faire voir': Chains of Circulation and Recommendations around the Microscope in the Early 1620s." *Nuncius* (2024): 1–26.

Valleriani, Matteo. "Galileo's Abandoned Project on Acoustic Instruments at the Medici Court." *History of Science* 50, no. 1 (2012): 1–31.

Van der Miesen, Leendert. "Studying the Echo in the Early Modern Period: Between the Academy and the Natural World." *Sound Studies* 6, no. 2 (2020): 196–214.

Van Orden Kate. "Descartes on Musical Training and the Body." In *Music, Sensation, and Sensuality*, edited by Linda Phyllis Austern, 17–38. New York/London: Routledge, 2002.

Van Orden, Kate. *Music, Discipline, and Arms in Early Modern France*. Chicago: University of Chicago Press, 2005.

Varwig, Bettina. *Music in the Flesh: An Early Modern Musical Physiology*. Chicago: University of Chicago Press, 2023.

Vendrix, Philippe. "L'augustinisme musical en France au XVII siècle." *Revue de musicologie* 78, no. 2 (1992): 237–55.

Verwaal, Ruben E. "Fluid Deafness: Earwax and Hardness of Hearing in Early Modern Europe." *Medical History* 65, no. 4 (2021): 366–83.

Wald-Fuhrmann, Melanie. *Welterkenntnis aus Musik: Athanasius Kircher's "Musurgia Universalis" und die Universalwissenschaft im 17. Jahrhundert*. Kassel: Bärenreiter, 2006.

Wardhaugh, Benjamin. *Music, Science, Experiment in England 1653–1705*. Farnham: Ashgate, 2008.

Weber, Joseph G. "Pascal and Music: World Harmony in Early Seventeenth-Century France." *Symposium: A Quarterly Journal in Modern Literatures* 30, no. 1 (1976): 75–91.

Wymeersch, Brigitte van. *Descartes et l'évolution de l'ésthetique musicale*. Paris: Mardaga, 1990.

2. Instruments between Art and Nature

Abstract: This chapter considers the role of musical instruments throughout Marin Mersenne's work on universal harmony. Although his interest in instruments and their makers is often mentioned, this encounter has received little attention. This chapter investigates what information Mersenne hoped to find at the maker's workshop and how he went about gathering such knowledge. After this investigation of the social world of instruments makers, I consider their use as experimental objects. Looking at Mersenne's efforts to demonstrate the naturalness of the consonances, I show that it was ultimately not the monochord but the trumpet that took on a central role for Mersenne.

Keywords: musical instruments, artisans, practical knowledge, trumpet, mechanics

"No sooner had he [Francesco da Milano] moved the air by three pinches ... he transported all those who were listening to him into such gracious melancholy," wrote the sixteenth-century author Pontus de Tyard, describing the powers of the sixteenth-century lute player Francesco da Milano.[1] Tyard here describes a scene at a banquet where the feasting immediately stopped when the lute player began playing. Da Milano had such power that all his listeners were enraptured by some melancholic frenzy. All those at the banket, he continues, had "gaping mouths and eyes more than half-closed" as the listeners "remained deprived of all feeling, except for their ears."[2] This description was cited throughout the seventeenth century. Étienne

Parts of this chapter have been reused with permission from: Leendert van der Miesen, "The Trumpet as Nature's Voice: Marin Mersenne and the Nature of Music," in *Music and Science between Leonardo and Galileo*, ed. Rudolf Rasch, 305–33 (Turnhout: Brepols, 2022).

1 "Il n'eut esmeu l'air de trois pinçades ... il transporte tous ceux qui l'escoutoient en une si graeieuse melancolie." Tyard, *Solitaire second*, 192–93.

2 "d'une bouche entr'ouverte et des yeux plus qu'à demy desclose ... demeuroient privez de tout sentiment, ormis de l'ouye." Tyard, *Solitaire second*, 193.

Van der Miesen, Leendert. *Marin Mersenne and the Study of Harmony: From Sound to Music.* Taylor & Francis Group, 2025.
DOI 10.5117/9789048564132_CH02

Binet described the control a good player had over his audience as follows: "one makes the lute sound however one wants, and makes the listeners do whatever one wants."[3] Mersenne's correspondent Pierre Trichet also quoted Tyard's statements and wrote that "If these examples did not suffice, I could produce here a few more similar ones to show how much power the gentle control of the lute governed by the modes of music has over the minds of those who with attention undertake to listen to it."[4] In her study on Claudio Monteverdi, Bonnie Gordon concludes that by the turn of the seventeenth century, "instruments became tools in an experimental practice in which new sonorities and effects could manipulate the ear and express emotions."[5]

Musical instruments took on a greater role in musical discourse in the early modern period. They were found in many homes, cabinets of curiosities, paintings, and in the hands of experimental scientists. A wider public could increasingly afford musical instruments and instruction books, or even hire a private teacher. Books such as Nicolas Faret's *L'honneste homme* (Paris, 1630) advised courtiers and noblemen to play the lute. Instruments with rare materials such as ivory or gold were collected within cabinets, cherished for their looks or expensive materials.[6] Louis XIII began lessons in the lute in 1612 with Robert Ballard, and Cardinal Richelieu took lessons from court lutenist Ennemond Gautier.[7] Homes and salons were used for concerts of both vocal and instrumental music.[8] Many inventories of the period list spinets, lutes, and guitars. According to Mersenne, "the lute has taken such an ascendancy over the other chordal instruments" that one hardly observes any other instruments, but according to Pierre Trichet, the spinet was in fact the most used instrument in France in the 1630s.[9]

3 "On fait dire au Luth tout ce qu'on veut, & fait-on des Auditeurs tout ce qu'on veut." Binet, *Essay*, 465.

4 "Si ces examples ne suffisoint, j'en pourrois produire ici quelques autres semblables pour faire voir combien le doux maniment du luth régi par les modes de musique a de pouvoir sur l'esprit de ceux qui avec attention s'engagent à l'escouter." Trichet, *Traite des instruments de musique*, 156.

5 Gordon "Orfeo's Machines," 211.

6 Faret, *L'honneste-homme*, 27. On the affordability of instruments and amateur engagement, see Lesure, *Musique et musiciens fraçais*, 68; Zecher, *Sounding Objects*, 8–9. The collecting of instruments is more extensively dealt with in chapter 3.

7 Anthony, *French Baroque Music*, 283–95.

8 Composer Jacques de Gouy, for example, mentions several such concert locations in the introduction to his collection of *airs*, see De Gouy, *Airs à quatre parties*, n.p. ("Preface").

9 "le Luth a pris un tel ascendant sur les autres instrumens à chorde." Mersenne, *Harmonie universelle*, "Livre second des Instrumens a chordes," 56; Trichet, *Traité des instruments de musique*, 171.

Instruments also took on a larger role in musical and acoustic investigations. They were objects to study, manipulate, and measure. The physical properties of lutes, spinets, organ pipes, and bells were studied in the hopes of learning something about the propagation of sound and the nature of music. The different aesthetic, social, and scientific spheres surrounding instruments were closely related, as instruments traveled easily through different realms, and Rebecca Cypess has noted that "musical performance could easily transition to scientific performance."[10] Academic societies such as the Royal Society in London and the Académie Royale des Sciences in Paris later invited musicians and makers, and instruments remained a central object of study in the subsequent history of acoustics. Previously shunned within music-theoretical works, artificial instruments became important tools for investigating and demonstrating the nature of music and sound at large.

This transition from performance to experiment is also evident in Mersenne's letters. In a late letter to the Dutch diplomat and poet Constantijn Huygens from 1647, he specified the instructions for hearing the overtones that accompany each musical tone: "Because when you sit in great silence and with great attention in your cabinet, plucking [a viola da gamba] softly or gently, hand lifted from the bow, you will always hear an echoing sound that accompanies the natural, the upper twelfth and often a seventeenth."[11] A year before, he had written to Constantijn's seventeen-year-old son Christiaan: "I forgot to find out from him [his father] if you know how to play the lute. If that is the case, I beg you to see if you can solve this beautiful problem well."[12] He then asked him to explain the creation of consonances in string lengths. It was only logical, Mersenne implied, to also perform an experiment on the instrument. The nobleman Nicolas-Claude Fabri de Peiresc similarly detailed experiments with lutes and guitars that he performed with others in his cabinet.[13] Many scholars in Mersenne's circle owned musical instruments, such as the mathematician Claude Mydorge with his six lutes.[14]

10 Cypess, "Giovanni Battista Della Porta's Experiments," 165. See also Cypess, *Curious and Modern Inventions*.

11 "Car si dans un grand silence vous vous mettes en vostre Cabinet avec grande attention, touchant doucement ou faiblement, main levée par l'archet, vous entendrez tousjours un son à l'écho qui accompagne le natural, à la douzieme en haut, et souvent un autre à la disseptiesme." Mersenne to Constantijn Huygens, 12 January 1647, CM 15, 44. I abbreviate the *Correspondance du P. Marin Mersenne* with CM and the number of the volume.

12 "J'ay oublié à sçavoir de luy si vous sçavez toucher le luth. Si cela est, je vous prie de voir si vous soudrez bien ce beau probleme." Mersenne to Christiaan Huygens, 16 November 1646, CM 14, 613.

13 Peiresc to Mersenne, 18 June 1634, CM 4, 176.

14 Jurgens, *Documents minutier central*, 893. For more on the spaces of music making in seventeenth-century France, see Berrada, "Music at Home."

The interest among early modern scholars in musical instruments was part of a larger culture of instruments in the early modern period. Telescopes, microscopes, and air pumps fundamentally altered the relation between making and knowing, as knowledge on lens grinding or glass blowing became an important part of the natural philosopher's tool kit. Books such as Salomon de Caus' *Les raisons des forces mouvantes* (Frankfurt, 1615) staged a wide variety of machinery, including acoustic devices that imitated birdsong.[15] Microscopes and telescopes uncovered worlds previously invisible to the human eye and raised questions about sensory processes and whether the senses could be trusted. As Jean-François Gauvin has noted, the notion of the instrument itself changed, increasingly viewed as an object for unlimited discovery rather than a simple tool.[16] The wide presence of instruments was also deeply felt in fields traditionally dominated by texts. As Jessica Wolfe has shown, machines and machinery became things to think with in ethics, diplomacy, politics, and pedagogy. Instruments and machines mediated intellectual problems but also human's subjective experience and the external world.[17] However, as much research in the last decades has shown, the scientific discourse adopted artisanal experience with great difficulty at the time. As Pamela Smith observes, "Artisanal knowledge is inherently particularistic" and as such often at odds with the universal claims early modern scientists wanted to make.[18]

This chapter focuses on how Mersenne approached musical instruments and the people who made them. It investigates the role of artisanal knowledge in Mersenne's work and how he utilized instruments as tools for the production of evidence in his investigation of harmony. This dual investigation allows one to be attentive to both the social boundaries of early modern knowledge production and the epistemological role of instruments within scientific contexts. The so-called "books on instruments" take up a large part of his *Harmonie universelle* and *Harmonicorum* (see appendices A and B) and have often been studied for their information on historical instruments and performance practice. As Gauvin has shown in his study on Mersenne's approach to musical instruments, these parts can also be read programmatically, with musical instruments as "powerful symbols of the emergent mechanical philosophy."[19] In Gauvin's words, the Minim's

15 On such depictions, see especially Popplow, "Why Draw Pictures of Machines?"; Lazardzig, "Die Maschine als Spektakel."

16 Gauvin, "Instruments of Knowledge."

17 Wolfe, *Humanism, Machinery, and Renaissance Literature.*

18 Smith, *From Lived Experience*, 88.

19 Gauvin, *Instruments of Knowledge*, 53.

work on instruments "epitomized like none other in the first half of the seventeenth century the new mechanical natural philosophy, which was ultimately grounded on theory, experiments, and a thorough understanding of the mechanical arts."[20] For Gauvin the organ in particular takes on this central role. Building on his insights, this chapter turns to an instrument of a very different nature, namely the trumpet. In Mersenne's hands, the trumpet became a powerful tool to distinguish between what is given by art and what is given by nature. Ultimately, I show that it was not the voice or string that could demonstrate the nature of the musical scale for Mersenne but the trumpet. His investigation of embouchure (the use of lips, muscles, and tongue) and the physical properties of the trumpet eventually led him to proclaim this instrument as proof for the naturalness of the diatonic series. By empirically investigating the trumpet to find the natural proportions of music, he arrived at a novel angle from which to approach an old problem. The individual skill of the players themselves remained unresolved and was ultimately rejected from his investigation.

The question of whether artificial objects such as instruments could demonstrate something about the natural world was felt deeply, pointing to the larger issue of which aspects of music might be considered natural. Within the realm of mechanics and engineering throughout the early modern period, certain musical practices and theories were considered more natural than others. Suzannah Clark and Alexander Rehding write that "the question was not so much *whether* music was connected with nature but *how* it was connected with it."[21] The early modern period saw a wide variety of efforts to solve musical questions by means of legitimation in nature, as seen birdsong and newly discovered animals, or through encounters with musicians from outside Europe. Historians have considered the variety of meanings of nature during this period. In addition to philosophers like René Descartes and Francis Bacon, who challenged the distinction between natural and mechanical movement, artists, engineers, chemists, and collectors all played with the boundaries between art and nature.[22]

Numerous scholars in Mersenne's vicinity worked hard to redefine the relationship between artificial objects and nature at large. Claude Picot, for example, paraphrased René Descartes in his translation of *Principia Philosophiae* (Amsterdam, 1644), stating that "all things that are artificial are like those natural" and comparing the movement of a clock to a tree

20 Gauvin, *Instruments of Knowledge*, 54.

21 Clark and Rehding, *Music Theory and Natural Order*, 1.

22 Daston, "The Nature of Nature."

producing fruits.[23] In the Aristotelian tradition, the internal motion of a tree or plant could not be compared to that of a mechanical device, but early modern approaches increasingly viewed each as a similar type of motion. Musical instruments were part of these shifting configurations in the early modern period. Although the voice had long been valued as a theoretical object over musical instruments, in the early seventeenth century instruments gained a higher status among scholars and experimenters. Their idiomatic possibilities were explored in instrumental music, and musical authors praised the versatility and sound of instruments like the violin. Some authors went as far as favoring instruments over the voice, which according to Bonnie Gordon reflected "a new emphasis on the possibility of mechanical instruments to both represent and deconstruct nature."[24] Mersenne's work, I argue, is especially suitable for investigating how instruments became primary tools for the production of evidence, but it shows at the same time how their epistemic potential was curtailed in his quest to limit his conception of harmony to the first six whole numbers.

"An abject mechanics"

The treatises or "books" on instruments of *Harmonie universelle* take up almost a third of the printed work. The work includes an entire book on the monochord, four books on string instruments, a book on wind instruments, a book on the organ, and a book on percussion instruments. In his *Harmonicorum libri*, this number is reduced to four parts: string instruments, wind instruments, the organ, and bells. In these texts, Mersenne writes on the smallest details of a spinet, offers one of the most elaborate descriptions of making gut strings in the early modern period, and reports on the expert hearing of instrument makers and tuners. Mersenne was a close observer of performance practices, incorporating the playing techniques of several Parisian musicians into his writings. Musical instruments were central to his acoustic investigations. In *Harmonie universelle*, he writes, "It is impossible to explain the diversity of sounds unless one speaks of the variety of instruments that produce them."[25] The parts on instruments are in many

23 "toutes les choses qui sont articielles, sont avec cele naturelles." Picot, *Les principes de la philosophie*, 480.

24 Gordon, *Monteverdi's Unruly Women*, 190.

25 "Il est impossible d'expliquer la diversité de tous les sons, si quant & quant l'on ne parle de la diversité des Instruments, qui les produisent." Mersenne, *Harmonie universelle*, "Traité des instrumens a chordes," 1.

ways a continuation of his previous books, as they put into action and make concrete what had been discussed in more abstract terms before.[26]

Mersenne's intent to include instrument making in his investigations is already evident in his 1627 *Traité de l'harmonie universelle*. Although the experimental content of the *Traité* itself is limited, Mersenne does hint at the inclusion of experimental practices in the study of music. Music, he advocates, is not only part of arithmetic or geometry but also of physics. It is here that Mersenne introduces the importance of understanding materials and instruments:

> I would just like add that one needs to consider the nature of materials and their properties, because the building of musical instruments belongs to the musician, who should know the wood and the other materials that are used to make musical instruments, such as skins, guts, and metals that make up the organs, viols, citterns, harps, flutes, trumpets, oboes, and other instruments, if he wants to be a perfect musician, otherwise one will not have the pleasure nor honor that one could receive from this knowledge and would be deprived from the great utility one could derive from it.[27]

Mersenne argues that building instruments belongs to the discipline of music and it is therefore necessary to investigate the skins, guts, woods, and metals of which they are made. This would not only bring pleasure and honor to the musician but also utility. In the outline for a future work, Mersenne promises to include musical instruments with "their materials, their making, their temperaments, their tuning, and their tablatures."[28]

26 Mersenne writes that one can bind the separate parts in any order one wants but suggests that "those who prefer harmony over physics" ("qui preferent l'harmonie à la Physique") start with the book on consonances and end with the books on sound. Those who already understand physics and mechanics should start the other way around and end with the books on instruments. See Mersenne, *Harmonie universelle*, "Livre premier des consonances," n.p. ("Preface & advertissement").

27 "J'ajouste seulement *qu'elle considere la nature des corps, & leurs proprietez*, parce que la fabrique des instrumens de Musique appartient au Musicien, qui doit connoistre le bois & les autres matieres qui servent pour faire les instrumens de Musique, comme sont les peaux, les intestins & les meteaux dont on fait les orgues, les luths, les violes, les cistres, les harpes, les flûtes, les trompettes, les hauts-bois, & les autres instrumens, s'il veut estre parfait Musicien, autrement il n'aura pas le plaisir ny l'honneur qu'il peut recevoir de cette connoissance, & sera privé de la grande utilité qu'il en pourroit tirer." Mersenne, *Traité*, 10–11.

28 "leur matiere, leur fabrique, leurs temperamens, leur accords, leur tablatures." Mersenne, *Traité*, n.p. ("Sommaire des seize livres de la musique").

More than his predecessors, Mersenne turned to the instrument shop as a source of knowledge. He was convinced that knowledge of instruments and material would strengthen his case for the harmonious composition of the world. In the same *Traité*, he criticizes authors like the alchemist Robert Fludd for not relying on measurements.[29] Mersenne wanted to avoid any mistakes like those of the famous anecdote of Pythagoras at the smithy. According to this legend, often reproduced by early modern authors, Pythagoras discovered the reasons of the consonances by measuring the weights of hammers producing different tones.[30] No wonder, Mersenne comments in his *Questions harmoniques*, that skeptics and non-believers deny the true principles of music if such myths continue to be repeated.[31] Music theorists like Gioseffo Zarlino and Pietro Cerone, he argues, spread "fables instead of histories."[32] According to Mersenne, new and better measurements of all the different sound objects were necessary to establish the true foundations of musical science.

Writing about instruments was not always a given. According to Boethius, the mythical Pythagoras did not devote himself to musical instruments, which he considered inconsistent and uncertain. In Boethius's words, Pythagoras considered all instruments "unreliable" and granted them "a minimum of trust."[33] Instead, he turned to the most unmusical of instruments: the monochord, which was a box with strings stretched over the top. This suspicion of instruments percolated into many Renaissance philosophical and theoretical texts. In Boethius's retelling of the myth of Pythagoras, instruments play an ambivalent role, simultaneously distrusted and investigated. This ambivalence was still very present among early modern scholars. In one of the first printed books solely devoted to instruments of the sixteenth century, Martin Agricola wrote, "several people who have shamefully scorned the Instrumental [Music] and me on account of it might almost have discouraged me from my intended and useful writing."[34] Somewhat earlier, Sebastian Virdung wrote in his book on instruments that he would not include any descriptions of how to build instruments, as this belonged more to architecture and woodworking than to music.[35] The sources in Mersenne's vicinity similarly point to a world

29 Mersenne, *Traité*, 352.

30 For this anecdote and its afterlife, see especially Heller-Roazen, *The Fifth Hammer*.

31 Mersenne, *Questions harmoniques*, 166.

32 "puis qu'ils nous ont donné des fables pour des histoires." Mersenne, *Traité*, 447.

33 Boethius, *Fundamentals of Music*, 18.

34 Agricola, *The "Musica Instrumentalis deudsch,"* 64. Translation by William E. Hettrick.

35 Virdung, *Musica getutscht*, 123.

where class differences between artisans and scholars matter. Lissa Roberts and Simon Schaffer conclude the following in their work on thinking and making in the early modern period: "In a society of orders, the steeply graded hierarchy of head and hand was vital to defining persons and their social places."[36]

Mersenne firmly planted musical instruments and their construction within the discipline of music and subsequently in his project of universal harmony. If one wants to understand music and sound in general, one must understand the objects that produce it, he argued. In *Harmonie universelle*, Mersenne emphasizes that the parts on instruments, in addition to the rest of the book, should inspire instrumentalists to consider the divine. If that happens, he will have "used the time as usefully for these treatises [on instruments] as if I had done the same for theology."[37] By reserving such an important part for the potentially dishonorable subject of building musical instruments, Mersenne knew that he was possibly treading on dangerous territory. In his dedication letter to Monsieur de Refuge, a government official to whom Mersenne dedicated numerous books, he wrote that would not have dared to dedicate these books on instruments to Refuge if he was not aware of Psalm 150 and its meaning:

> I know that you do not follow the advice of some of the ancients who said that one lowers the sciences when one reduces them to practice and use .... It is true that I would perhaps not have dared to offer you these books on instruments, which some people believe to be an abject mechanics [*une mechanique trop abjecte*], if I had not known the importance by which you hold the Royal Prophet's exhortations, or rather of the Holy Spirit who inspired him with these words *Praise him with sound of trumpet: praise him with psaltery and harp*, and what follows, to entice all men to publish the praises of God with all kinds of instruments.[38]

36 Roberts and Schaffer, "Preface," xiii.

37 "je croiray avoir employé le temps aussi utilement à ces Traitez, comme si j'en eusse fait autant de la Theologie." Mersenne, *Harmonie universelle*, "Traité de instrumens de chordes," n.p. ("Preface au lecteur").

38 "Je sçay que vous ne suivez pas l'avis de quelques Anciens qui disoient que l'on profane les sciences lors qu'on les reduit à la Pratique et à l'usage .... Il est vray que je n'eusse peut-estre pas osé vous offrir ces livres des Instrumens, que quelques-uns croyent appartenir à une mechanique trop abjecte, si je n'eusse sçeu l'estat que vous faites des Exhortations du Prophete Royal, ou plustost du S. Esprit qui luy inspiroit ces paroles, LAUDATE EUM IN SONO TUBAE, LAUDATE EUM IN PSALTERIO ET CITHARA, & ce qui s'ensuit, pour advertir tous les hommes de publier les loüanges de Dieu avec toutes sortes d'instrumens." Mersenne, *Harmonie universelle*, "Traité des instrumens a chordes," n.p. ("A Monsieur de Refuge").

Mersenne was well aware that ancient philosophers like Aristotle dismissed craft knowledge, who connected a hierarchy of knowledge forms to social class. Traditionally, theory was valued over practice, mind over hand. Perhaps Mersenne was also thinking of Boethius's warning in his *De musica*: "Now one should bear in mind that every art and also every discipline considers reason inherently more honorable than a skill which is practiced by the hand and the labor of an artisan."[39] Refuge, Mersenne suggested, did not follow this view, and neither did he. The legitimization of these books came straight out of the Holy Spirit, as God needed to be praised with all possible instruments. To those who thought that these matters were beyond the remit of a theologian like Mersenne, he replied that he had "forty-four thousand saints as my guardians, who sing every day new hymns, and delightful songs in honor of the immaculate lamb."[40] He hoped that all those who could play instruments would henceforth only use them to play psalms and hymns. Whether Refuge had a special interest in mechanics is difficult to say, but Mersenne also dedicated his *Traité de l'harmonie universelle* and translation of Galileo's work on mechanics to the government official. In the latter, he wrote in his dedication that "mechanics can teach how to live well," for example "by imitating heavy bodies which always seek their center in that of the earth, just as man's spirit must seek its own in the divine essence."[41] Both the study of mechanics and the study of artisanal objects like instruments thus played a central role in Mersenne's study of universal harmony and was similarly theologically motivated. The science of mechanics could be used to teach how to live well while also offering the knowledge to build musical instruments that could be used to play and sing praises to God, as divine harmony was present in the mechanical principles themselves.[42]

Although Mersenne valued musical instruments, instrumental music still held an ambiguous position for him. Vocal music had traditionally been considered more excellent than instrumental music, and this division

39 Boethius, *Fundamentals of Music*, 50 (I.34).

40 "si je n'avois quarante & quatre mille Saints pour mes garans, qui chantent tous les jours de nouveaux Cantiques, & des Airs ravissans à l'honneur de l'Agneau immaculé." Mersenne, *Harmonie universelle*, "Traité des instrumens a chordes," n.p. ("A Monsieur de Refuge").

41 "les Mechaniques peuvent enseigner à bien vivre, soit en imitant les corps pesans qui cherchent tousjours leur centre dans celuy de la terre comme l'esprit de l'homme doit chercher le sien dans l'essence divine." Mersenne, *Les Mechaniques de Galilée*, n.p. ("A Monsieur de Reffuge"). Henri de Refuge also took part in the astronomical observations of Mersenne and Gassendi, see Gassendi to Luillier, 2 November 1632, CM 3, 342n4; Mersenne to Gassendi, 5 January 1633, CM 3, 356.

42 Dear, *Mersenne*, 160.

was deeply present in musical scholarship, which mainly focused on how to sing. Numerous authors thought that music obtained its powers through words. Pontus de Tyard wrote that "the intention of music appears to give melody to words so that all the listeners will be moved and will understand the affect of the poet."[43] Around the same time, the poet Pierre de Ronsard wrote, "Poetry without instruments, or without the grace of a single line or several voices, is not at all agreeable, just as are instruments without being animated by the melody of a pleasant voice."[44] Poetry should be sung, and music should be sung. Mersenne's sixteenth-century predecessors thus aimed to reform music through a renewed union of music and poetry. In Howard Mayer Brown's words, "without them music becomes mere sound."[45] This bias toward vocal music seems to have still been present in early seventeenth-century France as the French viol player André Maugars marveled at the appreciation of instrumental music in Italy. In 1639 he wrote, "You would not believe, Monsieur, the esteem that the Italians hold for those who excel on instruments, and how much more they value they assign to instrumental music than to vocal, saying that one man alone can produce more beautiful inventions than four voices together and that it has charms and liberties that vocal music does not have."[46] For many seventeenth-century French observers, Italy was much more advanced in the cultivation and appreciation of instrumental music.

One instrumental genre that was especially popular in the first half of the seventeenth century was the instrumental fantasy. Several volumes have survived by, for example, Eustache du Caurroy, Claude le Leune, Charles Guillet, Etienne Moulinié, and Louis Couperin.[47] The fantasy was seen as a more learned genre of instrumental music, as opposed to instrumental

43 "l'intention de Musique semble estre de donner tel air à la parole que tout escoutant se sente passionné, et se laisse tirer à l'affection du Poëte." Tyard, *Solitaire second*, 214. Translation from Brown, "*Ut musica poesis*," 5.

44 "car la Poësie sans les instrumens, ou sans la grace d'une seule, ou plusieurs voix n'est nullement aggreable, non plus que les instrumens sans estre animez, de la melodie d'une plaisante voix." Ronsard, *Abbrege l'art poëtique françois*, 4. Translation from Brown, "*Ut musica poesis*," 7.

45 Brown, "*Ut musica poesis*," 7. This did not mean that instruments were not important for Ronsard and his circle; see Zecher, *Sounding Objects*.

46 "Vous ne sçauriez croire, Monsieur, l'estime que les Italiens font de ceux qui excellent sur les Instrumens, & combien ils prisent plus la Musique Instrumentale que la Vocale, disans qu'un homme seul peut produire de plus belles inventions que quatre voix ensemble, & qu'elle a des charmes & des licenses que la vocale n'a pas." Maugars, *Reponse faite à un curieux*, 17. On the epistemological importance of musical instruments in Italy, see especially Cypess, *Curious and Modern Inventions*.

47 Cohen, "The *Fantaisie* for Instrumental Ensemble."

dances. Some considered it an intellectual exercise or experiment. According to Mersenne, the fantasy is "enriched with all the subtleties of this science ... when the musician takes the liberty of using whatever comes to mind, without expressing the passion of any words, this composition is called *fantaisie*, or *recherche*."[48] Du Caurroy's fantasies published in 1610 are perhaps most well known. Du Caurroy's fantasies are contrapuntal exercises, at times taking their themes from psalms, hymns, and popular songs. However, instrumental music was not without its goals, as Du Caurroy's nephew André Pitard writes in the preface: "They may be called true fantasies, not of a fanciful or bizarre musician, but of a sharp mind, having only heaven and the holy image of truth as its object.[49] For Pitard, even instrumental music—or to be more precise, especially Du Caurroy's instrumental music—had the divine as its object. Rather than the ignoble music one could hear being played poorly in a tavern, Du Caurroy's fantasies were highly esteemed and intellectual.

Like his predecessors, Mersenne did seem to consider vocal music more excellent than instrumental music. His discussions of musical perfection often occur in the context of religious songs. He did exhort his contemporaries to familiarize themselves with instruments, as it could help composers to test out harmonies without having to rely on singers.[50] Another nod to instrumental music is found in the frontispiece of *Harmonie universelle*, which depicts a figure with a harp surrounded by animals. This figure sits somewhere between the mythical Orpheus and the biblical David, as Mersenne quotes Psalm 150 beneath the image: "I will also praise thee with the psaltery, even thy truth, O my God: unto thee will I sing with the harp, O thou Holy One of Israel."[51] But even the lyre is described as an instrument for accompanying the voice.[52]

48 "une pleine Musique figuree, & enrichie de toutes les subtilitez de cette science. ... Et lors que le Musicien prend la liberté d'y employer tout ce qui luy vient dans l'esprit sans y exprimer la passion d'aucune parole, cette composition est appellee *Fantaisie*, ou *Recherche*." Mersenne, *Harmonie universelle*, "Livre second, des chants," 164. "Recherche" refers to the musical term *ricercar*, most often used for an instrumental composition that could be homophonic or imitative.

49 "on les pourra nommer vraies Fantasies, non d'un fantasque ou bizarre Musicien, mais d'un esprit vif, qui n'ayant que le ciel & la saincte image de verité pour object." Du Caurroy, *Fantasies*, n.p. ("A monseigneur").

50 Mersenne, *Harmonie universelle*, "Livre sixiesme de l'art de bien chanter," 363.

51 "Nam & ego confitebor tibi in vasis psalmi veritatem tuam: Deus psallam tibi in Cithara, sanctus Israel." Mersenne, *Harmonie universelle*, frontispiece. Translation from the King James Bible. As *Harmonie universelle* was issued with the names of three different printers, there are slight differences between some copies. The version with only Pierre Ballard's name on the title page includes Psalm 150 here. In his *Quaestiones*, Mersenne included a lengthy discussion of David's lyre in the biblical story of Saul's madness and subsequent healing by means of music; see Mersenne, *Quaestiones*, col. 1708.

52 Mersenne, *Harmonie universelle*, "Livre quatriesme des instrumens a chordes," 206.

The tension between instruments and the human voice is most extensively discussed in one of the first propositions of his first treatise on instruments in *Harmonie universelle*. Here Mersenne discusses the late sixteenth-century debate between lutenist and composer Vincenzo Galilei and music theorist and composer Gioseffo Zarlino. Mersenne sought to answer the question of "whether musical instruments were made in imitation of the voice or if the intervals were governed by those of instruments" and, drawing a conclusion from this, "whether art perfects nature, or nature perfects art."[53] This debate between Zarlino and his former student Galilei centered around the use of either the syntonic Ptolemaic tuning or just intonation but quickly turned into a discussion of what could be called natural and what artificial. Throughout his writings, Zarlino emphasized the primacy of nature. Music, he declared, depends "first on nature, then on art."[54] Without nature, there would be no consonances and no harmony. Nature was for Zarlino a philosophical idea and norm to which all things should aspire. And it was the human voice that demonstrated this norm. In contrast to instruments like the lute and viol, the voice was given to humans by nature and should act as the guiding principle in music. The voice gave forth the perfect forms of the intervals and consonances, such as 2:1 for the octave, 3:2 for the fifth, 4:3 for fourth, 5:4 for major third, and 6:5 for minor third. As such, it held a special epistemological power for Zarlino. And consequently, musical instruments were considered artificial objects that could only be corrected by following the principles of nature. Instruments, Zarlino argued, were always made in imitation of nature and could only be corrected by imitating nature. Just as painting would always try to approach nature, artificial musical instruments would always be imperfect compared to music rooted in nature.[55] "One cannot conclude properly from the things of art about those of nature," Zarlino wrote.[56] And the voice naturally tended toward Ptolemy's syntonic diatonic scale. The consonances created by

53 "Determiner si l'on a fait les Instrumens de Musique à l'imitation des voix, ou si l'on a reglé les intervalles des voix par ceux des Instrumens; & consequemment si l'Art peut perfectionner la Nature, ou si la Nature perfectionne l'Art: et s'il faut juger des choses artificielles par les naturelles." Mersenne, *Harmonie universelle*, "Traité des instrumens a chordes," 7. The importance of this debate for Mersenne has already been pointed out by Claude Palisca and Jean-François Gauvin: Palisca, "Mersenne pro Galilei contra Zarlino"; Gauvin, *Instruments of Knowledge*, 43–49.

54 "che la Musica dipende prima dalla Natura che dall'Arte." Zarlino, *Sopplimenti musicali*, 18 (I.3).

55 Zarlino, *Sopplimenti*, 23 (I.4).

56 "& che non si può concluder bene dalle cose dell'Arte in quelle della Natura." Zarlino, *Sopplimenti*, 23 (I.4). Translation from Goldberg, *Where Nature and Art Adjoin*, 331.

this scale were not just "products of art nor inventions of man" but were enclosed in nature itself.[57]

Vincenzo Galilei had studied with Zarlino in Venice but came to question his former teacher's adherence to just intonation. His *Discorso* is a point-by-point rebuttal of Zarlino's *Sopplimenti*. In this book, Galilei criticizes Zarlino's use of the terminology "artifice" and "nature." Galilei was keen to point out that a tempered scale was no less natural than just intonation. He showed the impracticality and impossibility of using the system as it was calculated on the monochord. Musicians mixed their tuning systems, and no musical practice could be fully encapsulated in such a musical system. Galilei's work cites numerous examples in which musicians depart from the rules that Zarlino set. Lutenists, for example, often follow the diatonic tuning of Aristoxenus. Players of wind instruments like flutes or cornetts adjust to the situation, as do singers. As Daniel Walker has pointed out, their biggest disagreement was not the fact that the tuning of instruments would always remain imperfect; rather, they clashed on what was considered natural and what artificial.[58] Whereas Zarlino would always see the human voice as an ontologically different sound object than an instrument, Galilei rejected such classifications. The syntonic diatonic, hailed by Zarlino as the natural norm for all music, was according to Galilei "a wholly artificial thing, so made by artifice of Ptolemy."[59] If the human voice possessed the perfection that Zarlino ascribed to it, this perfection, Galilei argued, was acquired through practice and was not naturally given. Moreover, Galilei saw no difference between a sound from an instrument and a sound from a voice since, as both were composed of the material of sound. He argued, "Sound and voice, as the material of these intervals, are natural, as much in instruments as in voices."[60] The sound of a flute and the sound of the human voice were both sound. A man-made object like an instrument did not necessarily have to imitate nature but rather had to fulfill the function for which the artisan designed it. Ultimately, he concluded, "one can conclude properly from the things of art about those of nature."[61]

57 "non sono cosa dell'Arte, ne invention dell'Huomo." Zarlino, *Sopplimenti*, 8 (I.1). Translation from Palisca, *Humanism*, 272.

58 Walker, *Studies in Musical Science*, 14–26; Fix, "Esperienza," 552.

59 "il sintono è un solo tutto artifiziale, fatto cosi dall'artifizio di Tolomeo." Galilei, *Discorso intorno*, 31. Translation from Goldberg, *Where Nature and Art Adjoin*, 286.

60 "il suono, & la voce, come materia di essi intervalli é naturale, tanto in questi quanto in quelli." Galilei, *Discorso intorno all'opera*, 94. Translation from Goldberg, *Where Nature and Art Adjoin*, 351.

61 'Et che si può concluder bene dalle cose dell'arte in quelle della natura." Galilei, *Discorso intorno*, 72. Translation from Goldberg, *Where Nature and Art Adjoin*, 331. This belonged to a

When Mersenne wrote the parts on musical instruments for his *Harmonie universelle*, Zarlino's and Galilei's disagreement was very much on his mind. Although he would rely greatly on Zarlino for his writing on counterpoint and composition, Mersenne sided with Galilei on matters of musical instruments. In his dedication, he echoes Galilei (and prefigures Descartes's *Principia philosophiae*) by saying that the properties of air must be considered both "in the mechanics of art and nature."[62] Like Galilei, Mersenne emphasized that each object has a different goal. A flute has the goal of making sounds, not imitating nature. And like Galilei, he doubted that the voice possesses the perfection Zarlino ascribed to it: "There is no doubt that the degrees and intervals of voice take after art." If the intervals of the voice were natural, all children would naturally intone the exact intervals without having received previous education. Mersenne found that this was "against experience."[63] Moreover, there were a variety of fields that already demonstrated the perfection of art over nature, such as painting, medicine, agriculture, and fashion. A surgeon who set a dislocated bone did not learn this from nature; a painter did not solely imitate nature but rather painted everything he could imagine. Like Galilei, Mersenne saw musical instruments as especially suitable objects to learn about the nature of music. Instruments were even more revealing than human singers, since no element of custom or training was thought to be involved. Accordingly, they became central objects in Mersenne's investigation of music, capable of revealing music's nature and universal harmony at large.

"All sorts of persons:" Mersenne at the Workshop

Mersenne's attention to artisanal practices was no exception at the time. Pierre Gassendi, for example, described Mersenne's patron Peiresc as especially attentive to mechanics and wrote that he hosted all artisans who stopped by his house.[64] The early seventeenth century saw multiple

broader movement of reevaluating notions of art and nature. Bernard Palissy, for example, similarly argued that one can only learn about nature through "art"; see Smith and Lores-Chavez, "Counterfeiting Materials, Imitating Nature."

62 "dans les Mechaniques tant de l'art que de la nature." Mersenne, *Harmonie universelle*, "Traité des instrumens a chordes," n.p. ("A Monsieur de Refuge").

63 "Or il n'y a nul doute que les degrez, & les intervalles des voix tiennent de l'Art ... ce qui est contre l'experience." Mersenne, Harmonie *universelle*, "Traité des instrumens a chordes," 8.

64 Gassendi described Peiresc as follows: "He was studious of Mechanicks, or Handi-Crafts; for which cause, there was never any famous Workman that went that way, but he entertained him

scholarly projects dealing with crafts and artisanal practices, such as the endeavors formulated by the English philosopher and statesman Francis Bacon. Throughout his works, Bacon argues that scholars should investigate the machines and tools made by artisans. In *The Advancement of Learning* (London, 1605), for example, he writes that "the use of history mechanical is of all others the most radical and fundamental towards natural philosophy."[65] In his *Sylva Sylvarum* (London, 1626), he suggests that one could learn from builders about the operation and workings of musical instruments.[66] The investigation of crafts and mechanical objects was seen as a key to the transformation of natural philosophy. As others have shown, there were numerous problems with such projects. Artisans were generally reluctant to reveal the methods on which their livelihood depended.[67] Protective of their traditions and techniques, they were accused of working against an open and public science. Although they were interested in mechanics, many aristocratic scholars despised the workshop and the physicality of crafts, distinguishing their own endeavors from the handwork they observed.

Mersenne similarly hoped that artisans would share their work for the benefit of science, such as the chemists and metallurgists who could share the "thousand pleasant observations that they encounter while working" with scholars so that they could draw something useful out of them.[68] He envisioned great progress if artisans were to help to collect data and share their knowledge on materials but located the fault for absence of this exchange in the artisans themselves.[69] Mersenne agreed with later natural philosophers like Robert Boyle that artisans were better off providing their knowledge to scholars rather than writing their own accounts. The men and women who worked with their hands could in turn elevate their work by relying on his theoretical principles. This, Mersenne wrote, "would give them access to honorable company and a better reputation."[70] At other times,

at his House, and learned of him many works or mysteries of his Craft." Gassendi, *The Mirrour of True Nobility*, 186. During his time in Paris, Peiresc made observations of artisans and filled his notebooks with recipes; see Miller, "Peiresc."

65 Bacon, *Advancement of Learning*, 71. Bacon made similar remarks in his *Novum Organum* (London, 1620): Bacon, *The New Organon*, 152.

66 Bacon, *Sylva Sylvarum*, 44–45 (§ 186).

67 Eamon, "Markets, Piazzas and Villages," 222. See also the work of Pamela Long: Long, *Openness, Secrecy, Authorship*.

68 "mille gentiles observations qu'ils rencontrent en travaillant." Mersenne, *Questions theologiques*, 15.

69 Mersenne, *Questions théologiques*, 183.

70 "qui leur donnera encore plus d'accez dans les honorables compagnies, & beaucoup plus de reputation." Mersenne, *Harmonie universelle*, "Traité des instrumens a chordes," n.p. ("Preface

he seemed troubled by the attitude of artisans. In a letter to Peiresc from 1635, he wrote, "Craftsmen [*gens mecaniques*] fear everything, it seems to them that we will take away their livelihood, even when we do not think about it."[71] If science was to benefit greatly from artisanal experience, the exchange would not be without its difficulties.

The instrument shop or workshop attracted many different people. The Parisian harpsichord maker and organist Jean Denis, encountered in the previous chapter, formulated it as follows: "Since I am an instrument maker, I am obliged to receive all sorts of persons in my shop. Some come to see and hear my products, others come to buy."[72] Denis did not seem too keen on all the visitors, many of whom lacked musical or performance skills. As a salesman, however, he was obliged to receive them all, buyers, and non-buyers. Denis's shop was located at the corner of Rue des Arcis and Rue Jean-Pain-Mollet. Rue des Arcis, in the merchant district on the right bank, was the center of instrument makers in the seventeenth century; its narrow streets housed many skilled artisans and was surely a regular destination for any musically gifted Parisian. Denis lived with his family above the shop. All his three sons would become instrument makers themselves, like Denis's father and grandfather. According to the rules of the guild, he could have one apprentice and an extra worker, most likely a journeyman or one of his sons. It seems that he exploited those limits, as the shop's inventory indicates a lively workshop, including many unfinished spinets, harpsichords, benches, tools, and strings.[73]

Mersenne lived not far from Denis's shop and may have approached the instrument maker for information. That the two men knew each other is suggested by the many textual references to one another; the harpsichord maker even adopted certain parts of Mersenne's writings for his own *Traité de l'accord de l'epinette* (Paris, 1650).[74] Mersenne in turn mentioned Denis's expertise as a builder and tuner in his *Cogitata physico-mathematica*, where he described him as an "ingeniosissimus *lyropoios*," a luthier who brings his instruments to perfection.[75] He especially praised Denis's delicate ear. A few

au lecteur").

71 "Les gens mecaniques craignent tous; il leur semble qu'on leur vault oster leur gaigne-pain, quoyque l'on n'y songe seulement pas." Mersenne to Peiresc, 2 February 1635, CM 5, 47.

72 Denis, *Treatise on Harpsichord Tuning*, 10.

73 The inventory is reproduced in Dufourcq, "Une dynastie française," 153–55. A partial translation can be found in Hubbard, *Three Centuries of Harpsichord Making*, 286–87 (appendix C).

74 See the introduction to Denis, *Treatise of Harpsichord Tuning*.

75 "qualis est Joannes Dionysius praedictus, quo nullus clavicymbala perfectius construit, & ad concetum adducit." Mersenne, *Cogitata physico-mathematica*, 335–36.

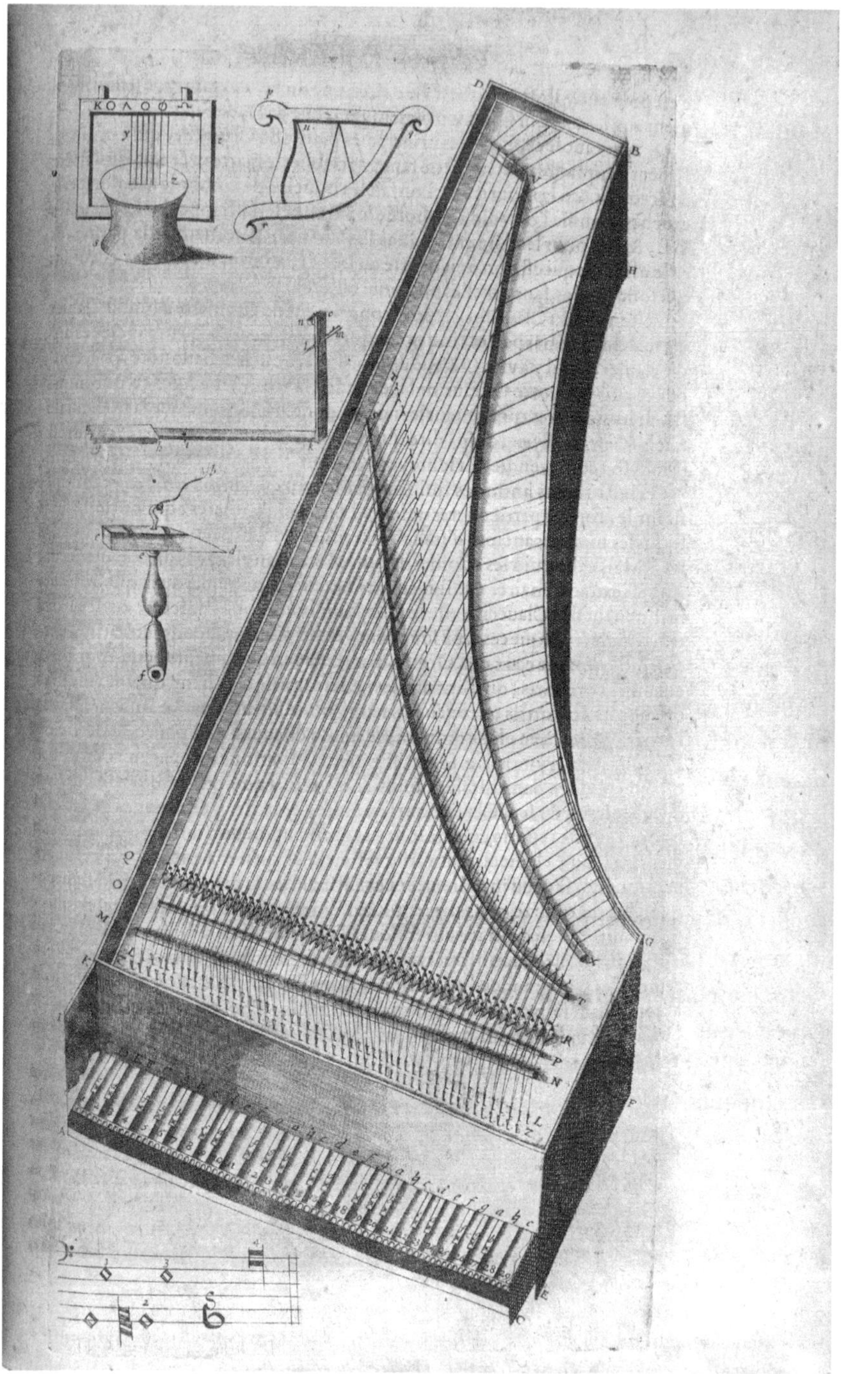

Figure 2.1: Depiction of a harpsichord. Mersenne, *Harmonie universelle*, "Livre troisiesme des instrumens a chordes," 111. Paris, Bibliothèque nationale de France, département Arsenal, FOL-S-1449.

years earlier in his *Harmonie universelle*, Mersenne had already mentioned a "Jean Denis" as one of the best instrument builders in France, but this might have been Denis's father, Jean Denis I. With all these biographical connections, it is likely that Denis was one of the builders whom Mersenne observed in preparation of his writings, but no further sources have survived. As is often the case with those working in Mersenne's vicinity, the closer the collaboration, the fewer sources seem to have survived.

As Mersenne himself observed, the harpsichord consists of countless parts, and he described the process of assembling the pegs, nails, strings, and mortises meticulously (fig. 2.1), suggesting that he had must have had some first-hand experience in workshops such as Denis's. He understood the details even in the choice of wood: beechwood, for example, was used for the soundboard, the sides were made from lighter woods such as linden or another white wood, and the table was made from cypress or cedar. This knowledge was essential, as a table made of oakwood would not sound. Mersenne described the process as a continuous balance between the resonance of the instrument, which requires delicateness, and solidity, which requires the instrument to be heavy and thick. These aspects of the building process could not be explained by reason but depended on experience that was only gained during the lengthy apprenticeship of instrument makers.

The sources that have survived are the long descriptions and images in *Harmonie universelle* and the references to artisans in Mersenne's letters. For example, he gave one of the first extensive descriptions of making gut string of the early modern period and seemed to have made special requests, asking for a string made of a single sheep gut and a very long string composed of a total of 144 guts.[76] The makers were able to make such as string by joining many guts together, which the makers "twist so well that they appear to be nothing but a single one."[77] He included a detailed account of what he encountered at the workshop, describing how the makers soak, clean, and scrape the gut. Not much is known about the men and women Mersenne observed, or how these contacts were established and mediated. The guild statutes that have survived contain hardly any information on their skills and social lives, instead detailing their formal organization.[78]

76 To compare, the largest strings of a theorbo required, according to Mersenne, was around fifty to sixty guts. However, this estimation is already a bit high according to present-day standards.

77 "qu'ils tordent si bien qu'ils ne paroissent que comme un seul." Mersenne, *Harmonie universelle*, "Traité des instrumens a chordes," 3.

78 For the guild statues of the Parisian string makers, see Haranger, *Articles, statuts, ordonnances*.

The instrument makers had their own guild starting in 1599.[79] The rules of the guild were strict: no one outside the guild was allowed to build or sell instruments; musical instruments that were built by anyone outside the guild could be confiscated; admission to the guild was only possible after the completion of a six-year apprenticeship in Paris; the sons of guild members could be excused from the apprenticeship if they proved themselves capable; widows could continue to operate a shop, but this right would cease when they remarried; no member of the guild was allowed to purchase wholesale materials from abroad without notifying the guild and offering the material to all the other members. Such rules were established to protect the position of the master builders and their investments.[80]

From the names mentioned in *Harmonie universelle* and the corresponding records in the Parisian archives, it is possible to identify some of the makers who moved in Mersenne's circles. For example, Mersenne mentioned the flute maker Thomas le Jeune Le Vacher several times, adopting his tablature for the flute and musette.[81] In the case of the flute, he included these examples to note the distances of the holes and to explain the fingering, i.e., which holes to close to get which pitch. In his list of makers, he also singled out Pierre Hubaut as the maker of a bowed clavier (*geigenwerk*) and the organ maker Valéran Héman, who worked on the organ for the Saint-Vivien Church in Rouen and the large organ at the Notre-Dame.[82] The fact that these conversations originated from Mersenne's engagement with makers becomes clear from the Latin *Harmonicorum*. In his Latin abridgement of the French *Harmonie universelle*, Mersenne goes out of his way to translate playing techniques and instrumental terminology in Latin, things "that can hardly be explained in Latin" in his own words, resorting to the French terminology used by builders.[83] In his preface, he writes that he did not want to alter the terms used by craftsmen "for fear of altering the practice of artisans."[84]

79 See the documents in Loubet de Sceaury, *Musiciens et facteurs*, 168–71. On Henri IV's approach to artisans, see Heller, *Labour, Science and Technology*, 184–91.

80 Hubbard, *Three Centuries of Harpsichord Making*, 85–86. On instrument makers in Paris, see Benoit, "L'apprentissage"; Hunt, "Jurors of the Guild"; Massip, "Facteurs d'instruments," Gétreau, "Entre l'oral et l'écrit." Whereas gut strings were imported from Italy in high numbers, international sale in instruments was rarer, see Massip, *La vie des musiciens*, 129.

81 Mersenne, *Harmonie universelle*, "Livre cinquiesme instrumens a vent," 233, 291.

82 Mersenne, *Harmonie universelle*, "Livre septiesme des instrumens de Percussion," 60; Mersenne, *Harmonie universelle*, "Livre sixiesme des orgues," 412.

83 "quae vix Latinem possunt explicari." Mersenne, *Harmonicorum*, "Liber primus de instrumentis," 13. For two such examples, see Mersenne, *Harmonicorum*, "De organis," 115, 123.

84 "de peur d'alterer la Pratique des artisans." Mersenne, *Harmonie universelle*, n.p. ("Preface au lecteur").

More information on contacts with practitioners comes from Mersenne's correspondence between 1623 and 1635. For example, Mersenne corresponded with a group of (amateur) scholars around the lawyer Robert Cornier and the organist Jean Titelouze in Rouen, urging them to visit instrument makers. He sent similar requests to his other correspondents, such as Claude Bredeau in Nevers, Christoph de Villiers in Sens, and Nicolas-Claude Fabri de Peiresc in Aix. But this endeavor was hardly a great success. The answers he received contain more annoyances than useful information. Giovanni Battista Doni, for example, complained that none of the harpsichord makers in Rome had even heard of a monochord.[85] Claude Bredeau wrote that none of the bell founders were willing to talk to him about their methods: "There is no master founder in this city who could tell us how much he reduces each bell for each tone."[86] Other letters displayed distrust toward artisans. Robert Cornier, for example, wrote to Mersenne that one needed to be present during the operations; with artisans, "there always remain some qualms about their certainty."[87] Peiresc complained of their secrecy, writing that there was a bell founder in his vicinity but that this maker protected his routines and rules and did not let anybody watch the process.[88] Christoph de Villiers wrote in several letters about an organ maker who repaired the organ at the Cathedral St. Estienne in Sens, describing him as "ignorant in theory, passable in practice."[89] Villiers later reported that the builder hardly knew how to tune an organ, spoke only of simple things, and was cruel toward his wife. Ultimately, he decided that the man was not worth Mersenne's time.[90]

A note among Peiresc's archival materials at the Bibliothèque Inguimbertine in Carpentras gives an example of what kinds of questions artisans were asked. The title reads: "If we can find some bellfounders in Aix or the surrounding area, and even if we could acquire some addresses in Italy, Venice, Spain, the Netherlands, and Germany, the following should be

85 Doni to Mersenne, 8 April, CM 4, 88–89.

86 "Il n'y a aucun maistre fondeur en cette ville, duquel on puisse sçavoir de combien il amoindrit chaque cloche pour chaque ton." Bredeau to Mersenne, 13 July 1628, CM 2, 99.

87 "il demeure tousjours quelque scrupule de leur certitude." Cornier to Mersenne, 16 March 1626, CM 1, 417.

88 Peiresc to Mersenne, 6 September 1633, CM 3, 475.

89 "ignorant dans la theorie, passable en la pratique." Villiers to Mersenne, 25 1635, CM 5, 53–54.

90 Villiers to Mersenne, 3 September 1635, CM 5, 382. Doni and Mersenne also corresponded about possible artisans that Mersenne could look up, see Doni to Mersenne, 8 November 1634, CM 4, 389–90; Mersenne to Doni, 2 February 1635, CM 5, 34–35; Doni to Mersenne, 8 September 1635, CM 5, 388.

known regarding to bells."[91] The list includes six questions regarding bell making, asking about the size, weight, proportion, mixture, and pitches of bells. The note even includes an ascending scale and asks for the width, height, and thickness of eight bells for these different pitches. Mersenne had inquired with Peiresc in 1634 on similar topics.[92] In light of the detailed questions regarding the measurements of a bell's shape and a request for a copy of their *brochette* (the *brochette* or *echele campanaire* was a wooden or bronze sphere indicating the proportions of bells of different weights), one can understand why artisans where hesitant to allow such visitors to witness the founding process.

Such sources indicate which questions were of interest to Mersenne but only provide the perspective of the musical scholar. There are no sources on the experience of the artisans questioned by Mersenne and his collaborators. Jean Denis's own treatise on harpsichord tuning does contain some references to the builder's perspective. The harpsichord maker mentioned musical gatherings in Paris where different tunings were presented, such as a tuning by the mathematician Jean Gallé, who became part of Mersenne's circle around 1635. Mersenne must have been present at these meetings or even organized them himself, since he also wrote on Gallé's method of dividing the octave into twelve equal parts, presented at one of these meetings.[93] Gallé's division assumes a 100,000,000,000 to 50,000,000,000 division of the octave, a number that did not have much practical use for tuners. When present at such musical gatherings, Denis found that he was ignored by others because he was an artisan: "When I have found myself in gatherings, people have not been inclined to listen to me, for I am considered a simple artisan."[94] Emphasizing his own experience as a master builder, Denis wrote that Gallé had only found his tuning "through numbers," not experience.[95] Denis's comments suggest that artisans were assigned a lower social position and that their testimony was considered less worthy because of this.

91 "Si on peut trouver quelque fondeur de cloches a Aix ou aux environs: & mesme si on pouvoit avoir quelque adresse en Italie, a Venise, en Hespagne, ou au Pays Bas & en Allemagne. Il faudroit sçavoir ce qui suit pour les cloches." Carpentras, Bibliothèque Inguimbertine, MS 1774, fol. 454r. Transcription from Scherpereel, "Peiresc et la musique," 162–63.

92 Mersenne to Peiresc, 14 May 1634, CM 4, 135.

93 Mersenne, *Harmonie universelle*, "Nouvelles observations," 21; Rasch, "Simon Stevin," 296.

94 Denis, *Treatise on Harpsichord Tuning*, 70.

95 Denis, *Treatise on Harpsichord Tuning*, 63. Mersenne himself wrote that Gallé told him that he could tune an organ without the ear. Mersenne, *Harmonie universelle*, "Livre sixiesme des orgues," n.p. ("Preface au lecteur").

Rationalizing Instruments

Throughout his work, Mersenne centered materials and artisanal knowledge. Commenting on the limits of human understanding, he wrote, "One is forced to admit that humans are not capable of knowing the reason for anything, other than that which they can make."[96] Only what could be made could be known. It seems that a special category was reserved for artisans, instrument makers, and those skilled in the crafts. In his *La verité*, he writes, "It is certain that the artisan must have an idea in order to carry out his work, otherwise it would never come about."[97] By making an object, artisans necessarily understand the constitution and materials of an artifact such as a musical instrument Mersenne saw this knowledge as a possible source of information for the musical scientist. Builders should share their knowledge of materials, and they would receive theoretical explanations and more prestige in return.

In his investigation of instruments and their makers, Mersenne was committed to finding perfect ratios, demonstrating Augustine's influence once again. Like Augustine, he maintained the idea that "reason is nothing other than number."[98] Reason and numbers were common to all things: they are universal. In his own writings, Augustine emphasized that the world of numbers is far removed from the bodily senses, but "all physical objects have their own numbers."[99] When humans judge on the properties of physical objects, they recognize the numbers that have also been implanted in themselves. Augustine specifically singled out the crafts:

> Craftsmen, who fashion all bodily forms, have numbers in their craft that they apply to their works. They use their hands and tools in designing until what is formed externally achieves its consummation when it conforms as much as possible to the inward light of numbers and, using sense as the go-between, it pleases the internal judge who looks upon the numbers above.[100]

96 "l'on trouve qu'ils ne sçavent rien, et l'on est constraint d'avoüer que l'homme n'est pas capable de sçavoir la raison d'autre chose que de ce qu'il peut faire." Mersenne, *Questions théologiques*, 111.

97 "il est certain qu'il faut que l'artisan ait une idee pour effectuer son ouvrage, autrement il n'en viendroit jamais about." Mersenne, *La verité*, 14. On this quote, see also Dear, *Mersenne*, 42. For this idea in relation to early modern skepticism, see Paganini, "Gassendi's Interplay."

98 Dear, *Mersenne*, 104.

99 Augustine, *On the Free Choice of the Will*, 55 (2.11.30.124).

100 Augustine, *On the Free Choice of the Will*, 63 (2.16.42.165).

According to Augustine, the hands of artisans during the process of work "are also full of numbers."[101] This idea also returned in Mersenne's writings. Solidifying music as the science of harmonies expressible in number, the scholar hoped to find these numbers at the workshop, but as Mersenne would soon learn, the messy world of making does not always comport with preconceived notion of harmonious numbers.[102]

This becomes clear from Mersenne's attention to wind instruments and bells. In contrast to a string instrument or monochord, the proportions of wind instruments are not immediately visible. Whereas the shortening of a string makes the immediate effect on its pitch comprehensible, wind instruments are an entirely different category of sound bodies. The same body of a flute could produce different pitches through air pressure, embouchure, and pressing any holes. The idea of an air column was not yet understood. In his *Harmonics*, Ptolemy advised the musical scholar not to rely on wind instruments in constructing the harmonic canon, since they were irregular, and it was not possible to find the same ratios as in string instruments.[103] Later texts proclaimed exactly this. In a famous image from Franchinus Gaffurius's *Theorica musicae* (Milan, 1492), for example, Pythagoras and Jubal are depicted measuring bells, glasses of water, pipes, and strings. In Gaffurius's description, the same numbers of the string experiment (2:4:8:12:16) are also used for other sounding bodies such as flutes, falsely connecting the two different acoustic procedures by Pythagorean numbers.[104]

Mersenne, seemingly unbothered by Ptolemy's advice, searched for the proportions in a wide variety of wind instruments, hoping to find a similar correlation between length and pitch, as in his work on strings. He knew that the placement of holes in wind instruments did not follow the proportions of strings, but he was ultimately unable to find a unifying theory.[105] Organ pipes posed a similar challenge. Throughout his correspondence, Mersenne urged his correspondents to visit and question organ makers regarding their rules for determining pitches. From his correspondent Robert Cornier, he received numerous letters anticipating the measurements of an

101 Augustine, *On the Free Choice of the Will*, 63 (2.16.42.166).

102 The continued emphasis on whole numbers had led Brandon Konoval to speak of a "Pythagorean disciplining of experience" in early modern musical science: Konoval, "Pythagorean Pipe Dreams," 46.

103 Ptolemy, *Harmonics*, 1.8, in Barker, *Greek Musical Writings*, 291.

104 Gaffurius, *Theorica musicae*, [18].

105 "Il est certain que les trous de ces Instrumens ne gardent pas la proportion des intervalles de Musique, & que nul n'a trouvé la theorie des distances que doivent garder lesdits trous entr'eux." Mersenne, *Harmonie universelle*, "Nouvelles observations," 18.

organ maker.[106] With the help of the engineer Cornu, Mersenne collected information on the cubing of pipes.[107] In Paris he observed organ builders, writing down description of the process of casting and flattening lead to make pipes. It seems he was aware of all the different parameters involved in making organs and hoped to find the ratios guiding this process. Not all his correspondents shared this belief. Christoph de Villiers, for example, doubted that any correspondence between size, shape, width, and sound could be found.[108]

The relation between the volume of pipes and their pitches comes up several times in Mersenne's writings. In the twelfth proposition in his treatise on the organ, he asks whether it is possible to make an organ with pipes of all the same length, only with differences in width. Here he recounts that he ordered several pipes to be made with the same length, with diameters proportioned with the numbers 1, 2, 4, 8, and 16. When sounding these pipes with the diameter doubled, the pitch only descended a minor third. This demonstrated to Mersenne that the proportions of strings could not be transferred to pipes.[109] He also experimented with pipes of the same width and with different lengths. But again, Mersenne struggled to come to any conclusive finding. His experimental results did not allow him to formulate a ratio of lengths as he could for strings.[110] Unable to find such a rule himself, Mersenne suggested that makers prepare catalogues of all the different sizes of pipes that bring forth the notes of an organ, but this remained an unfinished project.[111]

In a similar way, Mersenne and his collaborators turned to the investigation of bells, hoping that some rules about the relation between size, proportion, and sound could be discovered. The musical qualities of early modern bells were limited. Until the early seventeenth century, most bells were cast as ringing bells that did not require any elaborate musical function. With the rise of polyphonic musical functions for bells, such as the carillon, founders were more motivated to alter the internal structure of the bell and as

106 Cornier to Mersenne, 23 November 1625, CM 1, 310; Cornier to Mersenne, 16 March 1626, CM 1, 416–17; Cornier to Mersenne, 22 March 1626, CM 1, 427–28.

107 Mersenne, *Harmonie universelle*, "Livre sixiesme des orgues," 335.

108 Villiers to Mersenne, 15 May 1635, CM 5, 193.

109 Mersenne, *Harmonie universelle*, "Livre sixiesme des orgues," 331.

110 Mersenne, *Harmonie universelle*, "Livre sixiesme des orgues," 333. On these efforts, see also Dostrovsky, "Early Vibration Theory," 191.

111 Mersenne, *Harmonie universelle*, "Livre sixiesme des orgues," 346–47. For more on Mersenne's attention to organ pipes, see Konoval, "Pythagorean Pipe Dreams"; Gauvin, *Instruments of Knowledge*, 73–84.

such "tune" the overtones of bells.[112] With each bell different in size, metal mixture, and sound, Mersenne hoped that his questions would lead to an inductive investigation of measuring and observing bells to find an underlying principle. Although he included valuable information regarding the process of founding, Mersenne found that the makers did not have the knowledge he was looking for. When makers went outside their brochette, Mersenne argued, they relied on guessing. He maintained that founders did not have the "certain and infallible science" he presented in his book.[113] In his personal copy of *Harmonie universelle*, Mersenne included a more detailed drawing of a bell's shape (fig. 2.2). Within the history of campanology, this counts as a significant step up from Mersenne's image of the shape of a bell in the printed version of his *Harmonie universelle*, and it has been suggested that this was not his own construction but rather one obtained from a practitioner.[114] Bell founders, however, would not have searched for a perfect geometrical model but would manually adjust the shape of the bell. Mersenne himself remained unable to explain the irregular harmonics of bells, nor was he able to find out how the shape, weight, and thickness of a bell determines its pitch.

Like the other parts of *Harmonie universelle*, the books on instruments are filled with extensive tables. The table was an especially beloved method for Mersenne to communicate with his readers and were used to gather experimental results or quantify measurements. He suggested that practitioners use this information to standardize their craft and make their judgments more solid. In one table, for example, Mersenne gave the size and lengths of all the strings of the spinet. The first column contains the number of the string and the second the proportion that the strings have among one another (thus 10:9 for a major second, or 16:15 for a semitone). The rest of the rows contain the length, thickness, and diameter of the strings, coming to

112 The most famous example of this practice is given by the brothers Francois and Pieter Hemony in the Low Countries. Together with carillon player Jacob van Eyck, they improved bell founding by making the bell too thick in the beginning and removing the necessary amount by using tuning staves and sand to observe the overtones. The Jesuit Athanasius Kircher tried to gather more information by corresponding with one of the famous Hemony brothers. The letter is printed in Gaspar Schott's *Magiac universalis*: Schott, *Magiae universalis*, 357–60. Descartes would also write to Mersenne on Van Eyck, but it was too late to still include this in *Harmonie universelle*, see Descartes to Mersenne, 23 August 1638, CM 8, 57.

113 "parce qu'ils n'en sçavent pas la fabrique par une science certaine & infaillible que j'explique icy." Mersenne, *Harmonie universelle*, "Livre septiesme des instrumens de percussion," 11. The Dutch scientist Christiaan Huygens would later write that founders double the diameter of a bell in order to obtain an octave, but argued that this should not be taken as a theoretical principle. Huygens, *Oeuvres complètes*, XIII, 804.

114 Lehr, "Middeleeuwse klokkengietkunst," 97–101.

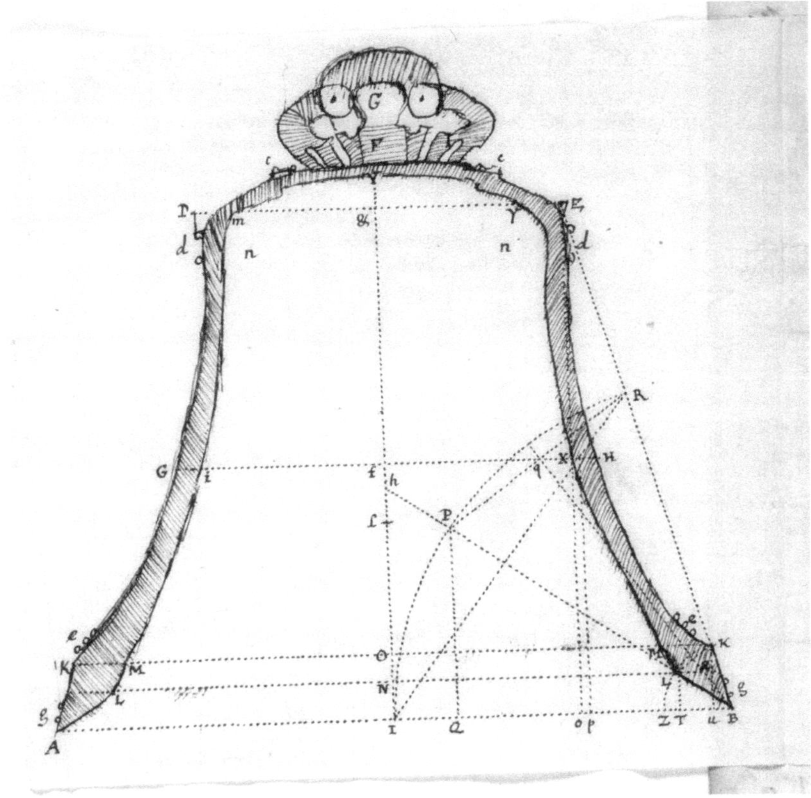

Figure 2.2: Drawing of a bell. Paris, Bibliothèque nationale de France, MS fr. 12357, fol. 16v.

an overwhelming result. Mersenne later provides a table of how many air vibrations each string of a spinet makes, trying to find another use for his method to count the vibrations of air and suggesting its use for practitioners.

One of Mersenne's most remarkable efforts to rationalize crafts is his so-called "tablature for the deaf" (fig. 2.4). With this table, one could tune a lute, viol, or spinet without hearing, provided one knows the material, the length, and thickness of the string. In the table and accompanying text, Mersenne presents nine hypothetical situations of two strings, each differing in length, thickness, or tension. The last rule deals with differences in material. The rules and corrections are "true and infallible," in Mersenne's words.[115] The first rule states that when two strings are equal in length and thickness, the necessary tension required to make an octave would be

115 "veritables & infaillibles." Mersenne, *Harmonie universelle*, "Livre troisiesme des instrumens a chordes," 123–26, here 123. See also Dostrovosky, "Early Vibration Theory," 187; Dear, *Mersenne*, 159–60. For some of the other occurrences, see chapter 1.

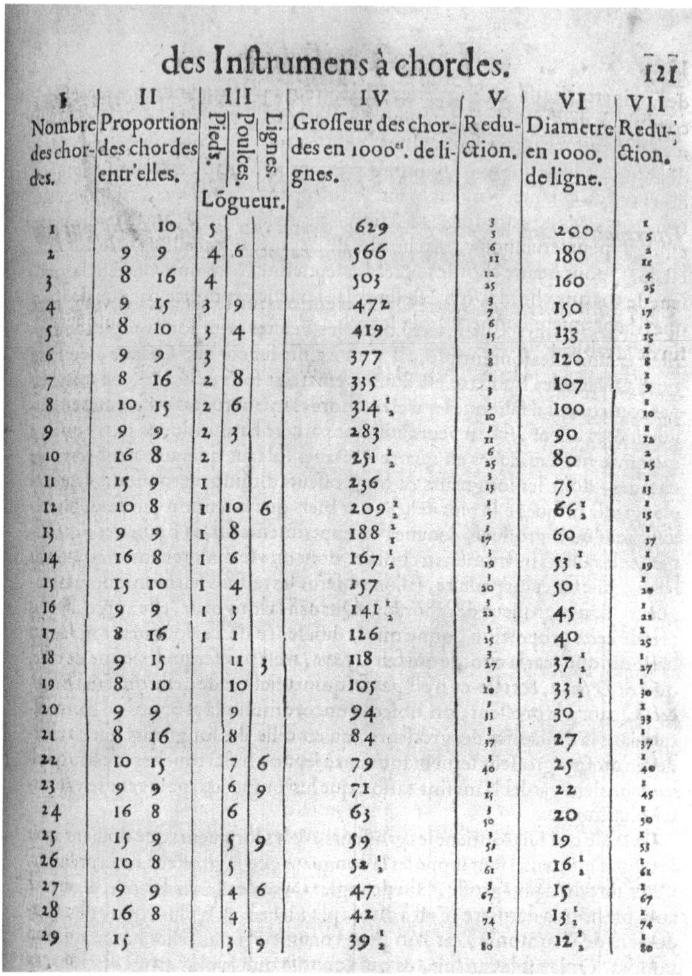

des Instrumens à chordes. 121

I Nombre des chordes.	II Proportion des chordes entr'elles.	III Lõgueur. Pieds.	Poulces.	Lignes.	IV Grosseur des chordes en 1000es. de lignes.	V Reduction.	VI Diametre en 1000. de ligne.	VII Reduction.
1	10	5			629	3/5	200	1/5
2	9 9	4	6		566	6/11	180	2/11
3	8 16	4			503	12/25	160	4/25
4	9 15	3	9		472	3/7	150	1/17
5	8 10	3	4		419	6/15	133	2/15
6	9 9	3			377	3/8	120	1/8
7	8 16	2	8		335	1/3	107	1/9
8	10 15	2	6		314 1/2	3/10	100	1/10
9	9 9	2	3		283	3/11	90	1/12
10	16 8	2			251 1/2	6/25	80	2/25
11	15 9	1			236	3/13	75	1/13
12	10 8	1	10	6	209 1/2	1/5	66 1/2	1/15
13	9 9	1	8		188 1/2	3/17	60	1/17
14	16 8	1	6		167 1/2	3/19	53 1/2	1/19
15	15 10	1	4		157	3/20	50	1/20
16	9 9	1	3		141 1/2	3/22	45	1/22
17	8 16	1	1		126	3/25	40	1/25
18	9 15		11	3	118	3/27	37 1/2	1/27
19	8 10		10		105	1/10	33 1/2	1/30
20	9 9		9		94	3/33	30	1/33
21	8 16		8		84	3/37	27	1/37
22	10 15		7	6	79	3/40	25	1/40
23	9 9		6	9	71	1/15	22	1/45
24	16 8		6		63	3/50	20	1/50
25	15 9		5	9	59	3/51	19	1/53
26	10 8		5		52 1/2	1/61	16 1/2	1/61
27	9 9		4	6	47	3/67	15	1/67
28	16 8		4		42	3/74	13 1/2	1/74
29	15		3	9	39 1/2	3/80	12 1/2	1/80

Figure 2.3: Table with the dimensions of the strings of a spinet. Mersenne, *Harmonie universelle*, "Livre troisiesme des instrumens a chordes," 121. Paris, Bibliothèque nationale de France, département Arsenal, FOL-S-1449.

4:1 (in contrast to what is stated in the depiction of Gaffurius mentioned before). The second rule is immediately a correction: in practice, one often needed a proportion of 4¼:1, Mersenne found. This might be due to the friction of the pulleys of the monochord that was used. As Peter Dear has argued, Mersenne divided his findings in two separate rules to maintain the perfect proportion of 4:1. By giving two different rules, Mersenne was able to keep the proportions of whole numbers that he was looking for.[116]

116 Dear, *Mersenne*, 159–60.

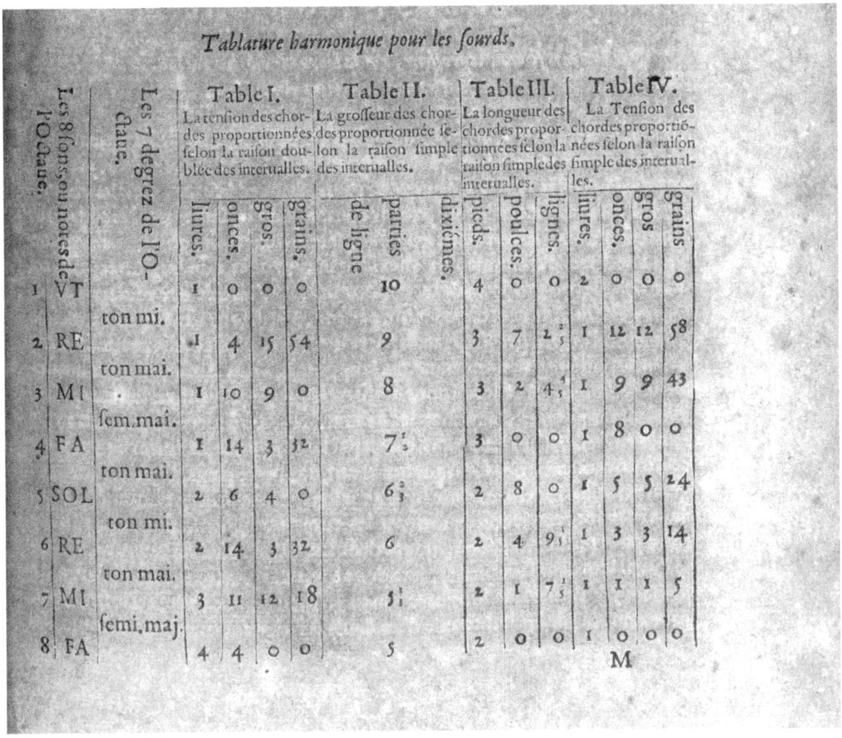

Tablature harmonique pour les sourds.

Les 8 sons, ou notes de l'Octaue.		Les 7 degrez de l'Octaue.	Table I. La tension des chordes proportionnées selon la raison doublée des interualles. livres.	onces.	gros.	grains.	Table II. La grosseur des chordes proportionnée selon la raison simple des interualles. parties de ligne	dixiémes.	Table III. La longueur des chordes proportionnées selon la raison simple des interualles. pieds.	poulces.	lignes.	Table IV. La Tension des chordes proportionées selon la raison simple des interualles. livres.	onces.	gros	grains
1	VT		1	0	0	0	10		4	0	0	2	0	0	0
		ton mi.													
2	RE		1	4	15	54	9		3	7	2 2/5	1	12	12	58
		ton mai.													
3	MI		1	10	9	0	8		3	2	4 4/5	1	9	9	43
		sem. mai.													
4	FA		1	14	3	32	7 1/2		3	0	0	1	8	0	0
		ton mai.													
5	SOL		2	6	4	0	6 2/3		2	8	0	1	5	5	24
		ton mi.													
6	RE		2	14	3	32	6		2	4	9 1/5	1	3	3	14
		ton mai.													
7	MI		3	11	12	18	5 1/3		2	1	7 1/5	1	1	1	5
		semi. maj.													
8	FA		4	4	0	0	5		2	0	0	1	0	0	0

M

Figure 2.4: "Harmonic table for the deaf." Mersenne, *Harmonie universelle*, "Livre troisiesme des instrumens a chordes," 125. Paris, Bibliothèque nationale de France, département Arsenal, FOL-S-1449.

With these rules, it would be possible to tune an instrument without resorting to the sense of hearing, instead solely relying on measurements. It is telling that Mersenne presented his experimental findings in such a form. He argued that one could bring experiential and aural forms of knowledge such as tuning outside the realm of experience and into the realm of science. Instrument makers, Mersenne argued, relied too often on their ears "without knowing the reason" for their operations.[117] With the help of rules and numbers, builders could learn the reasons behind their craft and make their work more certain and less prone to mistakes.

Ultimately, however, Mersenne realized that the ear could not be excluded. Since music was not a purely mathematical science, it had to account for the uncertainties that arose in dealing with materials. Writing on the tuning of organs, he stated,

117 "sans en sçavoir la raison." Mersenne, *Harmonie universelle*, "Livre premier des instrumens," 41.

> … if the material followed the correctness of the mind and if the manual operation corresponded completely with science. But since this surpasses the industry of all humans, who cannot foresee the great multitude of circumstances that accompany lead, tin, wood, and the other materials with which the pipes are made and which are met in the air, I consider the ears entirely necessary to tune the organ pipes.[118]

It was impossible to unite the realm of science completely with the realm of craftsmanship. A theme from the previous chapter reemerges here: when dealing with materials such as metal and wood, uncertainties surfaced that could only be approximated by the ears. The mind was not able to anticipate all the possible uncertainties, as this lied beyond the realm of what could be known. Ultimately, Mersenne did not seem convinced that "manual operation corresponded perfectly with science." Instruments could not be fully explained by reason, nor could they provide evidence in all circumstances. This helps to illustrate the ambivalent role of instruments and the world of craftmanship in Mersenne's musical epistemology. A successful interpretation of the workings of an instrument served to affirm the existence of whole numbers in nature. Bells, organ pipes, and flutes were investigated for their properties with the hope that their workings revealed the mathematical ratios that are beneath the sensory world. The ratios of the monochord still guided the investigation of organ pipes and bells. And although the image of Mersenne observing artisanal practices and transferring them to his writings is appealing, the sources tell a more complicated picture.[119]

The Trumpet as Nature's Voice

The example of the trumpet offers a clear case of how a musical instrument could offer evidence for Mersenne's harmonious universe. Like many music theorists of his day, Mersenne tried to determine which of the antique genera

118 "si la matiere suivoit la justesse de l'esprit, & si l'operation manuelle respondoit parfaitement à la science. Mais puis que cecy surpasse l'industrie des hommes, qui ne peuvent prevoir une grande multitude de circonstances qui accompagnent le plomb, l'estain, le bois & les autres matieres dont on fait les tuyaux, & qui se rencontrent mesme dans l'air, je suppose que les oreilles sont entierement necessaires pour accorder les tuyaux." Mersenne, *Harmonie universelle*, "Livre sixiesme des orgues," 363. On this passage, see also Gauvin, *Instruments of Knowledge*, 72.

119 For the wider project of rationalizing practices in the early modern period, see Vérin, "Rédiger et réduire en art."

could be called natural. Most texts from antiquity and the Middle Ages distinguished at least three musical genera: the diatonic, the chromatic, and the enharmonic. These genera consisted of different divisions of the tetrachord, all spanning a perfect fourth. The diatonic genus contains a major semitone and whole tones. The chromatic genus includes a major and minor semitone and a minor third, while the enharmonic genus includes a microtonal diesis. Although there were several attempts to revive the enharmonic genus in the sixteenth and seventeenth centuries—most famously by Nicola Vicentino—the diatonic was widely considered the most natural. For example, Zarlino argued that the voice naturally tended toward the diatonic genus and called the chromatic and enharmonic genera artificial ornaments.[120] Seventeenth-century French theorists argued similarly. According to engineer and music theorist Salomon de Caus, the voice was not able to make the small inflictions necessary for the chromatic and enharmonic genera and therefore must be regarded as unnatural.[121] Composer and music theorist Antoine Parran maintained that the pure chromatic genus was not singable and that the ear could not judge the distances of the enharmonic genus.[122] Echoing the widespread idea that the diatonic scale was natural, the Jesuit Athanasius Kircher, for example, argued that all cultures prefer the diatonic scale in their songs.[123]

Although Mersenne considered ethnographic evidence as well, as I will show in chapter 4, he ultimately rejected this argument, as there was always the chance that other cultures had learned these consonances by practice.[124] Here he turned to trumpets. Whereas humans and animals could have learned the consonances of the fifth or fourth, an "inanimate object" like the trumpet could not have done so.[125] And nonetheless, by changes in embouchure, the trumpet jumped to the octave, the fifth, the fourth, the major third, the minor third, the whole tone, and the semitone. Following Galilei's lead, Mersenne focused on instruments as demonstrating natural principles. But whereas Galilei had emphasized the difficulty of designating any consonance as natural, Mersenne's investigations into the trumpet led

120 Zarlino, *Sopplimenti*, 84 (3.1). For the revival of enharmonic and chromatic music in early seventeenth-century Italy in circles very close to Mersenne, see Kirnbauer, *Vieltönige Musik*. In France Mersenne's associate Titelouze had a spinet constructed that could play the enharmonic genus; see Cohen, "Jean Titelouze."

121 De Caus, *Institution harmonique*, 22v.

122 Parran, *Traité de la musique théorique et pratique*, 37.

123 Kircher, *Musurgia universalis*, VI, 568.

124 Mersenne, *Harmonie universelle*, "Livre troisiesme des genres," 148.

125 "les corps inanimez." Mersenne, *Harmonie universelle*, "Livre troisiesme des genres," 149.

him to a meditation on the divine, showing that nature was truly composed of harmony.

Players of the natural trumpet can play the notes of the harmonic series by means of embouchure. Some of these harmonics are quite flat and need to be adjusted. Players often skip the flatter notes, such as the seventh and fourteenth harmonic, which is a low B-flat for a trumpet in C; they go directly to the eighth harmonic, a C for a trumpet in C. By means of altering air pressure and lip embouchure, trumpeters can also play outside the series, making smaller intervals and altering jumps. Entire major and minor diatonic scales can be played in the higher registers, "tuning" the impure notes of the harmonic series. Although some of the trumpet notes confirmed the belief in an underlying diatonic system of music, difficulties remained in explaining the less harmonic notes.

Hardly any seventeenth-century scholar devoted as much space and effort to explaining the trumpet as Mersenne. The idea that the trumpet could demonstrate the universal distribution of intervals occupied Mersenne for a long time. He wrote on the phenomenon already in 1623 with his *Quaestiones*. There he recounted how the French composer Jean Titelouze observed someone playing through the different notes of the trumpet: first the octave and fifth, finally the whole and semitone.[126] In his unpublished *Livre de la nature des sons*, Mersenne continued this thought, writing that "the fifth is given to us by nature and is the first interval the trumpet makes."[127] After the fifth, the instrument would jump to the fourth, the major third, the minor third, the whole tone, and the semitone, only in that order. That is the proof that "these intervals were natural and proper for music, since nature gave them to us, and it was necessary to reject the other artificial ones."[128] The artificial consonances would be the sixth, the seventh, and anything smaller than a semitone. Similarly, in his *Les preludes*, Mersenne noted that the best reason to solely use the diatonic scale is that it was taken from nature. The trumpet did not emit the minor semitone or the enharmonic dieses, whereas it did emit the whole tone and the major semitone.[129]

That the trumpet occupied such a central role by Mersenne's considerations is remarkable. Historically, the trumpet received little attention from

126 Mersenne, *Quaestiones*, col. 1699.

127 "La quinte nous est enseignée par la nature, et est le premier intervalle que fait la trompette." Paris, Bibliothèque d'Arsenal, MS 2884, fol. 46r. Transcription from Taveau, "Le manuscrit 2884," 405.

128 "ces seuls intervalles étoient naturels, et propres pour la musique, puisque la nature nous les donne, et qu'il failloit rejetter les autres artificiels." Paris, Bibliothèque d'Arsenal, MS 2884, fol. 46r. Transcription from Taveau, "Le manuscrit 2884," 407.

129 Mersenne, *Les preludes*, 166.

music theorists. In sixteenth-century musical writings, such as Lodovico Zacconi's *Prattica di musica* (Venice, 1596), it is described as a signal instrument with few musical capabilities by itself. Zacconi excluded the trumpet from his overview of instruments because it is a field or battle instrument, one that could not play all the notes of the diatonic scale.[130] This reflected common musical practice, where trumpets indeed were most often used as a signal instrument for important events. It is notable that while Mersenne argued that the diatonic scale was natural because the trumpet was able to play it, Zacconi did not attribute the diatonic scale to the trumpet at all. This demonstrates the change the instrument underwent around 1600. Most early books of trumpet calls required no more than the first six or eight harmonics. Over the course of the late sixteenth and early seventeenth centuries, however, the range of the trumpet was expanded. Perhaps the most famous example is the opening toccata of Alessandro Striggio's and Claudio Monteverdi's *L'Orfeo* (first performed in Mantua, 1607), which requires harmonics up to the twelfth. A few decades later, players such as Girolamo Fantini were able to extend the range to non-harmonic notes and unlock the higher register of the harmonic series. Fantini's own instruction book, the *Modo per imparare a sonare di tromba* (Frankfurt, 1638) utilized the harmonic series as far as the eighteenth harmonic.[131] In addition to a wider range, he also used a wide variety of non-harmonic tones in his instruction book. Fantini and other trumpeters at the time transformed the signal instrument into a solo instrument, utilizing the full range of the trumpet and many notes outside the harmonic series.

Although Mersenne already announced in his earliest writings that the trumpet demonstrates that the diatonic intervals are given by nature, he continued to search for the reasons why. Between his first writings on the topic and the publication of *Harmonie universelle*, Mersenne extensively investigated the harmonic series and a range of wind instruments, experimenting with several musicians. According to Mersenne, the Parisian musician Jehan Price could play the range of a nineteenth on a flute with only three holes.[132] The trumpet was also the topic of numerous exchanges in his correspondence. In a list of questions Mersenne circulated among his correspondents, he asked why it is not possible to intone *ut, re, mi, fa,*

130 Singer, *Lodovico Zacconi's Treatment*, 59.

131 Wallace and McGrattan, *The Trumpet*, 98.

132 "Je viens ici ajouter une experience tres juste que j'ay fait faire à Jehan Price Anglois avec une fluste, de laquelle il joue excellemment." Paris, Bibliothèque d'Arsenal, MS 2884, fol. 46v. Transcription from Taveau, "Le manuscrit 2884," 411.

and *sol* in the lower range of the trumpet.[133] He also asked Nicolas-Claude Fabri de Peiresc to conduct an experiment himself if he were to encounter "an excellent trumpet player who knows a bit of music."[134] In the same letter, Mersenne wrote that he received the message that trumpeters play entire scales in Rome. And the physician Christoph de Villiers had asked a trumpeter to help him answer Mersenne's question, but the investigation was halted because the trumpeter could not read music notation.[135]

How he collaborated with individual musicians is difficult to tell, as only few sources have survived. Nor do Mersenne's printed books contain much information on the musicians with whom he experimented. Besides Thomas le Jeune Le Vacher, Mersenne also mentioned a "sieur Quiclet," "musician of the King," who played the cornet. Like Le Vacher, Quiclet provided Mersenne with examples of tablature and information on how to perform cadences and some of the higher notes.[136] In his introduction to *Harmonie universelle*, Mersenne mentioned a musician called "De Livet," most likely the court musician and valet Nicholas de Livet.[137] The trumpeter "l'Anglois" mentioned by Mersenne was most likely the court musician Claude Langlois. How the contact between Mersenne and these musicians was mediated remains unknown, but several of the musicians that Mersenne mentioned as his sources were employed at court. As Catherine Massip has shown, there was a pyramid of musical professions at the French court, from the superintendent at the top to the ordinary instrumentalist at the bottom.[138]

News of the star trumpeter Fantini also reached Paris. In his *Harmonicorum*, Mersenne cited a report from the French physician Pierre Michon Bourdelot from Rome about Fantini. According to Bourdelet, Fantini could perform all notes on his trumpet and play them in tune along with the notes

133 Deschamp to Trichet, January 1635, CM 5, 571.

134 " quelque excellent trompete qui sçache un peu de musique." Mersenne to Peiresc, 2 February 1635, CM 5, 48.

135 Villiers to Mersenne, mid-November 1633, CM 3, 548.

136 According to Mersenne, Quiclet could play eighty measures without taking a new breath. Mersenne, *Harmonie universelle*, "Livre cinquiesme instrumens a vent," 274–76. As Madeleine Jurgens writes, this is most likely Claude Quiquelet, who was also a singer at the court ("chantre ordinaire de la musique du roi"). Jurgens, *Documents du minutier central*, 115. For the singers associated with the court, see Bennett, *Sacred Repertories in Paris*, 286–90.

137 Mersenne, *Harmonie universelle*, n.p. ("Preface generale au lecteur"); Jurgens, *Documents du minutier*, 237.

138 Massip, *La vie des musiciens*, 89–90. Many city musicians tried to get a job at court, but this was difficult for those who did not belong to the musical dynasties such as the families La Barre, Champion, or Boësset.

that Girolamo Frescobaldi played on the organ.[139] Through such accounts and his observations of local trumpeters, Mersenne was well aware that expert players could manipulate the tones, departing from the harmonic series. For example, he writes in *Harmonie universelle*, "I have met someone who descends by a major third lower than the second sound [= second harmonic], instead of dropping an entire octave."[140] And in *Harmonicorum*, Mersenne similarly writes that expert trumpeters could adjust their breath in such a way that allows them to emit the tones between the standard series.[141] But he does not devote much attention to them and writes, "But, apart from the fact that this is not common, and that these notes are made with great difficulty, they are worth nothing and therefore they do not require any attention."[142] Rather, he was interested in what tones the instrument could "naturally" bring forth, without interference of the performer's skills.

Mersenne's depiction of the trumpet range in *Harmonie universelle* does not include the non-harmonic notes that expert players like Fantini were able to play (fig. 2.5). Rather, this description emphasizes that the trumpet naturally plays the diatonic scale, excluding the more dissonant seventh harmonic. He seems not to have observed that trumpet players exclude this note by choice. The fact that the eleventh and thirteenth harmonic are adjusted to confirm to the just intonation scale seems to have eluded him. Rather, Mersenne explains the trumpet tones by imagining that a single tone is caused by a single vibration of air. To this, another vibration of air is added, doubling the motion and sounding an octave higher. To this, another vibration of air is added, and the sound rises a fifth above the first octave. He explains the first series of trumpet tones with this principle (2:1, 3:2, 4:3, 5:4, 6:5; fig. 2.5b). But there is a problem: instead of a 7:6 (a flat minor third), the trumpet here sounds a fourth (the "seventh harmonic" in fig. 2.5). But this proportion that includes the number 7 is excluded, rejected by nature in Mersenne's words, because it is considered a dissonant:

> It seems that the sesquisexta [7:6] should follow the sesquiquinta [6:5], since it is the lesser interval or the lesser ratio that follows the minor

139 Mersenne, *Harmonicorum*, "Liber secundus, de instrumentis pneumaticis," 109. On this letter, see Downey, "Fantini and Mersenne."

140 "j'aye rencontré quelqu'un qui descend d'une Tierce majeure plus bas que le second son, au lieu de descendre d'une Octave entiere." Mersenne, *Harmonie universelle*, "Livre cinquiesme des instrumens a vent," 256.

141 Mersenne, *Harmonicorum*, "Liber secundus, de instrumentis pneumaticis," 109.

142 "mais outre que cela n'est pas ordinaire, & que ces tons se font avec une grande difficulté, ils ne valent rien, c'est pourquoy il n'en faut faire nul estat." Mersenne, *Harmonie universelle*, "Livre cinquiesme des instrumens a vent," 256.

> third. ... Because it is neither a consonance nor the difference of consonances, nature, which is harmonic, rejects it and prefers to interrupt the succession of its intervals and of its songs rather than to pass through an interval that is worth nothing except to offend the ear and the mind.[143]

Nature, not wanting "to offend the ear and mind," skips the seventh partial, which forms a flat seventh with the fundamental tone. The number 7 is "useless to harmony," according to Mersenne, and instead of this impure number, the trumpet jumps to the eight harmonic.[144] Similarly, the eleventh harmonic gets adjusted (so that it makes a semitone with the previous note, "F ut fa 96" and "E mi la 90" in fig. 2.5b). Following Mersenne's rule, one would expect here a 11:10 ratio, not a 16:15 (= 96:90). But here the semitone is a result of the difference between the previous series, between the fourth and the major third.[145] In his table of the trumpet notes, Mersenne gives the numbers 72, 81, 90, 96, 108, 120, 135, and 144 for the upper range. Together these numbers form the ratios 9:8, 10:10, 16:15, 9:8, 10:9; 9:8, and 16:15, resulting in a scale in just intonation. Combining different arguments, Mersenne's description goes a long way to explain how the trumpet might produce such a scale.

According to Mersenne, the notes that expert trumpeters like Fantini were able to play did not require the attention of the scholar, since they were not natural to the instrument and could therefore be considered artificial. Skilled performers could manipulate the pitches of the instrument, just as "the industry of man can trouble the movements of nature."[146] Nature appeared as a harmonious stable entity, emitting diatonic tones. He wrote, "So it seems that nature or its movements are nothing but a ravishing harmony, which invites us to consider the first sources, from which it takes its birth,

143 "La Sesquisexte devroit ce semble suivre la Sesquiquinte, puis qu'elle est le moindre intervalle, ou la moindre raison qui suit la Tierce mineure .... Mais parce qu'elle n'est ny consonance, ny difference des consonances, la nature qui est harmonique, la rejette & ayme mieux rompre la suite de ses intervalles & de ses chansons, que de passer par un intervalle qui ne vaut rien, que pour blesser l'oreille & l'esprit." Mersenne, *Harmonie universelle*, "Livre cinquiesme des instrumens a vent," 251.

144 "inutile à l'harmonie." Mersenne, *Harmonie universelle*, "Livre cinquiesme des instrumens a vent," 252.

145 "Ce qu'elle fait afin de donner la troisiesme difference des consonances, car le demiton majeur est la difference de la Tierce majeure, & de la Quarte, comme le ton majeur est la difference de la Quarte & de la Quinte, & le ton mineur celle de la Quinte & de la Sexte majeure." Mersenne, *Harmonie universelle*, "Livre cinquiesme des instrumens a vent," 253.

146 "l'industrie des hommes peut troubler les mouvemens de la nature." Mersenne, *Harmonie universelle*, "Livre cinquiesme des instrumens a vent," 256.

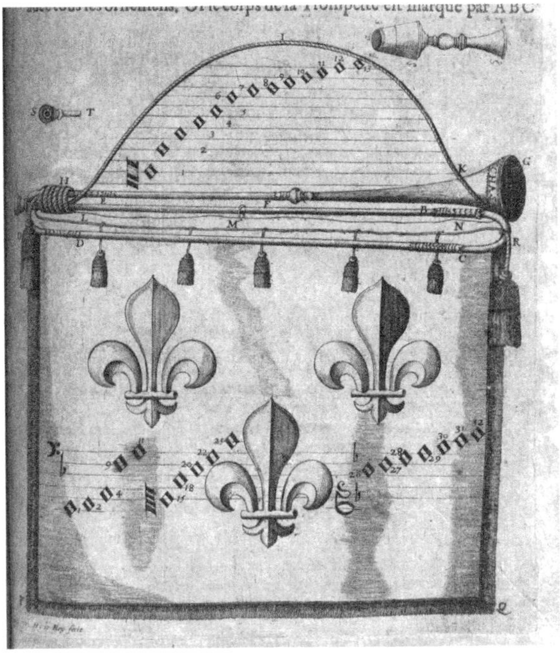

Eſtenduë de la Trompette.

16	C ſol vt fa	144
	demiton maieur	
15	♮ mi	135
	ton maieur	
	A mi la re	120
	ton mineur	
12	G re ſol vt	108
	ton maieur	
	F vt fa	96
	demiton maieur	
10	E mi la	90
	ton mineur	
9	D la re ſol	81
	ton maieur	
8	C ſol vt	72
	Quarte	
6	G re ſol vt	54
	Tierce mineure	
5	E mi la	45
	Tierce maieure	
4	C ſol vt	36
	Quarte	
3	G re ſol vt	27
	Quinte	
2	C ſol vt fa	18
	Octaue	
1	C ſol vt fa	9

Figure 2.5a and 2.5b: The range of the trumpet. Mersenne, *Harmonie universelle*, "Livre cinquiesme Instrumens a vent," 249 and 267. Paris, Bibliothèque nationale de France, département Arsenal, FOL-S-1449.

that is to say contemplate God, who is the Author, and to love Him above all things."[147] The trumpet shows the order of the natural intervals, starting with the octave and moving to the smaller intervals, rejecting artificial and dissonant notes.

After the publication of *Harmonie universelle*, the trumpet remained a topic of discussion in Mersenne's correspondence. For example, Mersenne received an explanation of the trumpet that suggested that trumpeters avoided the notes that were not part of the diatonic scale.[148] Musicians—rather than nature—rejected these notes. Mersenne shared this explanation with René Descartes, who found sound propagation in trumpets obvious:

147 "Par où l'on void que le progrez de la nature est amy de l'harmonie, qu'elle gouverne ou dont elle depend: de sorte qu'il semble que la nature ou ses mouvemens ne soient autre chose qu'une ravissante harmonie, qui nous invite à consider la premiere source, dont elle prend sa naissance, c'est à dire contempler Dieu, qui en est l'Autheur, & à l'aymer sur toutes choses." Mersenne, *Harmonie universelle*, "Livre cinquiesme des instrumens a vent," 250–51.

148 Gaudais to Mersenne, December 1638?, CM 8, 252.

sound is dependent on vibrations of air and their different proportions make the different intervals, the consonances before the dissonances: "When one blows harder, by means of which it makes the higher sounds, but which are all in tune with the first and therefore also with each other."[149] Why the divisions of 7:6 and 8:7 were skipped remains unknown. Theodor Deschamp similarly wrote that the numbers 7 and 11 are skipped because they are "unusual and unknown to the ear," suggesting that the breath of the player cannot accommodate it.[150] Whether Mersenne reconsidered his conclusion remains uncertain. Although he disagreed with Zarlino in his debate with Galilei, he arrived at a remarkably similar conclusion to that of the Italian composer: the first six numbers seem to be engraved in nature.[151] Whereas Mersenne was not opposed to using irrational fractions in his work on the consonances discussed in the previous chapter, his writings on instruments demonstrate his preference for whole numbers once again. And although it was considered an "inanimate object," the trumpet received a larger epistemological role than the human voice, since the human voice could be culturally dependent. Expert musicians like Fantini could only trouble this harmony, since certain skills of performers were considered artificial and rejected.

Mechanical and Natural Forces

Mersenne's views on the naturalness of the musical system remained conflicted. Later in his *Harmonie universelle*, he speaks of "natural and artificial intervals" and explains,

> we call intervals natural when they are made by everyone, by the shepherd in the woods or in the country as well as by musicians, these are the intervals of the diatonic: but musicians have used artificial ones to

149 "lorsqu'on souffle plus fort, au moyen de quoy elle fait des sons plus aigus, mais qui sont tous accordans avec le 1er, et par consequent aussy entre eux." Descartes to Mersenne, 9 February 1639, CM 8, 296.

150 "inusité, et incogneu à l'ouyë." Deschamp to Mersenne, 29 August 1640, CM 10, 72. See also the letter Deschamp to Mersenne, 26 March 1642, CM 11, 84–85.

151 Mersenne, *Harmonie universelle*, "Livre cinquiesme des instrumens a vent," 251. For Zarlino's arguments on why the first six numbers in nature were privileged, see Palisca, *Humanism*, 244–50; Moyer, *Musica scientia*, 208–9. By limiting the number to six, the minor sixth (created by an 8:5 proportion) was not included among the consonances. In his own work, Salomon de Caus therefore extended the harmonious number to eight with an "unused" ("inutile") number 7. De Caus, *Institution harmonique*, 11r.

> embellish their art and to enrich their songs, such as the minor semitone, the *diesis* [quarter-tone], etc., which are not practiced in music, except by chance.[152]

It is thus clear that the diatonic scale was considered natural. In chapter 4, I look more closely at Mersenne's use of non-European sources for this argument, but it is clear that he found such evidence inconclusive, as the role of learning could not be eliminated. But he continued to distinguish what he considered "natural" intervals like the tone, third, or fourth and "artificial" intervals such as the dieses. As he writes further on, "Nature seems to give us the intervals of the diatonic [genus]."[153] At the same time, this work emphasizes the role of art and practice in music: "For one cannot learn it [music] unless one has learned it by science, or by exercise, and by practice."[154] He did not believe that some were simply musical by nature, or at least he had not yet encountered them.

Mersenne's harmonious vision of nature was not incongruous with his interest in mechanics and artificial instruments. These were in fact deeply entangled. His interest in the world of mechanical objects has long been noted. Robert Lenoble's landmark study of Mersenne has emphasized his formation of a mechanistic philosophy. According to Lenoble, Mersenne's idea of nature "is reduced to a play of mechanical forces," codifying mechanistic philosophy before the work of Descartes.[155] In his own study of the scholar, Peter Dear has emphasized that Mersenne adhered less to a particular methodological position but that he gathered arguments from different traditions to argue for the harmonious creation of the world and the existence of certainties.[156] Mersenne's use of the trumpet shows that the "new" empiricism did not negate older theories but rather worked closely together with them. Practical and theoretical knowledge here operated hand in hand. The example of the trumpet demonstrates that Mersenne

152 "nous appellons les intervalles naturels, qui sont faits par tout le monde, c'est à dire aussi bien par le Berger qui est au bois, ou à la campagne, comme par les Musiciens, tels que sont les intervalles de la Diatonique: mais les artificiels ont esté inventez par les Musiciens pour embellir leur art, & pour enrichir leurs chants, comme sont le demiton mineur, la diese, &c. qui ne se pratiquent point hors de la Musique, si ce n'est par hazard. Mersenne, *Harmonie universelle*, "Livre second, des chants," 90.

153 "la nature semble nous donner les intervalles de la Diatonique." Mersenne, *Harmonie universelle*, "Livre second, des chants," 92.

154 "car l'on ne peut s'en servir si on ne l'a apprise par science, ou par exercice, & par la pratique." Mersenne, *Harmonie universelle*, "Livre second, des chants," 93.

155 "La nature réduite à un jeu de force mécaniques." Lenoble, *Mersenne*, 381.

156 Dear, "Method and the Study of Nature," 161–62.

indeed gathered arguments from a wide variety of fields, balancing on a tightrope regarding what is and what is not considered natural. Although he studied the movement in air in both artificial and natural objects, the idea of nature as a harmonious whole remained an important force in his work.

Although crafts like instrument building received a central role in many scientific projects of the early modern period, the distinction between those who were considered to work with their head and those who worked with their hands remained. Many of Mersenne's correspondents argued exactly this, describing artisans as ignorant, only concerned with protecting their trade, and generally deplorable characters. From his *Harmonie universelle*, it is clear Mersenne had an intimate knowledge of artisanal processes, and perhaps even tinkered with instruments himself.[157] His writings do not, however, dissolve the boundaries between theory and practice, or mind and hand. Rather, Mersenne aimed to establish theoretical standards for practice by means of reason. Builders could adopt his theoretical standards in order to elevate their social position, develop more reliable methods, and ultimately perfect their instruments. Arguing that builders relied mostly on their experience, Mersenne hoped to bring artisanal practices into the realm of science through the introduction of theoretical standards.

Ultimately, Mersenne's work helps to contextualize the turn to musical instruments by early modern scientists. As Rebecca Cypess recently emphasized in her study on musical instruments as experimental objects, someone like Giovanni Battista Della Porta showed "the beginnings of a rigorous, empirical method to the study of the natural world" centered around instruments.[158] Although the interest in instruments was indeed part of this empirical turn, for somebody like Mersenne, empiricism was still deeply mathematical and connected to his Augustinian beliefs. It is thus possible to reevaluate Mersenne's relation to his predecessors. In an important article on Mersenne, Claude Palisca already pointed to the centrality of the debate between Zarlino and Galilei for the French scholar. Galilei's and Mersenne's contributions are explained by "accurate observation and experimental evidence" rather than relying on "myth, metaphysics, or legend."[159] As this

157 Mersenne himself at least suggested that he could show some parts of the building process; see Mersenne to Haak, 20 January 1640, CM 9, 43.

158 Cypess, "Giovanni Battista Della Porta's Experiments," 160.

159 Palisca, "Mersenne pro Galilei contra Zarlino," 61. As Palisca points out, Mersenne would write late in his life that Zarlino and Salinas would write "true and beautiful" ("vera ... pulchra") things concerning numbers but that he speaks of "physical reasons" ("rationibus physicis"). Mersenne to Hevelius, 1 January 1647, CM 15, 10.

chapter has aimed to demonstrate, Zarlino's and Mersenne's goals were closer than one might expect, as both theorists mobilized different forms of evidence to assert the universality of a certain set of proportions. At the same time, Mersenne's discussion of the trumpet to demonstrate the nature of music would be unacceptable for Zarlino; to use an artificial object to say something about the natural world was mixing analytical categories. That does not mean that Aristotelian thought had disappeared. In his discussion of the trumpet, Mersenne still cited the Aristotelian maxims that nature takes the shortest routes and does nothing in vain.[160] Although ambivalent, the experimental investigation of instruments was a central part of his project of universal harmony, anticipating the rise of acoustic investigations of instruments in the later seventeenth and eighteenth centuries.

Bibliography

Manuscript sources

Paris, Bibliothèque nationale de France, MS fr. 12357. Marin Mersenne's annotations in his *Harmonie universelle.*

Carpentras, Bibliothèque Inguimbertine, MS 1774. Papers of Nicolas-Claude Fabri de Peiresc.

Printed primary sources

Agricola, Martin. *The "Musica instrumentalis deudsch" of Martin Agricola: A Treatise on Musical Instruments, 1529 and 1545*. Translated and edited by William E. Hettrick. Cambridge: Cambridge University Press, 1994.

Augustine. *On the Free Choice of the Will, On Grace, and Free Choice*. Edited and translated by Peter King. Cambridge: Cambridge University Press, 2010.

Bacon, Francis. *Advancement of Learning and New Atlantis*. Edited by Arthur Johnston. Oxford: Clarendon Press, 1980.

Barker, Andrew. *Greek Musical Writings II: Harmonic and Acoustic Theory*. Vol. 2 of 2 vols. Cambridge: Cambridge University Press, 1989.

Bergier, Nicolas. *La musique speculative*. Edited and translated by Ekkehard Jost. Cologne: Arno Volk Verlag, 1970.

Binet, Étienne [Rene Francois]. *Essay des merveilles de nature, et des plus nobles artifices*. Rouen: Osmon, 1623.

Boethius, Anicius Manlius Severinus. *Fundamentals of Music*. Translated by Calvin M. Bower, edited by Claude V. Palisca. New Haven: Yale University Press, 1989.

160 Mersenne, *Harmonie universelle*, "Livre cinquiesme des instrumens a vent," 250.

De Caus, Salomon. *Institution harmonique*. Frankfurt: Jan Norton, 1615. Reprint, Geneva: Minkoff, 1980.

Du Caurroy, Eustache. *Fantasies a III. IIII. V. et VI. Parties*. Paris: Ballard, 1610.

De Gouy, Jacques. *Airs à quatre parties, sur la paraphrase des pseaumes de Messire Antoine Godeau, évesque de Grasse*. Paris: Ballard, 1650.

Faret, Nicolas. *L'honneste-homme ou, l'art de plaire a la court*. Paris: Du Bray, 1630.

Gaffurius, Franchinus. *Theorica musicae*. Milan: Mantegatium, 1492.

Galilei, Vincenzo. *Discorso intorno all'opera di* [...] *Zarlino*. Florence: Mareschotti, 1589.

Gassendi, Pierre. *The Mirrour of True Nobility and Gentility Being the Life of the Renowned Nicolaus Claudius Fabricius, Lord of Pieresk, Englished by W. Rand*. London: Streater for Moseley, 1657.

Haranger, René. *Articles, statuts, ordonnances et réglemens de la Communauté des Vergetiers, Raquetiers et Brossiers* [...]. Paris: Grou, 1659.

Huygens, Christiaan. *Oeuvres complètes. Tome XIII: Dioptrique*. Edited by Diederik Johannes Korteweg. The Hague: Nijhoff, 1916.

Jurgens, Madeleine. *Documents du minutier central concernant l'histoire de la musique (1600–1650), Volume 1*. Vol. 1 of 2 vols. Paris: S.E.V.P.E.N., 1967.

Kircher, Athanasius. *Musurgia universalis sive ars magna consoni et dissoni in X libros digesta*. Rome: Corbelletti, 1650.

Maugars, André. *Response faite à un curieux, sur le sentiment de la musique d'Italie. Escrite à Rome le premier octobre 1639*. N.p.: s.n., ca. 1639.

Mersenne, Marin. *Correspondance du P. M. Mersenne*. Edited by Paul Tannery, Cornelis de Waard, Bernard Rochot, Armand Beaulieu, and René Pintard. Paris: Centre National de la Recherche Scientifique et al., 1932–88.

Mersenne, Marin. *Harmonicorum libri XII*. Paris: Baudry, 1648. Reprint, Geneva: Minkoff, 1972.

Mersenne, Marin. *Harmonie universelle contenant la théorie et la pratique de la musique (Paris, 1636). Édition fascimilé de l'exemplaire conserve à la Bibliothèque des Arts et Métiers et annoté par l'auteur*. Introduction by François Lesure. Paris: Éditions du Centre National de la Recherche Scientifique, 1963.

Mersenne, Marin. *Les mechaniques de Galilée: mathematicien & ingenieur du duc de Florence*. Paris: Guenon, 1634.

Mersenne, Marin. *Les preludes de l'harmonie universelle ou questions curieuses. Utiles aux predicateurs, aux theologiens, aux astrologues, aux medicins & aux philosophes*. Paris: Guenon, 1634.

Mersenne, Marin. *Quaestiones celeberrimae in Genesim, cum accurata textus explicatione: in hoc volumine athei, et deistae impugnantur & expugnantur, & vulgata edition ab hereticorum calumnijs vindicator. Graecorum, & Hebraeorum musica instauratur*. Paris: Cramoisy, 1623.

Mersenne, Marin. *Questions harmoniques. Dans lesquelles sont contenuës plusieurs choses remarquables pour la physique, pour la morale, et pour les autres sciences.* Paris: Villery, 1634.

Mersenne, Marin. *Les questions theologiques, physiques, morales, et mathematiques. Où chacun trouvera du contentement ou de l'exercice.* Paris: Guenon, 1634.

Mersenne, Marin. *Traité de l'harmonie universelle.* Paris: Baudry, 1627.

Mersenne, Marin. *La verité des sciences, contre les sceptiques ou pyrrhoniens.* Paris: Du Bray, 1625.

Parran, Antoine. *Traité de la musique théorique et pratique, contenant les préceptes de la composition.* Paris: Ballard, 1639. Reprint, Geneva: Minkoff, 1972.

Picot, Claude. *Les principes de la philosophie escrits en Latin, par René Descartes. Et traduits en François par un des Amis.* Paris: Pierre Des-Hayes, 1647.

Ronsard, Pierre de. *Abbrège de l'art poëtique françois.* Paris: Buon, 1565.

Schott, Gaspar. *Magiae universalis naturae et artis, pars II: Acustica, in vii libros digesta* […]. Würzburg: Hertz, 1657.

Taveau, Laurence. "Le manuscrit 2884 du P. Mersenne à la bibliothèque de l'Arsenal: Étude et édition critique." PhD diss., Université de Paris-Sorbonne, Laboratoire IREMUS, 2018.

Trichet. Pierre. *Traité des instruments de musique (vers 1640).* Edited by François Lesure. Geneva: Minkoff, 1978.

Tyard, Pontus de. *Solitaire second.* Edited by Cathy M. Yandell. Geneva: Droz, 1980.

Virdung, Sebastian. *Musica getutscht: A Treatise on Musical Instruments (1511) by Sebastian Virdung.* Translated and edited by Beth Bullard. Cambridge: Cambridge University Press, 1993.

Zarlino, Gioseffo. *Sopplimenti musicali del Rev. M. Gioseffo Zarlino da Chioggia: Terzo volume.* Venice: Sanese, 1588.

Secondary literature

Anthony, James R. *French Baroque Music: From Beaujoyeulx to Rameau.* Portland: Amadeus Press, 1997.

Bennett, Peter. *Sacred Repertories in Paris under Louis XIII: Bibliothèque nationale de France MS vm res.571.* Farnham: Ashgate, 2009.

Benoit, Marcelle. "L'apprentissage chez les facteurs d'instruments de Musique a Paris, 1600–1661, 1715–1774." *Recherches sur la Musique Française Classique* 24 (1986): 5–106.

Berrada. Tarek. "Music at Home: Spaces for Music in French Seventeenth-Century Residential Architecture." In *The Music Room in Early Modern France and Italy*, edited by Deborah Howard and Laura Moretti, 290–307. Oxford: Oxford University Press, 2012.

Brown, Howard Mayer. "*Ut musica poesis:* Music and Poetry in France in the Late Sixteenth Century." Early Music History 13 (1994): 1–63.

Clark, Suzannah and Alexander Rehding. "Introduction." In *Music Theory and Natural Order from the Renaissance to the Early Twentieth Century*, edited by Suzannah Clark and Alexander Rehding, 1–13. Cambridge: Cambridge University Press, 2001.

Cohen, Albert. "The *Fantaisie* for Instrumental Ensemble in 17th-Century France: Its Origin and Significance." *The Musical Quarterly* 48, no. 2 (1962): 234–43.

Cohen, Albert. "Jehan Titelouze as Music Theorist." In *Festa Musicologica: Essays in Honor of George J. Buelow*, edited by Thomas J. Mathiesen and Benito V. Rivera, 391–406. Stuyvesant, NY: Pendragon Press, 1995.

Cypess, Rebecca. *Curious and Modern Inventions: Instrumental Music as Discovery in Galileo's Italy*. Chicago: University of Chicago Press, 2016.

Cypess, Rebecca. "Giovanni Battista Della Porta's Experiments with Musical Instruments." *Journal of Musicological Research* 35, no. 5 (2016): 159–75.

Daston, Lorraine. "The Nature of Nature in Early Modern Europe." *Configurations* 7, no. 2 (1998): 149–72.

Dear, Peter. *Mersenne and the Learning of the Schools*. Ithaca: Cornell University Press, 1988.

Dear, Peter. "Method and the Study of Nature." In *The Cambridge History of Seventeenth-Century Philosophy*, edited by Daniel Garber and Michael Ayers, vol. 1 of 2 vols., 147–77. Cambridge: Cambridge University Press, 2000.

Dostrovsky, Sigalia. "Early Vibration Theory: Physics and Music in the Seventeenth Century." *Archive for History of Exact Sciences* 14, no. 3 (1975): 169–218.

Downey, Peter. "Fantini and Mersenne: Some Additions to Recent Controversies." *Historic Brass Society Journal* 6 (1994): 355–62.

Dufourcq, Norbert. "Une dynastie française: Les Denis." *Revue de Musicologie* 38, no. 114 (1956): 151–55

Eamon, William. "Markets, Piazzas and Villages." In *The Cambridge History of Science. Volume 3: Early Modern Science*, edited by Katharine Park and Lorraine Daston, 206–23. Cambridge: Cambridge University Press, 2006.

Fix, Adam. "Esperienza, Teacher of All Things: Vincenzo Galilei's Music as Artisanal Epistemology." *Nuncius* 34 (2019): 535–74.

Gauvin, Jean François. "Instruments of Knowledge." In *The Oxford Handbook of Philosophy in Early Modern Europe*, edited by Desmond M. Clarke and Catherine Wilson, 315–37. Oxford: Oxford University Press, 2011.

Gauvin, Jean François. *Instruments of Knowledge: Finding Meaning in Objects, Habits, and Museums*. Leiden: Brill, 2023.

Gétreau, Florence. "Entre l'oral et l'écrit: Pratique, transmission et théorie du métier de facteur d'instruments de musique." *Ethnologie française* 26, no. 3 (1996): 504–19.

Goldberg, Randall. *Where Nature and Art Adjoin: Investigations into the Zarlino-Galilei Dispute, Including an Annotated Translation of Vincenzo Galilei's Discorso intorno all'opera Di Messer Gioseffo Zarlino*. Ann Arbor: UMI Research Press, 2011.

Gordon, Bonnie. *Monteverdi's Unruly Women: The Power of Song in Early Modern Italy*. Cambridge: Cambridge University Press, 2004.

Gordon, Bonnie. "Orfeo's Machines." *The Opera Quarterly* 24, no. 3–4 (2008): 200–222.

Heller, Henry. *Labour, Science and Technology in France, 1500–1620*. Cambridge: Cambridge University Press, 1996.

Heller-Roazen, Daniel. *The Fifth Hammer: Pythagoras and the Disharmony of the World*. New York: Zone Books, 2011.

Hubbard, Frank. *Three Centuries of Harpsichord Making*. Cambridge: Harvard University Press, 1965.

Hunt, John N. "Jurors of the Guild of Musical Instrument Makers of Paris." *The Galpin Society Journal* 51 (1998): 110–13.

Kirnbauer, Martin. *Vieltönige Musik: Spielarten chromatischer und enharmonischer Musik in Rome in der ersten Hälfte des 17. Jahrhunderts*. Basel: Schwabe, 2013.

Konoval, Brandon. "Pythagorean Pipe Dreams? Vincenzo Galilei, Marin Mersenne, and the Pneumatic Mysteries of the Pipe Organ." *Perspectives on Science* 26, no. 1 (2018): 1–51.

Lazardzig, Jan. "Die Maschine als Spektakel: Funktion und Admiration im Maschinendenken des 17. Jahrhunderts." In *Instrumente in Wissenschaft und Kunst – Zu Architektonik kultureller Grenzen im 17. Jahrhundert*, edited by Helmar Schramm, Ludger Schwarte, and Jan Lazardzig, 167–93. Berlin: De Gruyter, 2006.

Lehr, André. "Middeleeuwse klokkengietkunst." In *Klokken en Klokkengieters: Bijdragen tot de Campanologie*, 13–158. Culemborg: Historische Commissie van Nederlandse Klokkenspel-Vereniging, 1963.

Lenoble, Robert. *Mersenne ou la naissance du mécanisme*. Paris: Vrin, 1943.

Lesure, François. *Musique et musiciens français du XVI siècle*. Geneva: Minkoff, 1976.

Long, Pamela O. *Openness, Secrecy, Authorship. Technical Arts and the Culture of Knowledge from Antiquity to the Renaissance*. Baltimore: The Johns Hopkins University Press, 2001.

Loubet de Sceaury, Paul. *Musiciens et facteurs d'instruments de musique sous l'Ancien Régime: Status corporatifs*. Paris: Pedone, 1949.

Massip, Catherine. *La vie des musiciens de Paris au temps de Mazarin: 1643–1661: essai d'études sociale*. Paris: Picard, 1976.

Massip, Catherine. "Facteurs d'instruments et maîtres à danser parisiens au XVIIe siècle." In *Instrumentistes et luthiers parisiens, XVIIe–XIX siècles*, edited by Florence Gétreau, 17–34. Alençon: Impremerie Alençonnaise, 1988.

Miller, Peter. "Peiresc in the Parisian 'Jewel House.'" In *Knowledge and Discernment in the Early Modern Arts*, edited by Sven Dupré and Christine Göttler, 213–35. London/New York: Routledge, 2017.

Moyer, Ann E. *Musica Scientia: Musical Scholarship in the Italian* Renaissance. Ithaca: Cornell University Press, 1992.

Paganini, Gianni. "Gassendi's Interplay between Skepticism and Empiricism." In *Pierre Gassendi: Humanism, Science, and the Birth of Modern Philosophy*, edited by Delphine Bellis, Daniel Garber, and Carla Rita Palmerino, 75–105. New York: Routledge, 2023.

Palisca, Claude V. *Humanism in Italian Renaissance Musical Thought*. New Haven: Yale University Press, 1985.

Palisca, Claude V. "Mersenne pro Galilei contra Zarlino." In *Essays in honour of David Evatt Tunley*, edited by Frank Callaway, 61–72. Nedlands: Callaway International Resource Center for Music Education, 1995.

Popplow, Marcus. "Why Draw Pictures of Maschines? The Social Contexts of Early Modern Machine Drawings." In *Picturing Maschines 1400–1700*, edited by Wolfang Lefèvre, 17–48. Cambridge, MA: MIT Press, 2004.

Rasch, Rolf. "Simon Stevin and the Calculation of Equal Temperament." In *Music and Mathematics in Late Medieval and Early Modern* Europe, edited by Philippe Vendrix, 253–320. Turnhout: Brepols, 2008.

Roberts, Lissa, and Simon Schaffer. "Preface." In *The Mindful Hand: Inquiry and Invention from the Late Renaissance to Early Industrialisation*, edited by Lissa Roberts, Simon Schaffer, and Peter Dear, xiii–xxvii. Amsterdam: Royal Netherlands Academy of Arts and Sciences, 2007.

Scherpereel, Joseph. "Peiresc et la musique." In *Peiresc ou la passion de connaître: colloque de Carpentras, novembre 1987*, edited by Anne Reinbold, 153–204. Paris: Vrin, 1990.

Singer, Gerhard. "Lodovico Zacconi's Treatment of the Suitability and Classification of All Musical Instruments in the *Prattica di musica* of 1592." PhD diss., University of Southern California, 1969.

Smith, Pamela H. *From Lived Experience to the Written Word: Reconstructing Practical Knowledge in the Early Modern World*. Chicago: University of Chicago Press, 2022.

Smith, Pamela H., and Isabella Lores-Chavez. "Counterfeiting Materials, Imitating Nature." In *The Matter of Mimesis: Studies of Mimesis and Materials in Nature, Art and Science*, edited by Marjolijn Bol and Emma C. Spary, 27–53. Leiden: Brill, 2023.

Vérin, Hélène. "Rédiger et réduire en art: un projet de rationalisation des pratiques." In *Réduire en art. La technologie de la Renaissance aux Lumières*, edited by Pascal Dubourg Glatigny and Hélène Vérin, 17–58. Paris: Éditions de la Maison des sciences de l'homme, 2008.

Wallace, Johan, and Alexander McGrattan. *The Trumpet*. New Haven: Yale University Press, 2011.

Wolfe, Jessica. *Humanism, Machinery, and Renaissance Literature*. Cambridge: Cambridge University Press, 2004.

Zecher, Carla. *Sounding Objects: Musical Instruments, Poetry, and Art in Renaissance France*. Toronto: University of Toronto Press, 2007.

3. Collecting Curiosities: Images, Observations, and Earwitnesses

Abstract: This chapter turns to methods of collecting and corresponding in Mersenne's work on music. Offering a possible way to read his desire for ever more observations and examples, I emphasize the role of curiosity, the term by which many of Mersenne's contemporaries described his state of mind and scope of interest. The chapter first looks at Mersenne's notion of curiosity and its relation to collecting practices, before considering some of the ways in which physical collections, letters, and images were utilized to gather new examples and experiences.

Keywords: Republic of Letters, correspondence, curiosity, drawings, transcribing

In 1634 Mersenne's patron Nicolas-Claude Fabri de Peiresc wrote a letter to the Capuchin missionary Gilles de Loches, who had just returned to France from Cairo one year earlier. In the letter, Peiresc asks de Loches if he could share with Mersenne what he knew about music in the eastern Mediterranean. If De Loches could show his notes on the Islamic prayers he might have heard, Mersenne would be "spared a great deal of trouble," Peiresc suggests. He then goes on to give the following description of the music theorist and his ambitions:

> For he [Mersenne] has brought to the printing press a great work on ancient and modern harmony—in which he has discovered very beautiful curiosities—and in which he wanted to include something on the music and the way of singing of all the religious sects and all the nations, and he has persecuted all his friends to educate him [*avoir des instructions*], [a project] that we have not yet ended.[1]

1 "tireroient d'une grande peine .... Car il a soubz la presse un grand ouvrage de toute l'harmonie ancienne et modern, où il a descouvert de tres belles curiosités, où il desiroit mettre quelque

Van der Miesen, Leendert. *Marin Mersenne and the Study of Harmony: From Sound to Music.* Taylor & Francis Group, 2025.
DOI 10.5117/9789048564132_CH03

According to Peiresc, Mersenne ventured where only few scholars had gone before; he wanted to know the music and ways of singing of all religious groups and nations, "persecuting" his friends by asking for materials and giving instructions. The list of subjects on which Mersenne requested information is long: possible instruments and their way of playing, styles of singing, images, tablatures, and the origins of instruments. His letters are filled with references to the singing in a Basilian monastery near Aleppo, a man with two different voices in Nevers, the trumpet marine, and the playing of traveling musicians in the Low Countries. In the printed *Harmonie universelle*, Mersenne includes several of such "curiosities," but many topics did not find their way into his printed works and can now only be retrieved from his letters and marginal notes. Peiresc's characterization is of special interest, since it highlights two central aspects of Mersenne's work that have received little attention: the notion of curiosity and his use of "friends" or correspondents to collect materials for his book. This chapter turns to the relation between these two particular observations. The second part of Peiresc's description—Mersenne's interest in the music of different nations—is explored in the subsequent chapter.

Mersenne's interest in collecting "beautiful curiosities" has already sparked scholarly interest. Katharine Park and Lorraine Daston characterize Mersenne as a curious and restless inquirer, pursuing "endlessly what is just out of reach."[2] In a similar way, Robert Lenoble describes Mersenne as "curious without measure."[3] His contemporaries already made similar observations. The Dutch diplomat and poet Constantijn Huygens saw Mersenne as a collector who "amasses the good and the bad at all costs."[4] Like many others, Huygens received a stream of incessant queries, problems, and observations from Mersenne. From his cell in Paris, the friar circulated his questions, inquiring on a myriad of topics, from the bizarre

chose de la musique, ou façon de chanter, de toutes les sects et de toutes les nations, et a persecuté tous ses amis pour luy en faire avoir des instructions, dont nous n'avions encore peu venir à bout." Peiresc to De Loches, 29 August 1634, CM 4, 342. Gilles de Loches would eventually report to Mersenne the following year. See Mersenne to Peiresc, 25 May 1635, CM 5, 215; Mersenne to Peiresc, 15 July 1635, CM 5, 300, and chapter 4 of this book. I abbreviate the *Correspondance du P. Marin Mersenne* with CM and the number of the volume.

2 Daston and Park, *Wonders*, 308. For an evaluation of Daston and Park's treatment of Mersenne that was especially important for this chapter, see the introduction to Mueller, *Marin Mersenne's Questions theologiques*.

3 "curieux sans mesure." Lenoble, *Mersenne*, 80. Mersenne's epistemology of collecting is also explored in Belin "La notion de collection scientifique."

4 "Le Pere Mersenne, qui amasse du bon et du mauvais de tous costez." Constantijn Huygens to Descartes, 8 October 1640, CM 10, 160.

to the mundane. This restlessness could also be interpreted negatively as the mark of a superficial scholar who cannot commit to a single topic. His contemporary Claude Saumaise, for example, saw Mersenne as lacking fundamental judgment.[5] And when the physician Christoph de Villiers wrote that Mersenne's books contained "infinite matters for the curious [*curieux*] only," this meant that many things in his books were useless to the more practically minded readers.[6] The term *curieux* is central here. In his dictionary from 1690, Antoine Furetière defines such a person as "someone who has a desire to learn, to see the good things, the wonders of art and of nature."[7] Although Mersenne's correspondence and scientific activity more broadly were supported by Minim order, Odile Krakovitch has argued that his "curiosity was often incompatible with the demands of his religious life."[8]

This chapter argues that curiosity is an essential feature of Mersenne's work that has been overlooked in previous studies on the relationship between music and science. While his positive evaluation of curiosity did set him apart from some of his contemporaries who condemned curiosity and the continuous search for new experiences, Mersenne's order enabled and provided for his curious interests. As this chapter will show, Mersenne's curiosity was theologically motivated. The concept of curiosity as an investigative mode helps to understand why Mersenne collected so much. As this chapter will explore, amassing information was a central endeavor for Mersenne, but with very specific aims. Convinced that it was impossible to find any ultimate truths, he turned his eye and ear to what is useful and wonderful, for this life and the afterlife. His primary metaphor for the insatiably curious person was the ear, departing from a tradition of seeing curiosity as an ocular occupation.[9] As Peiresc makes clear in the aforementioned description, many of the things that Mersenne collected in preparation for his *Harmonie universelle* were considered "curious" and were not limited to the visual realm. Moreover, curiosity was fundamentally linked to the concept and practice of collecting, so much that historian Neil

5 Saumaise to Peiresc, 2 September 1634, CM 4, 343–44.

6 "infinies pour les curieux seullement." Villiers to Mersenne, 7 July 1640, CM 9, 463.

7 "luy qui a desir d'apprendre, de voir les bonnes choses, les merveilles de l'art et de la nature." Furetière, *Dictionnaire universel*, n.p. ("Curieux").

8 "La curiosité de Mersenne était cependant souvent incompatible avec les exigences de sa vie religieuse." Krakovitch, "La vie intellectuelle," 31.

9 Benedict, *Curiosity*, 25. The tradition of describing curiosity as an ocular activity already began with Augustine, who described curiosity as "lust of the eyes." See Augustine, *Confessions*, 10.54, 211.

Kenny has spoken of "the curiosity-collecting tendency."[10] Early modern authors frequently described their works as collections of "curiosities," meaning that it was made up of observations, singularities, and fragments that did not necessarily add up to a whole. The astronomer Jacques Gaffarel, for example, called the content of his 1629 *Curiositez inouyes* "curiosities" to avoid censorship and to make clear that there was a collaborative effort behind the book.[11] Mersenne's Minim student Jean-François Niceron titled his book *La perspective curieuse* (Paris, 1638) because he saw his book as not only useful but also delightful.[12] In order to illuminate Mersenne's specific conception of curiosity, the first part of the chapter turns to his 1634 *Questions theologiques, physiques, morales and mathematiques*, where it becomes clear that curiosity was one of the most admirable qualities that one could aspire to in life. Here it also becomes clear that Mersenne redefined the notion of curiosities as those things that are not essential to life, thus also encompassing many mundane and prosaic objects and activities. Curiosity is therefore less a quality of an object than an investigate mode; the curious scholar approaches more mundane objects as if they were curious.

Whereas the first part of the chapter focuses on Mersenne's validation of curiosity, the second part looks closer at the mechanics of collecting the "beautiful curiosities" described by Peiresc. To do so, it introduces the term "earwitness" and asks how observations, lists, and images circulated in Mersenne's correspondence. The term "earwitness" was used by one of Mersenne's own correspondents. In January 1646, the Dutch diplomat and poet Constantijn Huygens wrote a letter to Mersenne in Paris informing him that a musical ensemble with three excellent musicians would soon arrive in The Hague. Huygens was especially looking forward to the viola da gamba player, a viol player whose instrument had bronze strings, and a dulcian player. He promised to keep Mersenne informed of his findings: "You will hear what I will learn as an earwitness" [*auribus testis*].[13] Mersenne had to remind him, but Huygens then shared a short description of the qualities of

10 Kenny, *The Uses of Curiosity*, 161.

11 Kenny, *The Uses of Curiosity*, 290. Gaffarel spoke of a suspension of judgment akin to a skeptical position. See Gaffarel, *Curiositez inouyes*, 643.

12 Niceron, *La perspective curieuse*, n.p. ("Preface"). Niceron added that the phrase "curious perspective" (*perspective curieuse*) was less loaded than "artificial magic" (*magie artificiele*), which may have sounded negative to less informed readers.

13 "Vous aurez advis de ce que j'en apprendray *auribus testis*." Constantijn Huygens to Mersenne, 26 November 1646, CM 14, 638; Huygens, *Driehonderd brieven*, 795. The last two words are left out of the edition of Mersenne's correspondence, so here I follow the transcription made by Rudolf Rasch.

the musicians, concluding, "you have heard many good things, but this will surprise you."[14] Reports like Huygens's were of special interest to Mersenne. Since he could not encounter every instrument or player in person, he had to rely on his wide network of informants. He had, for example, already described the viol with metal strings (called a baryton) in his *Cogitata physico mathematica* without hearing the instrument himself.[15] Through accounts such as those from Huygens, Mersenne could stay informed on new developments, instruments, and players—in Peiresc's terms, "persecuting all his friends" to send whatever they could to him. When the garrisons in Sens broke off any concerts and he had nothing new to report, Villiers once wrote disappointingly to Mersenne, "As for music, I have observed nothing new."[16]

Huygens's description of himself as an earwitness is noteworthy. By using the term "auribus testis" (more often "auritus"), he defined himself as a witness called to testify and recount his musical experience to Mersenne. The term itself was often associated with hearsay, and such testimony was not always recognized by the law, or at least it was valued less highly than that of a direct eyewitness. The Roman author Plautus famously argued that "one witness furnished with eyes is worth more than ten witnesses furnished with ears."[17] Such statements were repeated throughout the early modern period. In both legal and scientific realms, earwitnesses were associated with secondhand and thus less valuable information.[18] Huygens used the term differently, namely to refer to the observation, sharing, and validation of an audible phenomenon, in this case through in a written form. The ear is described as the organ of discernment, capable of judging and transmitting musical experiences through a network of letters. Such examples also included secondhand witnesses. Villiers, for example, wrote Mersenne concerning a pitch of a string (whether it descends or ascends at the end of its vibrations) and had his experiments confirmed by the music master of the church in Sens, "gifted with a good ear and expert in the practice of composition and singing."[19] The sharing of experiences can be

14 "Vous avez ouy beaucoup de bonnes choses, mais celle-cy vous surprendoit." Constantijn Huygens to Mersenne, 14 January 1647, CM 15, 48; Huygens, *Driehonderd brieven*, 796.

15 Mersenne, *Cogitata physico mathematica*, 365.

16 "Quand est de la musique, je n'y ay rien observé de nouveau." Villiers to Mersenne, 14 July 1642, CM 11, 216.

17 Plautus, *Stichus*, 320–21 (*Truculentus*, II.6, 8).

18 Shapiro, *A Culture of Fact*; Frisch, *The Invention of the Eyewitness*.

19 "doué d'un bon ouye et tres expert en la prattique de la composition et du chant." Villiers to Mersenne, 15 October 1633, CM 3, 519.

seen in the same light as Mersenne's emphasis on cooperation and group observations in his experimental work.[20] Mersenne's letters were central for this process of earwitnessing. The letters contain many questions about aural phenomena that correspondents were asked to observe. Many of these anecdotes and experiences can be described as observations. As Gianna Pomata has shown, the genre of the observation was especially popular in scholarly networks like Mersenne's.[21] An increasing number of books declared their content to consist of "observations," as in Mersenne's *Novarum observationum physico-mathematicarum* (New physico-mathematical observations). Observations were understood to be the result of experience and empiricism and thus closely tied to the category of experiment. Mersenne asked his correspondents to go out and observe echo effects, sit down in their cabinets and listen carefully to the overtones of their lutes, and report to him on any instruments in their vicinity.

Highlighting the ways in which Mersenne's correspondents transmitted auditory experiences and musical materials in their letters, the second part of this chapter focuses on the observations, experiences, and images that Mersenne collected. Mersenne circulated a variety of lists with questions, collecting different experiences and observations.[22] By circulating images of instruments, he aimed at gathering new examples and specimens that were unknown in Paris. In turn, his correspondents made images, engaged local musicians for information, and sent reports to Paris. By focusing on singularities, Mersenne was able to unite a wide variety of theoretical and methodological approaches that would otherwise clash. Viewing Mersenne's writings as a heterogenous whole, composed of many different observations and methodologies, helps us to understand why present-day scholars find it difficult to assign the labels to Mersenne's work that are traditionally used for seventeenth-century science, such as mechanism, empiricism, or sensualism.

Within this chapter, collecting thus takes place on different levels, from the physical collecting of objects in cabinets to the metaphorical collecting of sensory impressions and ideas. Circulating letters, collecting experiences, and writing books were highly interrelated activities. Like many early modern scholars, Mersenne worked in a tradition of commonplace books.[23] Authors collected paper scraps and notes that were used during the

20 Dear, *Discipline and Experience*, 133; Licoppe, *La formation de la pratique scientifique*, 26–35.

21 Pomata, "Observation Rising," 48.

22 For an introduction to the correspondence of Mersenne, see Rochot, *La correspondance scientifique*; Ruellet, "'Le grand négocient'"; Grosslight, "Small Skills."

23 For this aspect of Mersenne's work, see in particular Raphael, *Reading Galileo*, 75–97.

process of writing. Books like *Harmonie universelle* can be seen as collections in themselves, filled with personal experiences and others' findings. Furthermore, natural and artificial objects, paintings, and rarities were avidly collected by his contemporaries. Although Mersenne was part of a mendicant order that took an oath of poverty, many of his associates had extravagant collections of curiosities. He himself pointed to the usefulness of such collections for the advancement of science. The insistence on curiosity allowed Mersenne to highlight diversity while at the same time stressing the order and regularity in nature.

Excessive Curiosity

Although Mersenne is at times depicted as an austere friar solely working on mathematical problems, curiosity was a central aspect of his output and daily research. His contemporaries also described his activities with the term "curiosity." His friend Pierre Gassendi described Mersenne as "an exceedingly good man, learned and curious."[24] René Descartes offered the following characterization: "the good father Mersenne is so curious and eager to hear a marvel, that he listens favorably to all those [wonders] which please him."[25] According to Descartes's first biographer, Adrien Baillet, there was never "anyone more curious" than Mersenne.[26] The astronomer Ismael Boulliau wrote that Mersenne continuously searches for "rare and new things."[27] The Italian writer Élie Diodati was more negative in his description of Mersenne's "excessive curiosity."[28] His correspondents in turn often commented upon the "curiosities" or "rarities" the Minim requested. As becomes clear from these different usages, "curious" could refer to a state of mind, the characteristic of an object, and a person who was especially inquisitive.[29] Curiosity was a central feature of much of early modern scholarship, but it also had its ambiguities. It was closely related

24 Gassendi, *The Mirrour of True Nobility*, 115.

25 "Mais le bon pere Mercenne est si curieux et si ayse d'entendre quelque merveille, qu'il escoutte favorablement tous ceux qui luy en content." Descartes to Constantijn Huygens, 12 March 1640, CM 9, 202.

26 "Jamais mortel ne fut plus curieux que luy pour pénétrer tous les secrets de la Nature." Baillet, *Vie de Monsieur Des-Cartes*, 353.

27 "ut ipsius cupiditati, que res novas ac non vulgares appetit." Cited in De Coste, *La vie*, 53.

28 "necnon a Mersenno, cujus nimia curiositas vobis debuit esse suspecta." Diodati to Van den Hove, 16 March 1637, CM 6, 221.

29 For a good overview on the literature on curiosity, see Evans and Marr, *Curiosity and Wonder*. The relation between collecting and curiosity is explored further in Swann, *Curiosities and Texts*.

to other terms, such as "wonder," "marvel," and "rarity." No unified idea of curiosity existed, but rather a plurality of meanings, associations, and practices. But for many natural philosophers, curiosity was central to their self-understanding and their position in the natural world.[30]

Traditionally, curiosity was considered more a vice than a virtue. Augustine, for example, regarded curiosity as a variant of lust. He described it as "a vain inquisitiveness dignified with the title of knowledge and science" and "a lust for experimenting and knowing."[31] In *De vera religione*, he contrasts curiosity with contemplation, which leads to true knowledge.[32] Isidore of Seville similarly dismissed it, as it could lead to heresy: "curiosity is a harmful science."[33] Many medieval authors continued to describe it as such. To be curious was considered a vice, as it could only lead to questioning the foundations of the faith and pursuing a futile kind of science. Early modern scholars like Francis Bacon followed Augustine's warnings, describing it as "fruitless speculation or controversy."[34] Cesare Ripa's *Iconologia* (Rome, 1593) depicted curiosity as an uncombed figure dressed in a garment covered with ears, which symbolizes, according to Ripa, "the Itch of knowing more than concerns her."[35] Descartes also dismissed curiosity and emphasized the necessity of scientific method. In his *Regulae*, he writes, "So blind is the curiosity with which mortals are possessed that they often direct their minds down untrodden paths, in the groundless hope that they will chance upon what they are seeking."[36] But over the course of the early modern period, this negative stance toward curiosity changed. Curiosity was seen more and more as an advantageous trait, and Mersenne himself contributed to this process. It was increasingly seen as an essential characteristic of humans, related to discovery and a divine quest for knowledge. Mersenne's associate Thomas Hobbes described it as the "perseverance of delight in the continual and indefatigable generation of knowledge."[37]

Most of Mersenne's own reflections on curiosity can be found in his *Questions theologiques* of 1634. In this work, he comments that there are

30 Marr, "Introduction."

31 Augustine, *Confessions*, 211 (10.54).

32 Augustine, *De Vera Religione*, xlix.94, in Augustine, *Earlier Writings*, 273–74.

33 Isidore of Seville, *Synonyma*, 2.71. Quoted and translated in Pomian, *Collectors and Curiosities*, 59.

34 Bacon, *Advancement of Learning*, 28 (IV.6).

35 Ripa, *Iconologica*, 20. For the juxtaposition of the figure of Curiosity and the figure of Science, see Keller, *The Interlopers*, 3–5.

36 Descartes, *The Philosophical Writings of Descartes*, "Rules for the Direction of the Mind," 15–16.

37 Hobbes, *Leviathan*, 37.

certain pleasures that are found in movement and discovery, and not in possession. The most fleeting or glancing contact with a thing could create a deep moment of joy, different from the pleasure felt by possession. The feeling Mersenne attempts to describe here is also evident in his scholarly activities. Never arriving at the ultimate truth, he writes, "We always desire to tread beyond, so that the acquired truths only serve as means to arrive at others."[38] Later on in the book, Mersenne compares this feeling to listening to music: "the continuation of different harmonies pleases us more than the repetition of the same harmony, even when this is the most pleasant one in all of music."[39] Similar to a listener enjoying sequence of harmonies rather than the repetition of a single one, the curious scholar is thus someone who always searches beyond what is already known.

Mersenne's *Questions theologiques* not only contains a defense of curiosity; the book also simultaneously re-defines what exactly should be considered a curiosity. He defines curiosities as non-necessities for life and sustenance, considering them "those [things] that are easy to do without."[40] Building upon the widespread idea that the Indigenous populations of the Americas are underdeveloped and need to be evangelized, he writes:

> Humans show clearly by their behavior and tasks that they devote more time to curiosities than to necessities, for the Canadians and many other nations demonstrate by their ways of life that reading, writing, and all the other arts of which I will speak later are not necessary and as such may be counted as curiosities.[41]

As in many colonial writings of the period, Indigenous peoples of the Americas are depicted as absent of any cultural or religious practices and characterized as favorable for colonization and evangelization. Mersenne may have taken this from Samuel de Champlain, one of France's leading

38 "Et puis l'on desire tousjours passer outre, de sorte que les veritez acquises ne servent que de degrez pour arriver à d'autres." Mersenne, *Questions theologiques*, 114. For more this passage, see Daston and Park, *Wonders*, 308.

39 "la suite de differents accords nous plaist davantage, que la continuation d'un mesme accord, encore que cettuy cy soit le plus agreable de toute la Musique." Mersenne, *Questions theologiques*, 232.

40 "ceux dont il est aysé de se passer." Mersenne, *Questions theologiques*, 1.

41 "Or les hommes monstrent evidemment par leur procedé, & par leurs exercices, qu'ils donnent plus de temps aux curiositez qu'aux choses necessaires, car les Canadois & plusieurs autres nations témoignent par leur façon de vivre que l'art de lire, & d'écrire, & que tous les Arts, dont je parleray apres, ne sont pas necessaires; & consequemment qu'ils peuvent estre mise au nombre des curiositez." Mersenne, *Questions theologiques*, 2.

colonizers of North America, who emphasized in his travelogue that the Indigenous populations he encountered did not have any laws or cultures of worship. André Thevet's *Cosmographie universelle* (Paris, 1575) likewise listed things that Amerindians lacked, something Michel de Montaigne turned into a positive trait in his essay "On cannibals."[42] Since Mersenne was convinced that reading, writing, and other arts were absent from North America, he considered them curiosities. They could not be necessities, since they were not shared by all people in the world. Sciences such as mathematics were also considered curiosities, as well as gardening and playing music. In *Questions theologiques*, Mersenne concludes that "the greater part of life and labor of humans is spent on curiosities."[43] Rather than being relegated to the realm of idleness or unnecessary luxuries, the concept of curiosities is expanded to include mathematics, flowers, paintings, and music. Only those activities as essential in order to survive are not curiosities.

The idea of curiosities as "non-necessities" was certainly no indictment of engaging with these matters. Mersenne considered it wrong to spent time on the sciences if it did not advance knowledge of God. He thus dismissed the efforts of natural philosophers to discover the true meaning of the universe, which he considered heretical. Only the maker of the universe could know this. When a chemist tried to find the philosopher's stone, Mersenne considered this immoral, since it is impossible to know how God would make such an object.[44] All scientific knowledge would remain incomplete in this life, Mersenne argued. Only in the afterlife would "the blessed" know the effects and workings of nature. Emphasizing the utility of many curiosities, he pointed out that the study of curiosities could be very useful for the advancement of science and technology in this life but also in preparation for the afterlife. Here he departed from the negative description of curiosity within Catholic philosophy and turned it into a

42 "no knowledge of letters, no science of numbers." Montaigne, *The Complete Essays*, 153. For such lists on what Amerindians lack written by French contemporaries, see Whatley, "Savage Hierarchies." The work by Champlain and many other travel accounts were present in the Minim library; see Paris, Bibliothèque Mazarine, Ms. 4146. I would like to thank Emily Kent for pointing this out to me. In his study of the library, Whitmore counts 234 volumes in this section of the catalogue, making it the second largest section. Whitmore, *The Order of Minims*, 124.

43 "D'où l'on peut aisément conclure que la plus grande partie de la vie & du labeur des hommes s'employe au curiositez." Mersenne, *Questions theologiques*, 5.

44 He singled out the English Robert Fludd and the German physician and alchemist Heinrich Khunrath. Mersenne also dismissed sciences that he thought had no hope of success, for example trying to find the quadrature of the circle and attempting to reproduce the musical effects of the ancient Greeks. For Mersenne's criticism on these philosophers, see Hine, "Mersenne and Vanini"; Hine, "Marin Mersenne"; Fabbri, "Questioning Fludd."

fundamentally positive trait, perhaps to legitimate his own scholarly choices and scholarly appetite.

Later in *Questions theologiques*, Mersenne turns curiosity into even more of a virtue. Here he describes human inquisitiveness as an imitation of the divine, reflecting God's delight in diversity: "if we speak of nature as the author of the universe, it is certain that it is pleased by diversity."[45] Since the pleasures that humans experience on earth are only ephemeral, human curiosity is not a perfect image of this divine curiosity, but it does follow the same principles. The fact that the senses take pleasure in variety is understood here as a preparatory step toward a more divine sense of curiosity. According to Mersenne, each sense aspires to "contain all things which it is capable of receiving."[46] So the ear strives toward containing all possible consonances and dissonances. This is understood as a prelude to the state of humans in paradise, where they will have complete understanding of the divine essence and where all the consonances are encountered in their ultimate perfection. He writes,

> It is enough to know and experience that the objects of our senses do not have their perfection here, or that the senses are deprived here [in this life] of what they will have after the resurrection. But speaking only now of the senses that we use now, we may say that they are pleased by the variety of their objects so that they are able to find elsewhere what they cannot find in one thing, and that they may form some idea of perfection to which they aspire.[47]

Emphasizing that life on earth is only a preparation for the afterlife, Mersenne sees the human mind as repository of sensory experiences, increasing its knowledge and coming closer to the perfection that will be found in paradise. Curiosity here acts as a guide, attempting to see and hear what could be useful. Knowledge would always remain imperfect in

45 "Si nous parlons de la nature qui signifie de l'Autheur de l'univers, c'est chose asseurée qu'il se plaist à la diversité." Mersenne, *Questions theologiques*, 229.

46 "chaque sens doit contenir toutes les choses dont il est capable." Mersenne, *Questions theologiques*, 233.

47 "Il suffit de sçavoir & d'experimenter que les objects de nos sens n'ont pas icy leur perfection, ou que les sens sont privez de celle qu'ils auront apres la resurrection: Mais parlant seulement des sens dont nous nous servons maintenant, nous pouvons dire qu'ils se plaisent à la varieté de leurs objects, afin qu'ils puissent trouver separément, ce qu'ils ne peuvent rencontrer en une chose, & qu'ils se forment quelque idée de la perfection à laquelle ils aspirent." Mersenne, *Questions theologiques*, 234.

life on earth, but by obtaining as many impressions as possible, humans could gain insight into the perfection of afterlife. Mersenne compares the human mind to "a king in a kingdom" who sends out officers and valets to stock up on everything in nature, not only out of necessity but also out of "pleasure and fulfillment."[48] In the realm of listening, this means that the ear seeks out the pleasures of the different intervals and melodies much as collectors gather objects in their cabinets. By redefining what exactly counts as a curiosity and the function curiosities have in advancing science, Mersenne provides a moral and theological foundation for being curious. It is not heretical, but the opposite: an imitation of the divine. Moreover, by utilizing the metaphor of musical listening, Mersenne makes clear that curiosity is not just something of the eye. The ear seeks out new experiences as well, and musical listening provides a key metaphor for understanding the curious mind.

As many historians have shown, the early modern period saw an intense debate on the uses and merits of scholarly collecting. The English natural philosopher Francis Bacon projected science "as a process of collecting," though he was certainly not the first to make this claim.[49] Like many of his contemporaries, Mersenne was intrigued by Bacon's proposals. In the 1620s, he corresponded with numerous scholars on the English statesman and began his own translation of the work on sound found in Bacon's *Sylva Sylvarum*.[50] Like Bacon, Mersenne pointed to the value of artisans, gardeners, and instrument makers for the sciences; many of Mersenne's early experiments can be linked in some way to Bacon's writings, though there were some key differences between the two scholars. In several of his books, the Minim was highly critical of Bacon and associated him with those who seek to establish apodictic truths. Unlike Bacon, Mersenne did not think that it was possible to reach to "the bottom of things."[51] Again

48 "comme un Roy dans un Royaume … non seulement en ses necessitez, mais aussi à ses plaisirs & contentemens." Mersenne, *Questions theologiques*, 236. On the idea of collecting as an imitation of the divine, see Bredekamp, *Antikensehnsucht und Maschinenglauben*, 70.

49 Swann, *Curiosities and Texts*, 55. On the collecting aspects of early modern science, see, among others, Findlen, *Possessing Nature*; Cook, *Matters of Exchange*; Smith and Findlen, *Merchants and Marvels*; Bleichmar and Mancall, *Collecting Across Cultures*.

50 The translation is found in Paris, Bibliothèque nationale, MS 2884, fols. 61r–68r. Mersenne was one of most important disseminators of Baconian philosophy in France, but not the only one. In 1631 the book *Histoire naturelle de Mre François Bacon* was published in Paris, containing an edited and translated version of Bacon's *Sylva Sylvarum* by Pierre Amboise. For more on this, see in particular the two articles by Claudio Buccolini: Buccolini, "Mersenne Translator of Bacon?"; Buccolini, "Mersenne et la philosophie baconienne."

51 Bacon, *The New Organon*, 65–66. For Mersenne's critique, see Mersenne, *La verité*, 212–13.

and again, he emphasized that humans can only experience surfaces. He wrote to Constantijn Huygens in 1640 that "theology teaches me that we are far beneath these celestial spirits, who see all things in God according to their essence and truth, instead of knowing only the effects and surface of things."[52] Ultimately, he estimated Bacon's designs as impossible and argued that they only serve to come up with new experiments. Because of this, Claudio Buccolini has stressed that Mersenne's interest lay "mostly in practical and experimental aspects" of the statesman's work.[53] Like Bacon and many of their contemporaries, Mersenne would stress the importance of collecting, both on paper and in person.

The Collecting Impulse

Mersenne's comparison between the senses and the king collecting for his cabinet is especially interesting since it connects the mental collection of auditory experiences to the physical collection of objects. The comparison between the inner and outer world was a popular theme in the early modern period. John Locke, for example, described the process of learning as follows: "the senses at first let in particular ideas, and furnish the yet empty Cabinet."[54] Allegorical depictions of the senses portrayed female figures within luxurious cabinets, surrounded by objects such as instruments, sheet music, and animals. Both the senses and the cabinet demonstrated the central role of variety. When collectors filled their cabinets, this was typically motivated by pleasure, but in Mersenne's interpretation it also served as a prelude to the possibilities in the afterlife. Not knowing what might be useful in the next life, humans should aspire to collect. Transgressing the walls of the cabinet, the interest in collecting was visible in a wide variety of realms, as the friar belonged to a generation that increasingly depicted scientific progress as a process of collecting and the cabinet as the ultimate embodiment of this spirit.

The early modern period saw an incredible number of cabinets, museums, and Kunst- and Wunderkammern. Across Europe collections were established to manage this explosion in materials. It was a fundamentally

52 "car la Theologie m'aprend que nous sommes bien au dessoubz de ces espritz celestes, qui voyent toutes choses en Dieu selon leur essence et leur verité, au lieu que nous ne cognoissons icy que les effectz et l'escorce des choses." Mersenne to Constantijn Huygens, 14 November 1640, CM 10, 237.

53 Buccolini, "Mersenne Translator of Bacon?," 44.

54 Locke, *The Locke Reader*, 126 (*An Essay*, 1.2.15); Bolzoni, *The Gallery of Memory*, 245.

social activity. Collecting, Paula Findlen writes, "had become an activity of choice among the social and educated elite .... Through the possession of objects, one physically acquired knowledge, and through their display, one symbolically acquired honor and reputation that all of men of learning cultivated."[55] The cabinet of curiosity was one of the central places where curious men and women gathered.[56] Collectors tried to create miniature versions of the larger universe in their cabinets and museums, a microcosm in which all objects could be gathered and categorized.

The cabinet of curiosities provided a site to play with the boundaries between art and nature and to experiment with order and disorder. In such spaces, early modern scholars performed chemical experiments, discussed art, investigated automata, and heard rare musical instruments. Like many others, Mersenne singled out the cabinet as especially valuable for the progress of science. He wrote that it would be desirable if the owners of cabinets could "take note of what is most exquisite in each kind" and make public what utility these objects could have "for the arts and the sciences."[57] Instead of hiding their collections from society, Mersenne encouraged collectors to contribute to the public good. As an example of how this could be done, he pointed to the work on rare shells by the artist Bernard Palissy. The work of the sixteenth-century craftsman was for Mersenne an example of how the workings of nature could be discovered through careful observation.[58] Although objects like shells or corals were collected for their value and appearance, they were also seen as central objects for understanding the underlying geometry of nature. Mersenne argued that curiosities within cabinets and museums could be especially important in scholarly efforts.

Musical instruments were esteemed possessions in such cabinets. They were often grouped with mechanical instruments such as telescopes, optical devices, or automata, for example in Samuel Quiccheberg's *Inscriptiones* (Munich, 1565).[59] In Italy and France, there were numerous extensive collections of instruments, for example that of Mersenne's associate Pierre Trichet. Trichet was a lawyer for the parliament in Bordeaux and well known for his

55 Findlen, *Possessing Nature*, 3.

56 See especially Pomian, *Collectors and Curiosities*, 45–64.

57 "Il seroit à desirer que ceux qui ont des cabinets tres-rares, remarquassent ce qu'il y a de plus exquis dans chaque genre, & qu'ils advertissent de l'utilité que l'on en peut retirer pour les Arts, & pour les sciences." Mersenne, *Questions theologiques*, 5–6.

58 Mersenne, *Questions theologiques*, 6.

59 Quiccheberg, *Inscriptiones*, n.p. ("Quarta Classis"). On musical instruments in early modern cabinets, see especially Seipel, *Für Aug' und Ohr*; Falletti, Meucci, and Rossi-Rognoni, *Marvels of Sound and Beauty*; Cavicchi and Vendrix, "L'érudit et l'amateur."

cabinet of curiosities. In 1635 he published a description of his cabinet, which shows that he had an extensive collection of plants, minerals, fossils, corals, glasses, and mechanical objects. Many objects demonstrate the increasing number of voyages overseas and the influx of non-European materials in early modern France, such as Chinese combs, Indigenous pipes, and rare animals. His cabinet also contained more than 4,000 volumes, a collection of medals, and an extensive collection of musical instruments, including harpsichords, flutes, and string instruments.[60] The interplay between sonic and visual curiosities in Trichet's cabinet must have been especially impressive and makes clear that the cabinet of curiosity addressed not only the eye but also the ear. A contemporary description mentions the sound of the instruments, animating the many dead animals such as crocodiles, snakes, and dragons, turning the collector in a present-day Orpheus.[61] Trichet himself worked on a treatise on musical instruments, which in many ways functioned as the written counterpart for his physical collection.[62]

Much of the evidence on a visit to such a collection comes from Mersenne's journey to Italy in 1644 when he visited the cabinet of Antonio Goretti in Ferrara. Goretti had a large musical collection, widely described by contemporaries. One earlier visitor described it as follows: "one room with all sorts of ancients and modern instruments, both winds and strings, of exquisite beauty and quality."[63] In 1594 Hercole Bottrigari described this collection as well and observed that whereas some instruments were still playable, others were just admired for their ingenuity or rarity.[64] In his own description, Mersenne mentioned three spinets, one strung with gold strings, one with silver, and one with steel, each of them playable.[65] Reflecting on the collection, the Italian author Giovanni-Andrea Spinola wrote to Mersenne that the collector possessed "everything concerning music and sounds, that which would appeal to every king and especially

60 Trichet, *Synopsis rerum*. The musical instruments are mentioned under "Instrumenta Mechanica."

61 Godefroy, *Ode sur les raretés*, 1.

62 Paris, Bibliothèque Sainte-Geneviève, MS 1070. A partial transcription is published by François Lesure in Trichet, *Traite des instruments*.

63 "ha in una camera ogni sorte di stromenti Antichi e Moderni ... ma tiene ancora con ordine bellissimo in un'altra Stanza tutta la Musica Antica, e Moderna, così da Camera come a Chiesa, che sia possibili tirovarsi." Piccini, *Intavolatura*, 8. Translation by Flora Dennis in Dennis, "When is a Room a Music Room," 41–42.

64 Goretti was believed to have acquired several instruments of the Estense court after the death of Alfonso II, including Nicola Vicentino's archicembalo. Bottrigari, *Il Desiderio*, 50.

65 "triplice spinetam habet, quarum prima chordis aureis, secunda argentis, tertia chalybeis instruitur." Mersenne, *Novarum Observationum*, 165.

Figure 3.1: The cabinet of Manfredo Settala in Milan. Paolo-Maria Terzago, *Museo o galeria adunata dal sapere* [...] *di Manfredo Settala* (Tortona: Viola, 1666), n.p. (pasted at end of the book). Paris, Bibliothèque nationale de France, 16-S-492.

the most illustrious professors in the science of music."[66] In his description, he specifically singles out the instruments using sympathetic vibration, the collection of music treatises, and a collection of portraits of musicians and composers. Drawing a connection to his own efforts, Mersenne wrote that Goretti had physically collected almost all instruments that he himself had tried to describe in his *Harmonie universelle* a decade earlier.[67]

Goretti's spinets with silver and golden strings demonstrate that rare and valuable instruments were of special interest for collectors and visitors. This is confirmed by the catalogues of many other cabinets. Manfredo Settala's museum in Milan included many musical instruments described as "rare" and "curious."[68] Between astrolabes, paintings, vases, crocodiles, and animal horns, musical instruments were displayed, often with extravagant materials and decorations (fig. 3.1). Rare, (pseudo-) ancient, and exotic instruments were eagerly collected. Of special interest were instruments from outside Europe.[69] The German music theorist Michael Praetorius depicted several African instruments in his *Syntagma musicum* (Wolfenbüttel, 1619), most likely drawn from physical collections.[70] Interestingly, the catalogue of Settala's cabinet makes an explicit reference to Mersenne's *Harmonie universelle* and *Harmonicorum*

66 "qui habet domi quae ad Musicam et sonos spectantis posset ostendere viris ad addica ... Regibus omnibus grata, sed gratissima Professoribus celeberrimis in musicae scientia." Spinula to Mersenne, 10 April 1645, CM 13, 433–34.

67 Mersenne, *Novarum observationum*, 165.

68 Terzago, *Museo ò galeria*, 363.

69 Terzago, *Museo ò galeria*, 367–68.

70 Meyer, "In Sound Similar to the Harps."

as books that taught musical discourse.[71] Much like Mersenne's book, Settala's cabinet aimed for a certain completeness, although common European instruments are not described in the catalogue so as not to bore the reader.

Throughout Mersenne's letters, the cabinet appears as an object of interest and source of materials. In the process of writing *Harmonie universelle*, Mersenne got in touch with several such collections, for example through the French author and classicist Jean-Jacques Bouchard. In 1635 Bouchard wrote to Mersenne that he had shown parts of *Harmonie universelle* to Cardinal Francesco Barberini, who especially approved of the images. According to Bouchard, the cardinal would send someone to the cabinet of Francesco Gualdi to copy "certain very rare and extravagant instruments" that he had in his possession.[72] Gualdi's cabinet of curiosities in Rome was well known, and the collector had at least one Egyptian sistrum in his possession.[73] Giovanni Battista Doni later reported to Mersenne that Gualdi only had a couple instruments, such as a psalterium and an Ottoman string instrument. But according to Doni, the collector was not eager to share his collection.[74] Gualdi was hardly the only Italian collector approached on behalf of Mersenne. Mersenne also received drawings of coins from the collection of the Italian collector Cassiano Dal Pozzo, which was also studied by Doni in preparation for his work on ancient stringed instruments.[75] The French librarian Gabriel Naudé and astrologer Jacques Gaffarel, both active in Italy, were also asked to find depictions of instruments. Several of the images that Naudé and Gaffarel collected were ultimately included in *Harmonie universelle*.[76] Late in his life, Mersenne still inquired with Constantijn Huygens on getting help to obtain a drawing of a certain type of lute that the French statesman Pierre Séguier had in his possession.[77]

71 "ch'hanno insegnato con le stampe il discorrerne con termini adattati all'arte musicale." Terzago, *Museo ò galeria*, 364.

72 "pour copier certains instruments fort rares et extravagans." Bouchard to Mersenne, 1 January 1635, CM 5, 2.

73 The Englishman John Evelyn described the cabinet in his diary, but the only musical instrument that Evelyn mentioned was an antique Roman horn. Evelyn, *Diary*, 165. A plate of the sistrum has survived among Peiresc's papers. See Paris, Bibliothèque nationale, MS fr. 9530, fol. 176r.

74 Doni to Mersenne, 8 September 1635, CM 5, 392.

75 Palisca, *G.B. Doni's Lyra Barberina*, 47–61.

76 Mersenne's reliance on Gaffarel is remarkable, since they were in a vehement discussion a decade before. When Gaffarel had published a counterattack on Mersenne's criticisms on Kabbalah in his *Quaestiones*, Mersenne retaliated by calling Gaffarel ignorant and crazy. For this episode, see Taussig, "La réception des curiositez inouyes."

77 Mersenne to Constantijn Huygens, 17 April 1648, CM 16, 231.

The person who did the most to provide Mersenne with rare materials was Peiresc, who possessed several exotic and rare musical objects, including horns and plaques with depictions of musical instruments. Peiresc's cabinet in Aix was well known across Europe, containing Egyptian mummies, antique tripods, a large collection of coins, chameleons, and many drawings.[78] The published inventory of Peiresc's cabinet is divided into manuscripts and books, stones, sculptures, and metal works such as coins.[79] How many musical instruments Peiresc possessed remains unsure. From the correspondence with Mersenne, it is certain that he had many decorative objects with musical depictions, such as vases and medals. Several of these objects were discussed extensively in his correspondence with Mersenne, as he offered his own collection to Mersenne and guided him to other cabinets. For example, he connected Mersenne with a Parisian collector (again described as "curieux") who owned an Egyptian musical instrument (a sistrum).[80]

Mersenne got into contact with Peiresc in early 1633 through their mutual friend Pierre Gassendi. Gassendi, who knew both men well and resided in southern France, pitched Mersenne's project to Peiresc, who seemed to have been pleasantly surprised: "as for father Mersenne's [book on] music, the account you give me of it makes me think more highly of it than I could possibly have imagined."[81] The social hurdles to address Peiresc directly were possibly too high for Mersenne and instead required Gassendi's mediation. It most likely also did not help that Peiresc was involved with Mersenne's argument with Gaffarel several years earlier. But Gassendi's proposal of his project on universal harmony seemed to have been convincing, and Peiresc was eager to help Mersenne from the start and provide him with materials on ancient and non-European music. Peiresc immediately mentioned drawings that he had in his possession and instruments that he had heard about, such as the recently invented "Sambuca lincea" by the naturalist Fabio Colonna, a keyboard instrument that could play the diatonic, chromatic, and enharmonic tetrachord species. He also mentioned that he had seen a great number of ancient statues and depictions that included musical instruments but that he had neither the determination nor time to study them in depth.[82] Over the coming years, he would continue to seek out

78 Jaffé, "Peiresc," 307; Schnapper, *Le géant, la licorne, la tulipe*, 139–40, 161.

79 Chapard, *Fabriciani*.

80 Peiresc to Mersenne, 13–15 October 1633, CM 3, 504.

81 "Quant à la musique du P. Mercenne, la relation que vous m'en faictes, m'en faict prendre meilleure opinion que je ne l'eusse possible peu concepvoir." Peiresc to Gassendi, 12 December 1632, CM 3, 351.

82 Peiresc to Gassendi, 12 December 1632, CM 3, 351–52. Colonna published his description in 1618; see Colonna, *La sambuca lincea*.

materials for Mersenne and connect him with other collectors. Peiresc did in turn also desire things from Mersenne, such as a list of all ancient authors on music that he had encountered. The Minim was also frequently asked about items in Peiresc's possession, such as a Chinese plaque or the Delphic tripod found on an excavation in southern France in 1629.[83] In 1634 Peiresc sent a drawing of the tripod's structure to Paris and asked what it could mean and if there were any hidden meanings in the shapes, perhaps related to music or proportions: "I ask you to examine the proportions and dimensions of the attached figure ... to see if you can find anything in there which has a relation to your tones and harmonic proportions, or some other mystery of your philosophy."[84]

The cabinet of curiosity thus not only provided a model for scholarship and the human senses; it was also a repository that could be studied and investigated. It was the ultimate embodiment of the collecting spirit that captured so many of his contemporaries. As Horst Bredekamp has argued, the visual culture of the cabinet of curiosity deeply influenced the thinking of (natural) philosophers like Kepler and Locke. In this vein, the aural components of the cabinet seemed to have played a key role in Mersenne's work.[85] Many cabinets contained musical instruments, such as lutes made from ivory and spinets with golden strings. In his *Questions theologiques*, hearing is imagined as a process of collecting, filling a cabinet with sounds and experiences.[86] That process is materially enacted in this book. Leafing through the pages, readers would have encountered flutes played by peasants in the French countryside, lutes and harpsichords played at the court, and instruments such as a Chinese sheng stored in a cabinet. To see how these objects, experiences, and observations were shared, I turn now to Mersenne's correspondence.

A Paper World

There was a lively culture of sharing musical observations in the early seventeenth century. Many scholars and *honnête hommes* shared their

83 Peiresc to Mersenne, 13–15 October, CM 3, 504.

84 "Je vous prie d'examiner les proportions et dimensions de la figure cy-jointe ... pour voir si vous y trouveriez rien qui eusse du rapport à voz tons et proportions harmoniques ou à quelque autre mystere de vostre philosophie." Peiresc to Mersenne, 18 June 1634, CM 4, 183. On this anecdote see also Miller, *Peiresc's Europe*, 25–26; Ossa-Richardson, "Nicolas Peiresc."

85 Bredekamp, *Antikensehnsucht und Maschinenglauben*, 41.

86 On aural collecting and its relation to the cabinet of curiosity, see especially Dell'Antonio, *Spiritual Listening*; Cypess, *Curious and Modern Inventions*, 148–56.

musical adventures through letters. The astronomer Ismaël Boulliau reported on Venetian music in an extensive letter currently in among the papers of the Dupuy brothers, and in 1639 the viola da gamba player André Maugars wrote a response to what he heard in Italy. Such reports on music were part of a larger culture geared toward sharing observations. Travelers on the Grand Tour were asked to take note of not only what they saw but also what they heard. As Andrew Dell'Antonio writes in his study on early modern Italian listening practices, "collecting, recollecting, parsing/categorizing, and sharing experiences were seen as essential activities of the sophisticated individual."[87] Travel was an important factor in how this musical discourse developed. Musical difference were articulated through travel and the subsequent reflection on touring experiences. As soon as he entered the city of Turin, Boulliau wrote that he realized the distinctions between French and Italian music, noting that the singer's voices were "quite different from ours."[88] Mersenne frequently received information from his correspondents in Italy on Italian composers such as Claudio Monteverdi, Girolamo Frescobaldi, and Stefano Landi. According to the French author Jean-Jacques Bouchard, Landi took six months to give him the rules of music for composing in the Italian style, which he forwarded to Paris. Mersenne described the infatuation for the Italian style of making music as follows in 1635: "From Rome I am told that their music is much better, more studied, and more knowledgeable than ours."[89]

Through the postal networks of early modern Europe, a wide variety of individuals shared their observations and theories with Mersenne. Emphasizing the importance of letters for scholarly conduct, Mersenne imagined a European academy that would be conducted solely by letters. This was preferable to conversations in person because debates could become too heated in cabinets and academies.[90] Although the names of Gassendi, Peiresc, and Descartes are the most prominent, his correspondence contains many regional scholars who are little known—men such as Claude Bredeau, Robert Cornier, Christoph de Villiers, and Gabriel de la Charloyne. Throughout

87 Dell'Antonio, *Spiritual Listening*, 57.

88 "bien differénte des nostres." Transcribed in Launay, "La musique à Venise vers 1645," 271; Maugars, *Response faite à un curieux*.

89 "De Rome on me mande que leur musique vault bien mieux, et plus estudiee et plus sçavante que la nostre." Mersenne to Peiresc, 2 February 1635, CM 5, 47; Bouchard to Mersenne, 1 January 1635, CM 5, 1–2. As discussed in chapter 2, Mersenne received a report from the French physician Bourdelot on a concert by the trumpeter Fantini and Frescobaldi, which he cited in his *Harmonicorum*.

90 Mersenne to Peiresc, 15 July 1635, CM 5, 301–2.

his life, Mersenne corresponded with many of these (amateur) scholars, who were lawyers and physicians by profession and identified themselves as *honnête hommes* or *curieux*. Mersenne especially admired his regional correspondents, writing that "there are often men in small towns who surpass almost all those that are esteemed knowledgeable, particularly in certain parts of the arts and science, which they have found through fine genius rather than acquired by books."[91] He found that although they might lack bookish knowledge, provincial scholars were often inventive and unconventional in their approach. These correspondents reported to Mersenne with answers to his questions, observations on instruments, and data from experiments. In turn, they were kept up to date on the latest developments in the French capital and had an outlet for their interests. The physician Christoph de Villiers, for example, complained to Mersenne that there were few men in his area who were interested in the subjects that he and Mersenne discussed.[92] He described himself as being on an island "where one does not learn about praiseworthy curiosities."[93] Théodore Deschamps similarly complained of the lack of *hommes curieux* in Bergerac (southern France) and was thrilled to hear about "the progress of the sciences by the efforts of so many good spirits of Paris."[94]

One of Mersenne's first regular correspondents was the lawyer Claude Bredeau from Nevers, where Mersenne had taught at the Minim convent and where he most likely encountered the lawyer. After the friar returned to Paris, Mersenne and Bredeau would correspond for over a decade on biblical, classical, and musical matters. Although music featured prominently in their discussions, Bredeau emphasized multiple times to Mersenne that he was not a musician and had been taught "little more than *ut, re, mi, fa, sol*, and *la*."[95] At one point, Bredeau felt uncomfortable with the many questions on music and wrote in 1627, "If so many learned musicians are unable to agree on your questions, how is it possible that someone who is

91 "qu'il y a souvent es petites villes des gens qui surpassent quasi tous ceux qu'on estime sçavoir particulierement en de certaines parties des arts ou des sciences qu'ils ont plustost trouvees par leur bon genie qu'apprises dans les livres." Mersenne to Peiresc 15 July 1635, CM 5, 301.

92 Villiers to Mersenne, 1 February 1641, CM 10, 458–59.

93 "où l'on aprend point les curiositez louables." Villiers to Mersenne, 17 August 1634, CM 4, 318.

94 Deschamp to Mersenne, 14 June 1642, CM 11, 182; "Je suis bien aise du progrès des sciences par les labeurs de tant de bons esprits qui resident à Paris." Deschamp to Mersenne, 8 July 1642, CM 11, 204.

95 "Je vous ay jà escry sur ce sujet que n'estant musicien, il estoit difficile que je peusse satisfaire à vostre affection. Car je n'ay jamais esté enseigné plus avant que *ut, re, mi, fa, sol, la*." Bredeau to Mersenne, 12 December 1625, CM 1, 317.

not a musician could solve anything?"[96] Bredeau had to admit that he did not understand large parts of *La verité*, even the parts on music, as he was not versed in mathematics.[97] Eventually their exchange was limited by Bredeau's bad health. His son would continue to write some of the letters and read Mersenne's *Traité* to his father, but according to Bredeau, this son could not do the research Mersenne wanted, since he knew even less about music.[98]

Although Bredeau was unable to answer the many questions on music, it continued to stimulate a great deal of discussion over the ten years of their correspondence. Like Mersenne, Bredeau was able to move quickly between topics such as ancient musical meters and the physics of instruments. In a letter from 1625, Bredeau presented his ideas on why an organ pipe jumps to one of the harmonics when air pressure is increased. According to Bredeau, the proportion of the notes could be related to the blower's level of strength, with the fault being in the man, not the pipe. He also reported on different instruments and corrected Mersenne when he wrote on the range of the sackbut.[99] The lawyer received multiple requests from Mersenne to inquire with local musicians and builders, such as to investigate the size and tones of the church bells in Nevers. Bredeau obtained this information from the local cantor of the church and forwarded it to Mersenne.[100]

Bredeau's letters demonstrate that circulating questions was a central method for Mersenne. The list of questions was one of the most inconspicuous research technologies of the early modern period. A large part of Mersenne's correspondence consisted of questions, making him famous for his many queries. According to Blaise Pascal, Mersenne had a "particular talent for forming beautiful questions," although not everyone appreciated the stream of inquiries that Mersenne sent from Paris.[101] Constantijn Huygens cautioned Anna Maria van Schurman against entering into correspondence with Mersenne on the grounds that she would be soon deluged and fatigued

96 "Car si tant de doctes musiciens se trouvent empeschez à s'accorder de vos questions, comment est-il possible qu'un qui n'est pas musicien en puisse rien resoudre?" Bredeau to Mersenne, 15 September 1626, CM 1, 500.

97 Bredeau to Mersenne, 15 September 1626, CM 1, 496.

98 Bredeau to Mersenne, 31 August 1627, CM 1, 567; Bredeau to Mersenne, 2 November 1627, CM 1, 588.

99 Bredeau to Mersenne, 16 May 1625, CM 1, 217; Bredeau to Mersenne, 23 February 1626, CM 1, 387.

100 "Monsieur le chantre de l'Ecclise." Bredeau to Mersenne, April 1628, CM 2, 76; Bredeau to Mersenne, 13 July 1628, CM 2, 97–98.

101 "Il [Mersenne] avoit un talent tout particulier pour former de belles questions." Pascal, *Histoire de la roulette*, 1.

with "letters, questions, and problems" on a daily basis.[102] The theologian André Rivet complained that Mersenne's questions were "without end."[103] Many of the letters Mersenne himself sent are unfortunately lost, but it is possible to partially reconstruct the questions he circulated from the answers he received. The letters he received often referred to the questions of the previous letter, containing phrases such as "I now come to your questions" and "but let us return to your questions."[104]

An example of such a collection of questions can be found in a letter from Theodore Deschamps to Pierre Trichet dated January 1635. Mersenne had sent these questions in April 1632 to the physician and chemist Jean Rey. Rey most likely sent them to Deschamps, who forwarded his answers to Pierre Trichet in Bordeaux. Three years after Mersenne had sent his letter to Rey, he was still receiving answers on this list. The questions are as follow:

> Is sound something different than percussions of the air? ... Why does one string of an instrument make several sounds, without being pressed by the finger to shorten it, and why do wind instrument do the same, without opening or closing any holes? ... Why do both string and wind instruments, without pressing the string on the neck or opening or closing any holes, skip from the lowest tone first to the octave, then to the twelfth, and then to the fifteenth? ... Why is it that the trumpet cannot intone *ut*, *re*, *mi*, *fa*, and *sol* in the bass or lower register?[105]

Mersenne shared similar lists with Christophe de Villiers, René Descartes, Isaac Beeckman, and Ismael Boulliau, asking about the production of sound, the creation of overtones, and the pitches of wind instruments like trumpets. Questions were often grouped together around a particular topic, for example those dealing with overtones or the human voice. In one letter to Trichet, Mersenne asked several questions about the extremities of the human voice

102 "epistolis, quaestionibus, problematis." Huygens to Van Schurman, 26 August 1639, CM 8, 484. No letters between Mersenne and Van Schurman have survived.

103 "questions du P. Mersenne qui sont sans fin." Rivet to Saumaise, 25 June 1640, CM 9, 436

104 "Je viens à vos questions." Villiers to Mersenne, 25 February 1635, CM 5, 59. "Mais retournons à vos questions." Villiers to Mersenne, 12 March 1635, CM 5, 98.

105 "A sçavoir si le son est quelque chose de distinct des percussion de l'air? ... Pourquoy une corde d'instrument sans estre pressee du doigt pour l'accourcir, fait divers sons, et les instruments à vent sans ouvrir ni fermer aucun trou? ... Pouryouy tant les instruments à corde que ceux à vent, sans presser la corde sur le manche, ni ouvrir ou fermer aucun trou, sautent du son grave à l'octave premierement, puis apprès à la douziesme, puis à la quinsiesme? ... Pourqouy en la trompette vers le bas ou son grave, on ne peut entonner ut, re, mi, fa sol?" Deschamps to Trichet, January 1635, CM 5, 566–71.

and the movement of the vocal cords.[106] These letters with questions were shared by correspondents themselves and seem to have functioned as a quick device for Mersenne to gauge the knowledge and ideas of new contacts.

It becomes clear that the answers to such questions were used in the process of writing his *Harmonie universelle* through the example of two of Mersenne's most famous correspondents: René Descartes and the atomist schoolteacher and candlemaker Isaac Beeckman. Mersenne rejected atomism as a philosophical principle, but throughout his writings he relied on excerpts from Beeckman's letters, such as his proof of the inverse proportionality of length and frequency.[107] When writing the Latin *Harmonicorum*, Mersenne often quoted their letters, for example when paraphrasing Beeckman's theories on the consonances and citing his assumptions on laypeople singing psalms.[108] At other times, material from the letters was incorporated into the running text without any indication of their origin. Mersenne, for example, cited Beeckman's explanation of overtones word for word without mentioning him.[109] Descartes's letters demonstrate that this excerpting of letters was no exception. Perhaps anticipating this, he asked Mersenne not to include his name in *Harmonie universelle* when he used material from their letters.[110] In the manuscript held at the Bibliothèque de l'Arsenal, Mersenne cites "the reasoning of an excellent mathematician," most likely referring to Descartes.[111] And he continued to anonymize the philosopher, even though portions of the letters are quoted in *Harmonie universelle*, for example in his discussion of the consonances in a letter from December 1629. A comparison between the letter and Mersenne's printed book seven years later demonstrates that he utilized materials from their correspondence word for word without any reference to their origin.[112] Other correspondents

106 Trichet to Mersenne, 27 April 1631, CM 3, 157.

107 Beeckman had notated this proof in his unpublished journal in 1614 and 1615 and communicated it to Mersenne in 1629. Beeckman, *Journal*, vol. 1, 54–55; Mersenne, *Harmonie universelle*, "Livre troisiesme du mouvement," 157–58.

108 For example: "Ita siquidem mihi respondet." Mersenne, *Harmonicorum*, "De sonis consonis," 67; Beeckman to Mersenne, 7 October 1631, CM 3, 201–2. On the singing of psalms, see Mersenne, *Harmonicorum*, "De generibus," 108; Beeckman to Mersenne, March 1629, CM 2, 219–20; Beeckman to Mersenne, June 1629, CM 2, 230–31.

109 Mersenne, *Harmonicorum*, "De organis," 140; Beeckman to Mersenne, 30 May 1633, CM 3, 403–4. Mersenne mentions Beeckman's explanation again at the end of *Harmonie universelle*. See Mersenne, *Harmonie universelle*, "Nouvelles observations," 16.

110 Descartes to Mersenne, 15 May 1634, CM 4, 145.

111 "la raison d'un excellent mathematicien." Paris, Bibliothèque nationale de France, MS 2884, fol. 48r. Transcription from Taveau, "Le manuscrit 2884," 421.

112 Descartes to Mersenne, 18 December 1629, CM 2, 350–51; Mersenne, *Harmonie universelle*, "Livre premier des consonances," 65.

and associates like Villiers, Bredeau, Titelouze, Cornu, and Jean le Maire were similarly quoted anonymously or by name, making clear that these examples were no exception of the extensive use of materials from letters in the printed books. Multiple paragraphs of a letter by René Moreau were reprinted as Mersenne's own words, without any mention of their author.[113] These examples are less conspicuous than the entire treatises presented in his books, such as La Mothe le Vayer's essay on music, Roberval's work on mechanics, Desargues instructions for music, or the anonymous essay on echo research, made by "an excellent mind" in Marne around 1625.[114] But these make clear that Mersenne's process of writing was very much driven by the collective nature of his scientific approach.

The range of topics that Mersenne covered with his correspondents is extensive. Many correspondents were asked about specific acoustic phenomena such as overtones, sympathetic vibration, organ pipes, or echoes. The physicians, lawyers, and collectors frequently wrote on practitioners they encountered or observations they made. These letters thus allowed Mersenne to extend his networks to other localities not within his own reach, such as an organ maker in another town or a book that was hard to find in Paris, such as Praetorius's *Syntagma musicum*. Mersenne's letters also helped to find possible sources in the capital itself. Mersenne was advised by the music master Bomot to look up "Monsieur Vincent," living in the Rue Saint-Jean in Paris, who might know something about sympathetic vibration. And from Villiers, he heard about the organ stops of the instrument in the Parisian church Saint-Jean-en-Grève potentially worth investigating.[115]

The epistolary network here functioned as a virtual community of observers that could replicate experiments and find new examples. After being questioned by Mersenne on echo effects, Villiers wrote to Mersenne that he hoped "to go for a walk three short leagues away when the weather is better" to study reflected sounds more extensively.[116] The Minim friar Gabriel Thibaut described an echo that responded seven or eight times when there was no wind and wrote on a whispering gallery, inquiring with Mersenne on what the physical origins of these acoustic phenomena might be.[117] Descartes also shared his observations on an echo in his own

113 Moreau to Mersenne, date unknown, CM 1, 634–37, lines 32–140; Mersenne, *Harmonie universelle*, "Livre second, des chants," 105–6.

114 Mersenne, *Harmonie universelle*, "Livre premier de la nature," 50.

115 Bomot to Mersenne, 29 June 1642, CM 11, 192; Villiers to Mersenne, 3 September 1635, CM 5, 382.

116 "j'espere faire une pourmenade à troys petites lieues quand il fera plus beau." Villiers to Mersenne, mid-November 1633, CM 3, 546.

117 Thibaut to Mersenne, 18 December 1646, CM 14, 680.

garden. This echo, Descartes wrote, always rendered the same sound, no matter what sound one made, whether speaking high or low or clapping the hands. Namely, the returned sound always sounded like the sound of a hen.[118] When the grass was cut, the echo almost disappeared. Mersenne reacted with skepticism to the description and thought that a simple "Jean des Vignes" tricked the philosopher, but Descartes promised that he himself had observed it, in his own garden, with nobody around. He assured Mersenne that it could also happen with other tall grasses. In the margins of his *Harmonie universelle*, Mersenne describes the whole affair as "strange."[119]

Between the discussions of the mechanics of musical strings and the physics of sound propagation, many of these exchanges can indeed by characterized as "strange." Mersenne's curiosity seemed to have few boundaries and he continued to inquire with his correspondents about rarities and strange occurrences. Claude Bredeau, for example, reported to Mersenne about certain men who can "form two voices with their mouths." He was also told about a man who made marvelous sounds with a small leaf in his mouth. According to the account, he "learned it from some of the Indian peoples."[120] Coincidentally, Peiresc had met the same man in Aix, whom he called "La Feuille," and described him five years later:

> On which point I must tell you that we once had a man called La Feuille, who is employed in the church of Beauvais and who has traveled all over the world, even to the Indies, he claims. He put a small ivy leaf [*feuille de liairre*] in his mouth and imitated the cornet as admirably as one could imagine, going up or down two or three octaves. He said that he had learned this from some very barbaric people in the East and had seen a number of very strange instruments. ... You must see this man, who will give you great satisfaction. He escaped from me so quickly that I was unable to arrange it, and he had promised to return to me on his return from the trip he made to the archbishop's house [archbishop of Aix, Louis de Bretel] in the countryside, but he was so badly received there that he left, much to my regret. You will be able to see him from there and will no doubt get some very rare reports from him, by flattering him a little

118 Descartes to Mersenne 23 August 1638, CM 8, 58.

119 "ce qui est estrange." Marginal note in Mersenne, *Harmonie universelle*, "Livre premier de la nature," 50. Transcribed in CM 8, 69.

120 "qui formoient de leur bouche deux voyx. ... disoit l'avoir appris parmy quelques peuples Indiens." Bredeau to Mersenne, 13 July 1628, CM 2, 97.

> and recommending him to your friends, as I believe he deserves, more for your pleasure than for any other.[121]

Apparently, "La Feuille" had traveled the world and picked up the skill of playing numerous exotic instruments, including an ivory horn in Peiresc's collection. Peiresc hoped that Mersenne could see him soon and obtain rare information from him; if anybody would be interested in this man, it would be Mersenne. To his contemporaries, Mersenne seemed a scholar in desire of wonders. Peiresc himself would later write that he could "easily excuse the credulity of the good father Mersenne, even though it is not free of vice, and do not blame it as much as the incredulity of the others, who neglect all things and want to mock everything."[122] Topics alternated quickly. In a letter from 1635 to Pierre Gassendi, Mersenne recounted the topics he discussed in person with Jean le Maire, which ranged from the effect of wind on the sound of a cannon to two men, "one of whom let himself be pinched and his ear and nose pulled as hard as he wanted without pain, and the other could be hit with all kinds of force on the buttocks without any pain."[123]

After his publications on music in the 1630s, Mersenne remained interested in collecting musical and acoustic observations. He continued to receive descriptions of new instruments, rare acoustic occurrences, and scientific explanations for sympathetic vibration, resonance, and echoes. Many letters were in fact stimulated by his books, as correspondents asked for explanations of what they had read or requested parts of *Harmonie universelle* that their copy did not include. Mersenne also continued to receive and engage in discussions on many things that he had already

121 "Sur quoy il fault que je vous dise que nous avons eu icy un nommé la Feuille qui est employé en l'église de Beauvais, lequel a bien couru le monde et jusques dans les Indes, ce dit-il. Il mettoit une petite feuille de liairre dans la bouche et contrafaisoit le cornet à bouqin le plus admirablement qu'on sçauroit imaginer, montant ou descendant de deux ou trois octaves. Il disoit avoir apprins cela de gents de fort barbares en cet Orient et avoir veu quantité d'instruments bien estrange. … Il faut que vous voyiez cet homme, qui vous donnera de grandes satisfactions. Il m'eschappa si tost que je n'eu pas le moyen de le gouverner, et m'avoit promis de revenir au retour du voyage qu'il fit vers Mr l'Archevesque à sa maison des champs, mais il y fut si mal recue qu'il s'en alla tout despité à mon grand regret. Vous le pourrés voir de par de là et en tirerés sans doubte de tres rares notices, en le flattant un peu et le recommandant à voz amys, comme je croys qu'il merite, plus pour vostre goust que pour aultre quelconque." Peiresc to Mersenne, 13–15 October 1633, CM 3, 503–4.

122 "C'est pourquoy j'excuse facilement la credulité du bon P. Mercene quoyque non exempte de vice." Peiresc to Jacques Dupuy, 20 May 1636, CM 6, 68.

123 "l'un se laissoit pinser et tirer l'oreille et le nez tant fort que l'on vouloit sans douleur, et l'autre frapper de toute sorte de force sur les fesses sans aucune douleur." Mersenne to Gassendi, 17 November 1635, CM 5, 486.

dismissed in his published writings, suggesting an open-ended science that was already described in his *Questions theologiques*. For example, in 1642 Baptiste Masoyer Deshommeaux send him an explanation of the human voice stating that it is caused by the soul emanating from the lungs, an Aristotelian notion that Mersenne had already dismissed several years earlier.[124] The same author gave Mersenne an explanation of the interval division based on Plato's *Timaeus*, arguing that this should still be the foundation for the art of composition.[125] It is difficult to say what Mersenne would have found useful from these observations, as often only the letters to Mersenne have survived and not his answers. It is, however, clear that Mersenne was convinced of the epistemological strength of gathering experiences and observations from a wide of group as people. In the words of William Ashworth, "Mersenne concluded that the problems of science and faith could be worked out by gathering the opinions of intelligent men."[126] Although any ultimate truths were out of reach, some sort of consensus was necessary. The correspondence had a central role in this respect. It allowed Mersenne to gather many observations and theoretical explanations, always moving onward to the next subject.

Circulating Notes and Images

The letters that were circulated frequently contain a variety of visual formats, be it diagrams, drawings, engravings, and musical notation depicting a scale or cadence to entire compositions. Mersenne seemed to have frequently circulated such materials together with his letters, but since many letters have been lost or only the main text has survived, this is difficult to reconstruct. In numerous letters, notes are included within the text, such as in Christoph de Villiers's description of *musica ficta*, where a small staff is interspersed between the lines of text. Villiers also included entire solmization systems in his letters, carefully writing around the tables, to which Mersenne refers in the Latin version of his book.[127] Mersenne himself

124 Deshommeaux to Mersenne, 10 September 1642, CM 11, 262.

125 Deshommeaux to Mersenne, 7 July 1642, CM 11, 200.

126 Ashworth, "Catholicism and Early Modern Science," 139.

127 Villiers to Mersenne, 15 October 1633, CM 3, 517; Paris, Bibliothèque nationale, MS 6205, fol. 369r; Mersenne, *Harmonicorum*, "De generibus," 102–3. Solmization is a method for attributing distinct syllables to each note. Traditionally, the system included six syllables, but over the course of the seventeenth century, a seventh syllable was included. The topic was discussed at length in Mersenne's correspondence, especially in the letters exchanged with the physician

included notes, diagrams, and compositions to solicit musical opinions and theories. When the astronomer Boulliau was in Venice, Mersenne asked him to show a musical diagram to "the most excellent musicians, whether practitioners or theorists, in Venice or elsewhere."[128] With the provincial nobleman Gabriel de La Charlonye, Mersenne frequently circulated musical compositions. For example, for a musical contest in Angoulême, Mersenne asked Peiresc to circulate this musical problem for a prize contest with any musicians in his vicinity: "If you have some excellent musicians (as I have no doubt there are many good masters in the Provence), you could suggest to them this piece [*desfi*] that has been sent to me from Angoulesme, where someone believes that there is no one in Paris who can manage it."[129] Mersenne frequently asked La Charlonye for musical advice, for example about cadences that he sent in anonymized form to the scholar.[130] On one occasion, Mersenne forwarded a canon, which La Charlonye heavily criticized, writing that Mersenne could have already experienced this on the spinet if he had made the effort to play its parts.[131] La Charlonye was not a professional musician, but, perhaps partially because of this, Mersenne seemed to have regularly sought out his advice on musical matters, even printing one of his compositions in *Harmonie universelle*.[132]

Such exchange about musical compositions or specific musical techniques was a regular activity; many of Mersenne's correspondents discussed modes, cadences, and solmization techniques between other topics such as natural history of mathematics. This correspondence was actively put to musical use. Most famously, Mersenne staged a musical competition between the Dutch composer Joan Albert Ban and the Parisian court composer Anthoine Boësset. This matter began in 1639 when Mersenne received several analyses by Ban of compositions by the French composers Fleury, Jacques Mauduit, and Claude Du Chemin. The Dutch composer

Christoph de Villiers: Villiers to Mersenne, Mid-September 1733, CM 3, 484–87; Villiers to Mersenne, 15 October 1633, CM 3, 517–18; Villiers to Mersenne, 10 June 1635, CM 5, 229–30.

128 "les plus excellens Musiciens, soit praticiens, soit theoriciens de Venise ou d'ailleurs." Mersenne to Boulliau, 16 March 1646, CM 14, 136.

129 "Si vous avez quelques excellents musiciens (comme je ne doute pas qu'il y ait quantité de bons maistres dans la Provence) vous pourriez leur faire proposer ce desfi qu'on m'envoye d'Angoulesme, où quelqu'un croit qu'il n'y a personne dans Paris qui en puisse venir à bout." Mersenne to Peiresc, 15 July 1635, CM 5, 305. On La Charlonye, see Wymeersch, "Gabriel de la Charlonye."

130 La Charlonye to Mersenne, 30 June 1635, CM 5, 260–61.

131 "Celuy qui l'a faict, s'il m'eut demandé mon advis, je ne luy eusse pas conseillé de l'offrir à V.R." La Charlonye to Mersenne, 2 July 1634, CM 4, 217.

132 Mersenne, *Harmonie universelle*, "Livre cinquiesme de la composition de musique," 318–19.

argued that the compositions he discussed were fundamentally flawed, as they went again the rules that he himself proposed. Only his own compositions were deemed satisfactory. Ban seemed to have struck a nerve with his critique of beloved composers such as Mauduit en Le Jeune, and Mersenne circulated a French text to Ban with the request to set it to music. Ban's resulting composition was deemed inferior by Mersenne's correspondents, and a lengthy, heated exchange ensued despite Mersenne's observation that debates in correspondence get less heated than those in person.[133] Interestingly, the matter of performance was hardly raised at all, and the musical merit was discussed on paper alone.

Although the controversy surrounding Ban's compositions is the most famous, it is hardly the only discussion and exchange about musical merit. Theodor Deschamps, for example, solicited Mersenne's opinion about the lute music of Ennemond or Denis Gaultier in 1642 with the following words:

> I am also surprised that you have not given me your opinion, as I requested, of my lute tablature pieces in plagal mode. It may be that you found them rough and ungracious. However, I did not send them to you for their beauty but for their novelty, and by disapproving of these compositions, you would not have done anything against me, as I am not the author, but Gaultier, and I have only changed them and transposed them into their opposite mode. However, you will have seen how consonances can be suitably used differently from what is customary.[134]

These discussions between lawyers, physicians, and *curieux* were not isolated but rather were connected to circles of musicians in the specific local towns. Mersenne seemed to have sought out the advice of the musicians in Paris about compositions he received. When he received Stefano Landi's opera *Sant'Alessio* (performed 1632, possibly at Palazzo Barberini, Rome), he wrote to Peiresc, "You must excuse me regarding the music of *Sant'Alessio* if I do not state my opinion so fast, since the King's greatest music masters take

133 Ban's critique was entitled "Anatomia quarundam cantionum musicarum Parisiensium." It is printed in CM 8, 387–96. See also Rasch, "*Me veux-tu voir mourir.*"

134 "Je m'estonne aussi que vous ne m'ayés donné vostre jugement selon ma prière de mes pieces de tablature de luth en Mode plagale. C'est peut estre que vous les avez trouvées rudes et de mauvaise grace. Or je ne vous les envoyay pas pour leur beauté, mais pour leur nouveauté, et en desapprouvant ces compositions, vous n'eussiés rien fait contre moy qui n'en suis pas l'autheur, mais Gaultier, et je n'ay fait que les changer et transposer en leur Mode contraire. Touttesfois vous y avés peu voir comme les consonances se peuvent convenablement employer autrement qu'on n'a accoustumé de faire." Deschamp to Mersenne, 14 June 1642, CM 11, 182.

a good three months to give their opinion on a simple piece of music."[135] Through Peiresc's mediation, Landi's opera (and piece of cultural propaganda for Cardinal Francesco Barberini) reached the composers of the French capital, whose opinion Mersenne solicited. The correspondence also includes many references to unknown "maîtres de musique," for example Monsieur Le Riche, a local composer from Sense whom Christoph de Villiers knew personally. Le Riche would also participate in the contest with Ban and Boësset, but as Villiers wrote, he experienced delay: "I had hoped to send you the musical compositions of Mr. Le Riche, master of music in our town, but the fever has taken hold of him, being quite common for the present climate, in addition to his usual laziness in composing and transcribing."[136] Mersenne's response to these pieces has not survived, but it is clear that aside from exchanges of new and curious examples, there was discussion of many more mundane compositions in an effort to find rules for composition and justify artistic licenses. Others also seemed to have sought methods to scientifically demonstrate what compositions could work and what could not. Baptiste Masoyer Deshommeaux, for example, wrote to Mersenne, "I am sending you the verse you want in four-part music, simply and in counterpoint, with the numbers I use. It seems to me that the science of music is entirely physical. It consists of the body of air sounding pleasant to human ears."[137] As in many other cases, this composition did not survive, but it is clear that the correspondence was central in gauging musical opinions, circulating new examples, and solving music theoretical problems.

These compositions were usually short excerpts or already existing pieces, but at times Mersenne's correspondents were also asked to transcribe music they encountered. As already mentioned before, Peiresc engaged musicians for this purpose, such as Monsieur Simian, "an old friend of our house," who made the music notation for the timbales that Peiresc sent to Paris.[138] On

135 "vous devez m'excuser de la musique de Saint-Alexis si je ne prononce pas si viste mon sentiment, puisque le plus grands maistres de musique du Roy prennent bien trois mois de temps pour dire leur avis d'une simple piece de musique." Mersenne to Peiresc, around 1 September 1635, CM 5, 373.

136 "J'esperois vous envoyer les compositions de musique de Mr Le Riche, maistre de musique de nostre ville, mais la fievre double tierce l'a pris, estant assez comune pour le present en ce climat, oultre la paresse à composer et transcrire qui luy est assez ordinaire." Villiers to Mersenne, 21 August 1640, CM 10, 47.

137 "Je vous envoye le verset que desirés en musique à quatres, simplement et à contrepoinct, avec les nombres desquels je me sers. Il me semble que la science de musique et toutte physicale. Elle consydere le corps de l'air sonnant aggreablement aux oreilles humaines." Deshommeaux to Mersenne, 7 July 1642, CM 11, 199.

138 "des anciens amys de nostre maison." Peiresc to Mersenne, 20 November 1636, CM 6, 154.

another occasion, Peiresc described getting help from a local musician in Marseille to obtain a transcription of music, making clear that this was no exception. Peiresc himself kept such notes in his extensive archive, including notes on the melody of one of his nightingales or the pitches of a bell.[139] When he wanted to send a song sung by a galley rower that he obtained in Marseille, Mersenne had to wait nine months for Peiresc to find the transcription. The difficulty of transcribing becomes clear when looking at the further correspondence on Mersenne's behalf. In 1633 Peiresc repeatedly reached out to several of his contacts in Tunis, Cairo, and Jerusalem to notate melodies. One correspondent was asked to "mark" the differences in singing among different Christian groups at the Church of the Holy Sepulchre.[140] Peiresc also wrote to the French ambassador in Istanbul Henri de Gournay de Marcheville requesting that the French musicians he had brought with him to Istanbul study the musical instruments and ways of singing in the city. According to the antiquarian, these musicians should have no trouble in notating "on paper" the different chants of the Orthodox Church and observing its differences.[141] These requests most likely intimidated even those informers who were trained in the field of music, and only a few answers have survived, perhaps partially due to the difficulty of putting unfamiliar sounds on paper.

Mersenne also experimented with circulating images of instruments. In preparing his *Harmonie universelle* for print, he circulated proofs of the eventual images of the book to his correspondents in Sens, Aix, and Rome. Copies of these images can still be found among Peiresc's papers at the Bibliothèque nationale de France (figs. 3.2 and 3.3). Some images include the range and tuning instruments, which would allow them to be compared with other instruments abroad. Responding to these images, Mersenne's correspondents commented upon proportion and shape or suggested instruments that were not found among Mersenne's examples. He was able to receive reports on any regional differences, for example comparisons of the tuning of the lute and theorbo in Rome with the tuning in use in France.[142] Mersenne received several images of non-European

139 Peiresc to Mersenne, 16 July 1634, CM 4, 242–43; Paris, Bibliothèque nationale de France, MS fr. 9531, fol. 270r; Carpentras, Bibliothèque Inguimbertine, MS 1774, fol. 454r.

140 "pour faire marquer la difference des chants des Grecs aux Cophtes, Armeniens, Maronites, Abyssins et autres." Peiresc to Minuti, 12 October 1633, CM 3, 495. For these and other requests, see Miller, *Peiresc's Mediterranean World*, 106–9.

141 "de la cotter sur le papier." Peiresc to Mersenne, 13–15 October 1633, CM 3, 499; Peiresc to De Gournay, 19 March 1634, CM 4, 80.

142 Villiers to Mersenne, 15 October 1633, CM 3, 518–19. Peiresc and Doni both made observations on the proportions of the instruments and pointed out the lute that was in reverse. Peiresc to

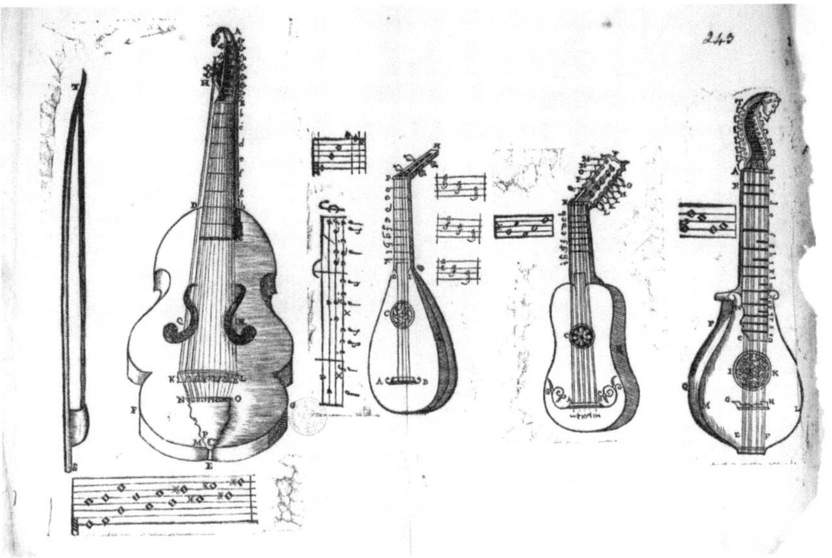

Figure 3.2: String instruments from *Harmonie universelle.* Paris, Bibliothèque Nationale, MS fr. 9531, fol. 243r.

instruments using this method. In his printed books, he even presents an image of a Chinese sheng and Indian veena.[143] He was first alerted to the Chinese sheng in 1634 and tried to locate a drawing of it, which he eventually obtained through Claude Hardy.[144] The circulation of images was a central method in obtaining more examples and observe regional differences. The librarian Gabriel Naudé, for example, circulated proofs of Mersenne's images "to some *curieux* among my friends" in Italy to see if they had any such instruments in their collections.[145] But many of such images were from marbles or coins, containing little information on their use, sound, or acoustic principles. When he received an image of an Egyptian sistrum from Rome, Mersenne wondered whether the animal on top was a cat, lion, or ox.[146] In response to the images he received from Doni in Rome, he expressed his disappointment that nobody seemed to know how to use these instruments.[147] Instruments on paper could only tell so much.

Mersenne, 18 June 1634, CM 4, 180; Doni to Mersenne, 8 November 1634, CM 4, 388–89.

143 Mersenne, *Harmonie universelle*, "Livre quatriesme des instrumens a chordes," 227–28; Mersenne, *Harmonie universelle,* "Livre cinquiesme des instrumens a vent," 308; Mersenne, *Harmonicorum*, "De instrumentis pneumaticis," 111–12.

144 Mersenne to Peiresc, 2 July 1634, CM 4, 230; Mersenne to Peiresc, 2 February 1635, CM 5, 46.

145 "à quelques curieux de mes amis." Naudé to Mersenne, 12 November 1633, CM 3, 534.

146 Mersenne to Peiresc, 26 May 1635, CM 5, 218.

147 Mersenne to Doni, 2 February 1635, CM 5, 35.

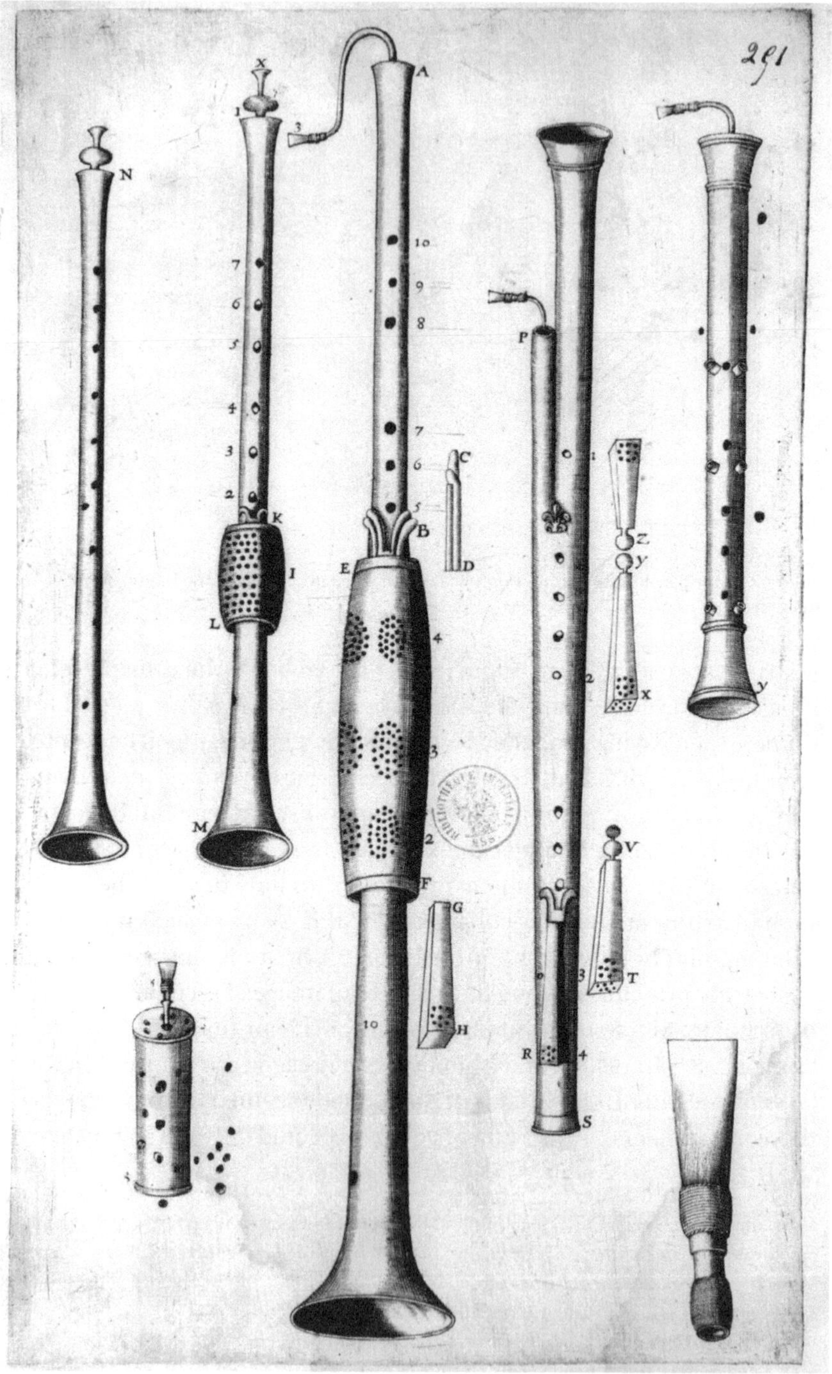

Figure 3.3: Wind instruments from *Harmonie universelle*. Paris, Bibliothèque Nationale, MS fr. 9531, fol. 251r.

Some of the most elaborate images that were obtained through the correspondence are still among Peiresc's papers at the Bibliothèque nationale in Paris such as these images of cymbals and kettledrums (fig. 3.4). The antiquarian had several scribes and artists at his disposal in Aix, who drew animals, buildings, and antiquities for his collection.[148] Stimulated by Mersenne's questions, Peiresc set out to provide the friar with several drawings of cymbals and drums in use in southern France. Peiresc wrote to Mersenne that it was difficult to find musicians who could help with the music examples.[149] Peiresc also wrote that he struggled to find an artist who could draw the instruments in "the right dimensions according to the rules of geometry."[150] When he did receive the images after a year of correspondence, Mersenne thanked the antiquarian, but his curious gaze had already moved onward as he wrote Peiresc with yet another request: he wished to know whether musicians in the Ottoman Empire, the Persian Safavid Empire, Japan, and Ethiopia had similar instruments.[151] And when Mersenne printed the images in his *Harmonie universelle* and *Harmonicorum*, he could only offer an anticlimactic discussion and depiction of this topic. These texts exclude the pitches that the local musician Peiresc had engaged had worked hard to notate and include the figures Peiresc had sent out of respect for his patron and his efforts in "aiding all those who work in the arts and sciences."[152]

Peiresc also promised to send the engraved and woodcut images of instruments along to other correspondents. Seeing the potential to gather examples of instruments in use beyond southern France or Italy, Peiresc had the idea to send the images even further than other correspondents had:

> I have received at the same time the letter which you wrote to me, with the depictions of your lutes, your theorboes, your flutes, and bagpipes, which I was very pleased to see, and I will send them to the Levant to show them what we would like to have of them. If you send me some more copies, it would not be bad to send them to multiple places.[153]

148 Miller, "Looking at the Past."

149 Mersenne to Gassendi, 5 January 1633, CM 3, 355; Peiresc to Mersenne, 13–15 October CM 3, 503.

150 "ses justes dimensions selon les reigles de la geometrie." Peiresc to Gassendi, 12 October 1632, CM 3, 351.

151 Mersenne to Peiresc, 14 May 1634, CM 4, 135.

152 "ayder tous ceux qui travaillent aux Arts & aux sciences." Mersenne, *Harmonie universelle*, "Livre septiesme des instrumens de percussion," 53; Mersenne, *Harmonicorum*, "De campanis," 165.

153 "J'ay recue en mesme temps la lettre qu'il vous a pleu m'escrire, avec les portraicts de voz lucts, de voz tiorbes, et de voz fluttes et cornemuses, que j'ay esté bien aise de voir, et les envoyeray

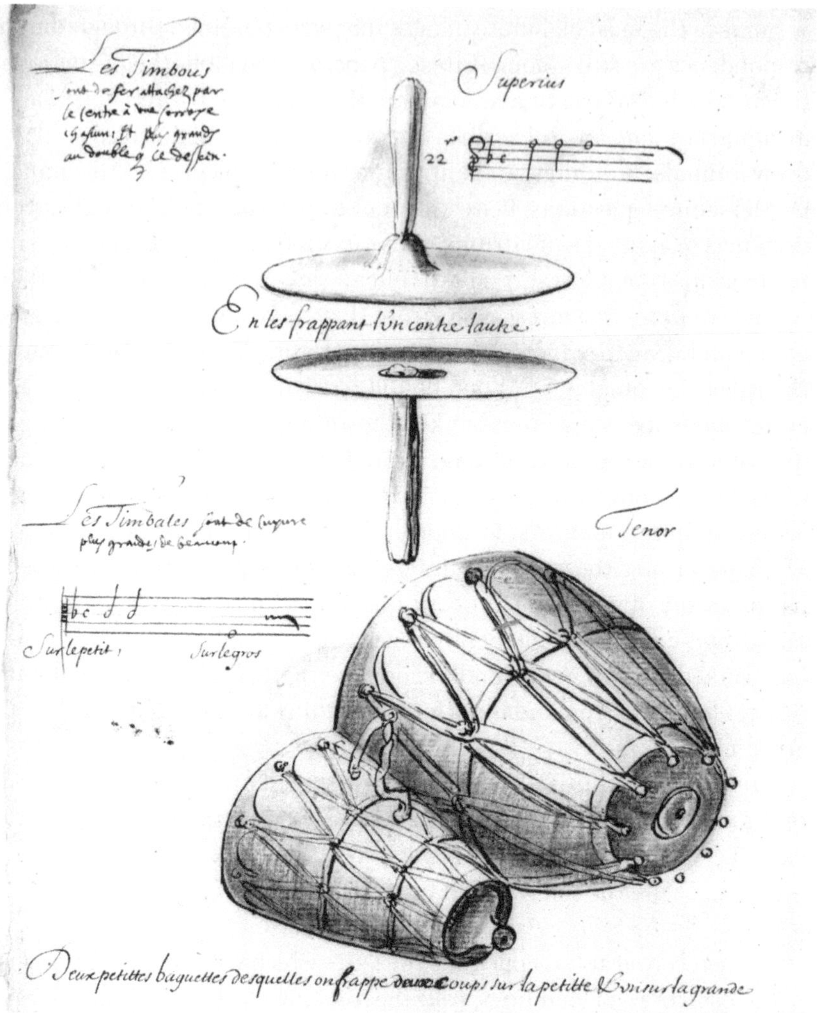

Figure 3.4: Drawings of instruments in use in southern France. Paris, Bibliothèque nationale de France, MS fr. 9531, fol. 272v.

It was thus considered useful to send Mersenne's images not only to correspondents in Sens and Rome but also to Peiresc's many contacts in the eastern Mediterranean. Peiresc was well connected and had his own personal agents in Tunis, Cairo, Jerusalem, and Aleppo. Within Peiresc's subsequent

en Levant pour leur faire voir ce qu'on desireroit avoir d'eux. Sy vous m'en envoyiez quelque autre exemplaire, il ne seroit pas inutile pour envoyer en divers lieux." Peiresc to Mersenne, 6 September 1633, CM 3, 474–75.

letters, the images are described as conversation starters, getting local musicians talking about their music and the different names of musical instruments, with the hope to glean information about the etymology and origins of instruments. As Peiresc wrote to the Minim Théophil Minuti, Mersenne's images should be used

> to show them to those interested in music, and to induce them to communicate the portraits and dimensions of theirs and, if possible, some relation or instruction of the proportions of their tones, notes and musical characters, and the way of singing and sounding of their instruments.[154]

A month later, Peiresc wrote several requests on music to his agents in Aleppo, Cairo, and Jerusalem, attaching Mersenne's images and several requests on music. He also asked for a complete piece of music, hymn, or psalm in their musical notation and perhaps an interpretation by "someone who knows a little of our music."[155] Minuti was asked to pay special attention to the use of musical notes and methods for singing and playing. When words failed, perhaps images could speak.

The correspondence with the physician Christoph de Villiers seemed to have been more successful. In 1633 Villiers and Mersenne started a correspondence that would last for at least nine years. Mersenne utilized several experiences and materials from Villiers in his printed books. Already early in their correspondence, Mersenne sent the images of the instruments in the *Harmonie universelle* to his correspondent.[156] Villiers commented upon the shape and proportion of the instruments, which he considered to be incorrect. Villiers also did not understand the differences between all the flutes and wrote that Mersenne was missing the flutes and recorders used by shepherds and those imitating bird calls.[157] Mersenne did receive new examples of instruments. For example, Villiers sent a description of the mirliton or eunuch flute to Mersenne, which he saw was missing among Mersenne's images. In his letter, he depicted the instrument in the margins and gave an explanation of the sounds produced by the instruments, which

154 "pour les faire voir aux curieux de la musique, et les induire à communiquer les portraits et dimensions des leurs et, s'il est possible, quelque relation ou instruction des proportions de leurs tons, notes et characteres de musique, et façon de chanter et sonner de leurs instrumens." Peiresc to Minuti, 12 October 1633, CM 3, 494.

155 "quelqu'un qui entendoit un peu de nostre musique." Peiresc to Minuti, 11 October 1633, quoted and translated in Miller, *Peiresc's Mediterranean World*, 107.

156 Villiers to Mersenne, 15 October 1633, CM 3, 518–19.

157 Villiers to Mersenne, mid-November 1633, CM 3, 542.

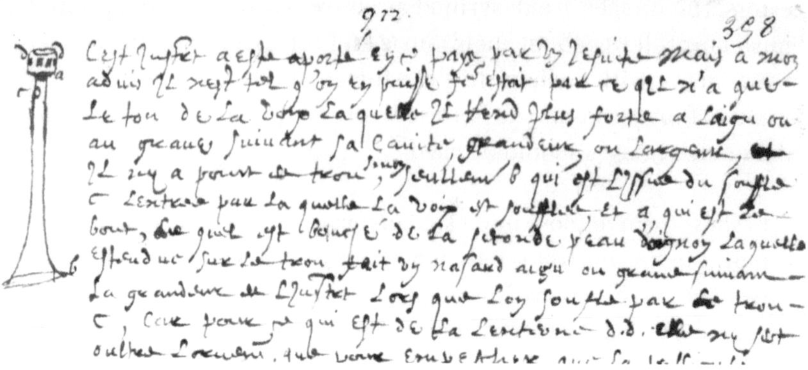

Figure 3.5: Marginal drawing of a mirliton in a letter from Christoph de Villiers to Mersenne. Paris, Bibliothèque nationale de France, MS NAF. 6205, fol. 358r.

he simultaneously dismissed, since the instrument could only alter the natural human voice (fig. 3.5). Mersenne included the eunuch flute in his survey of flutes, where he seems to have also used other sources beyond Villiers's description.[158]

Although Villiers argued that such instruments did not deserve scholarly attention, Mersenne would later explicitly write that he considered all sound-producing objects significant: "science does not pertain to the rich any more than the poor and there is nothing so base and vile in nature or the arts that is not worthy of consideration."[159] At the same time, as Mersenne writes in the introduction to *Harmonie universelle*, he spent more time on instruments like the lute, spinet, and trumpet than other instruments, since they are held in higher esteem than others.[160] The curious gaze of the collector was not only limited to expensive exotica but also included more mundane instruments, such as the mirliton, in the quest for totality. Some instruments were also especially noteworthy because of their method of sound production, such as the trumpet marine, which consists of long wooden body in the shape of a cone with only a single string. The string is not stopped but only lightly touched at the nodal points so that one hears the natural harmonics. Because of its use of the natural harmonics or flageolet tones, it was an

158 Mersenne, *Harmonie universelle*, "Livre cinquiesme des instrumens a vent," 230.

159 "la science n'appartient pas davantage aux riches qu'aux pauvres, & qu'il n'y rien de si bas ny de si vil dans la nature, ou dans les arts qui ne soit digne de consideration." Mersenne, *Harmonie universelle*, "Livre quatriesme des instrumens a chordes," 212.

160 Mersenne, *Harmonie universelle*, "Traité des instrumens a chordes," n.p. ("Preface au lecteur").

especially interesting instrument for later natural philosophers. While the instrument was not widely used at the time, Mersenne could have read about the instrument in several German treatises. In March 1634, Villiers wrote,

> I forgot to tell you that I have been told about an instrument that is not known called the trumpet marine. If you have not heard of it, I will send you an image, as it has been represented to me. And if I see it, as I hope I will, I will give you an even more particular understanding of it. It is not currently in the country. The person who told me about it heard it trumpeting and says that it is heard as far away as the trumpets of war. And yet it is only a wooden instrument, with a triangular body and a stretched string (like your viol), which is touched with the bow and makes the aforementioned sound. ... I have sent you this, because I have not seen any such instrument in the letters you sent me.[161]

At this moment, Villiers had not seen the instrument himself and thus could only give a secondhand account, but since it was missing from Mersenne's materials, he thought that it might be worth describing. By August of that year, Villiers had managed to see the instrument and was able to send Mersenne an image and description in which he describes the size, shape, and sound of the instrument (fig. 3.6). Villiers could not give much detail on why it could only play certain intervals and how its rattling sound is created.[162] Coincidentally or not, the next month Descartes also sent a description of the same instrument to Mersenne, explaining what he thought to be "the secrecy" of the instrument. Descartes, however, had to admit that he was reporting from "assumptions" [*conjecturer*] rather than "experience."[163] In October 1634, Villiers wrote once more about the instrument, happy that

161 "J'oubliois encor à vous dire que l'on m'a fait part d'un instrument qui n'est pas connu, qui s'apelle Trompette marine. Si n'en avez entendre parler, je vous en envoyeray un pourtraict, tel qu'on me l'a representee. Et s'il arrive que je la voye, comme j'espere, je vous en donneray encore plus particuliere intelligence. Elle n'est pas maintenant au pays. Celuy qui m'en a parlé, l'a ouy trompetter et dit qu'elle se fait entendre aussi loin que les trompettes de guerre. Et neantmoins n'est que boys, le corps triangulaire, avec une corde tendue (comme à vostre viole), laquelle se touche avecque l'arcelet et rend le son susdit. ... Je vous ay mandé cecy, parce que je n'ay veu pourtrait de tel instrument dans les vostres que vous m'avez envoyé." Villiers to Mersenne, 3 March 1634, CM 4, 59–60. For a comprehensive history of the trumpet marine and its sources, see Adkins, *A Trumpet by Any Other Name*.

162 Villiers to Mersenne, 17 August 1634, CM 4, 319–22; Paris, Bibliothèque nationale de France, MS fr. 6205, fol. 357r–v.

163 Descartes to Mersenne, around 15 September 1634, CM 4, 360 (date is inferred).

Mersenne had finally been able to see one in person.[164] For the physician, the instrument's sound and specific tones continued to puzzle, and he wrote that it would take "geniuses to give reasons to these secrets." In his eventual description of the instrument, Mersenne again seemed to use his own observations together with the several reports he received.[165]

At other moments, he seemed to prefer personal communication, such as when discussing experimental set-ups or artisanal practices. Such topics were more easily discussed face to face. In 1628 Titelouze expressed his regret that Mersenne did not live closer, as he could show him how to practice the musette in person; this was more difficult to express in words.[166] Similarly, Mersenne excused himself for the poorly drawn images of instruments in his letters, writing that the construction of instruments is more easily shown in person than described in words.[167] His prolific letter-writing had its limitations.

The next chapter will deal more extensively with the circumstances surrounding these letters on music sent to the eastern Mediterranean. For now, it is remarkable how far Mersenne's images circulated and how instruments printed on paper were used to aid correspondents in their own observations, specify what instruments had already been studied, and indicate what instruments were still missing from his investigation. The circulation of such materials makes clear that a wide range of scholars, antiquarians, and collectors were involved in Mersenne's musical project, contributing theories, observations, and images. His correspondence functioned as a repository for observations and experiences. The network was characterized by a variety of scholarly disciplines and motivations. There was a remarkable fluidity in topics discussed, ranging from mathematical problems to parlor tricks. A different kind of fluidity is visible in terms of scholarly background and scholarly aims. Although they frequently collaborated, Peiresc, Trichet, and Doni would sharply distinguish their own interests from Mersenne's. Doni contrasted his own work with Mersenne's "beautiful and curious investigations," emphasizing that he spent his time

164 Villiers to Mersenne, 20 October 1634, CM 4, 370.

165 "qu'il faut des genies pour donner raison de ces secrets." Villiers to Mersenne, 20 October 1634, CM 4, 371; Mersenne, *Harmonie universelle*, "Livre quatriesme des instrumens a chordes," 217–21. Later, in 1647, Gabriel Thibaut would send Mersenne another division of the trumpet marine, with the proportions necessary for each tone. Thibaut engaged a music master from Brioude, who designated all the distances on a large sheet of paper. Thibaut to Mersenne, 3 February 1647, CM 15, 77.

166 Titelouze to Mersenne, 26 March 1628, CM 2, 45–46.

167 Mersenne to Haak, 20 January 1640, CM 9, 43–44.

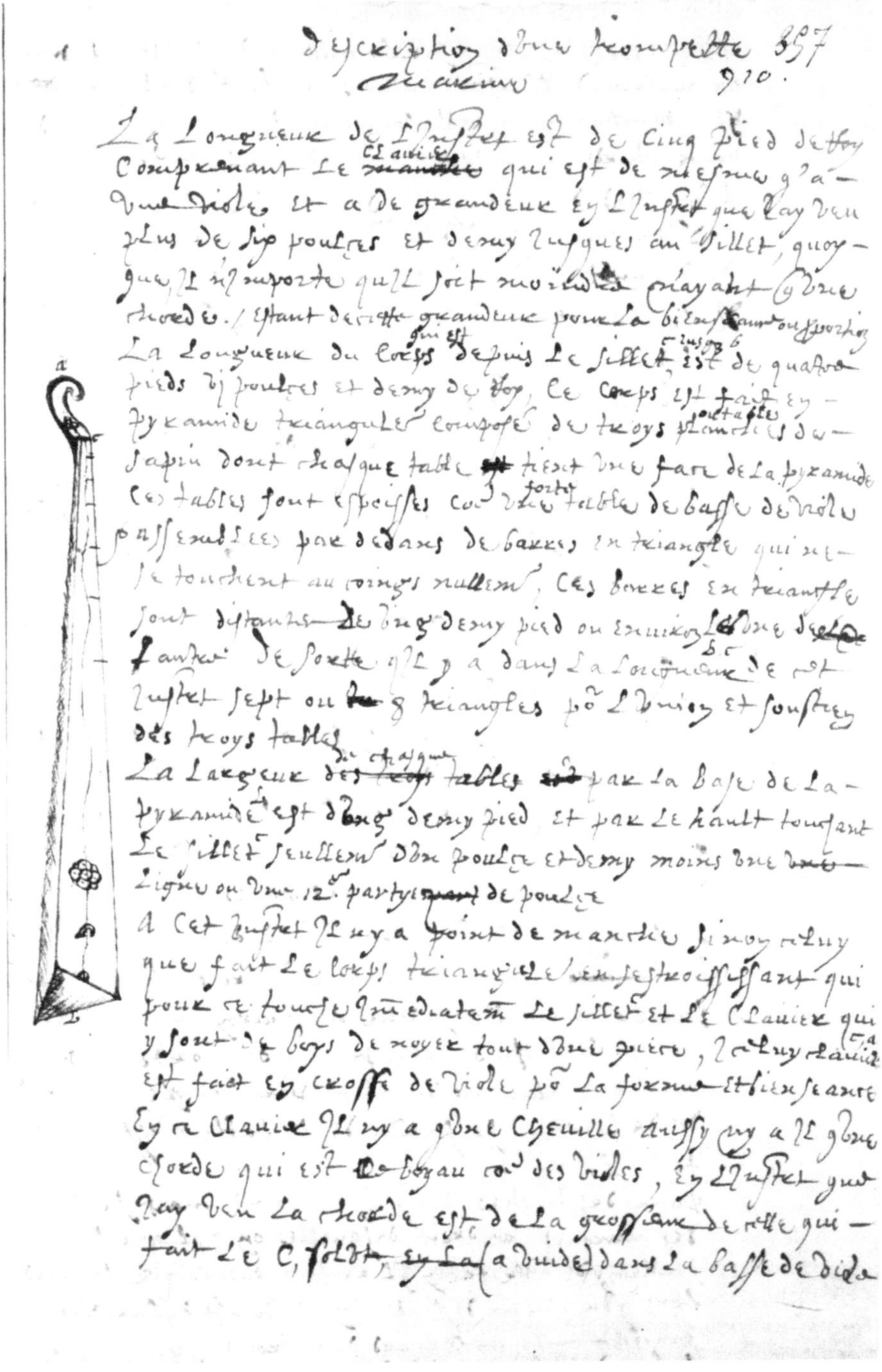
Description d'une trompette marine 357
910.

La longueur de l'instrument est de cinq pied de roy comprenant le clavier qui est de mesme qu'a une viole et a de grandeur en l'instrument que j'ay veu plus de six poulces et demy jusques au sillet, quoyque il importe qu'il soit moindre n'ayant qu'une chorde. Estant de cette grandeur pour la bienseance ou proportion
La longueur du corps qui est depuis le sillet jusqu'au b est de quatre pieds dix poulces et demy de roy, le corps est fait en pyramide triangulaire composée de troys planches ou tables de sapin dont chasque table tient une face de la pyramide ces tables sont espoisses comme une forte table de basse de viole assemblées par dedans de barres en triangle qui ne se touchent au coings nullement, ces barres en triangle sont distantes d'un demy pied ou environ l'une de l'autre de sorte qu'il y a dans la longueur de cet instrument sept ou 8 triangles pour l'union et soustien des troys tables
La largeur de chasque table par la base de la pyramide est d'un demy pied et par le hault touchant le sillet seullement d'un poulce et demy moins une ligne ou une 12e partye de poulce
A cet instrument il n'y a point de manche sinon celuy que fait le corps triangulaire en s'estrecissant qui pour ce touche immediatement le sillet et le clavier qui y sont de boys de noyer tout d'une piece, celuy la a est fait en crosse de viole pour la forme et bienseance. En ce clavier il n'y a qu'une cheville aussy n'y a il qu'une chorde qui est de boyau comme des violes, en l'instrument que j'ay veu la chorde est de la grosseur de celle qui fait le C, sol ut (a vuide) dans la basse de viole

Figure 3.6: Description and drawing of a trumpet marine in a letter from Christoph de Villiers to Mersenne. Paris, Bibliothèque Nationale de France, MS. Fr. 6205, fol. 357r.

"more willingly on those studies concerning musical practice."[168] Peiresc wrote, for example, to Mersenne, "All my studies have had a very different purpose than yours."[169] Trichet in turn wanted to deal "only historically" with musical instruments.[170] Befitting their status as *curieux*, they did not limit their interest to a single discipline or scholarly field. Archaeological excavations, mechanics, measurements of eclipses, and musical observations were all part of what Paula Findlen has called "the Janus faces" of early modern science. Within the fluid disciplinary boundaries of early modern science, there was a space and time for scholars like Mersenne to exploit different scholarly domains.[171] As Vera Keller has made clear in the context of Stuart England, rather than a secure order or disciplining of knowledge, there was "a loosening of knowledge structure."[172]

Many of the objects that Mersenne gathered were labeled "curious." Peiresc, for example, described the drawings of cymbals as "praiseworthy curiosities."[173] As Mersenne's discussion in his *Questions theologiques* makes clear, not only the rare and valuable were considered a curiosity; a musical instrument or scientific observation could be curious as well. Convinced that the ultimate truth would only be revealed in paradise, Mersenne argued that one should collect as much as possible as preparation for the afterlife. Through his emphasis on the theological dimension of collecting and curiosity, Mersenne provided a justification for his own activities and those of his contemporaries. This scholarly magpie became the archetypal collector, a practice that gestured toward collecting on earth as preparation for the riches of paradise. Between the discussions of overtones, sympathetic vibration, and instruments, the letters between Mersenne and Peiresc are filled with references to their endeavor to find materials on music from the Ottoman Empire, the subject of the next chapter. In many ways, it was their most curious effort, linking different scholarly interests, engaging a wide variety of actors, and attempting to find order among variety.

168 "Queste son belle e curiose ricerche, ma io per me più volentieri spendo il tempo in quelli studii che concernono la pratica." Doni to Mersenne, June 1647, CM 15, 280. For an insightful exploration of Doni's and Mersenne's differences, see Bianchi, "Bad Latin, Bad Manners."

169 "Mais toutes mes estudes ont eu un object trop different du vostre." Peiresc to Mersenne, 18 April 1634, CM 4, 176. When corresponding about a Delphic tripod, Peiresc became tired of Mersenne's unwillingness to make an attempt to uncover its symbols, writing: "I am only seeking to understand what the ancients believed" ("Mais je n'y cherche que de recongnoistre ce que les Anciens en ont creu."). Peiresc to Mersenne, 19 December 1634, CM 4, 415.

170 "seulement de traitter historiquement." Trichet to Mersenne, 9 January 1631, CM 3, 4.

171 Findlen, "The Janus Faces of Science," 246.

172 Keller, *The Interlopers*, 9.

173 "louables curiositez." Peiresc to Mersenne, 20 November 1636, CM 6, 154.

Bibliography

Manuscript sources

Carpentras, Bibliothèque Inguimbertine, MS 1774. Papers of Nicolas-Claude Fabri de Peiresc.

Paris, Bibliothèque de l'Arsenal, MS 2884. Marin Mersenne. "Livre de la nature des sons."

Paris, Bibliothèque Mazarine, MS 4146. Catalogue of the library of the Minim convent in Paris from the mid-seventeenth century.

Paris, Bibliothèque nationale de France, MS. fr. 6205. Letters of Marin Mersenne.

Paris, Bibliothèque nationale de France, MS fr. 9530. Papers of Nicolas-Claude Fabri de Peiresc.

Paris, Bibliothèque nationale de France, MS fr. 9531. Papers of Nicolas-Claude Fabri de Peiresc.

Printed primary sources

Augustine. *Confessions.* Translated by Henry Chadwick. Oxford: Oxford University Press, 1991.

Augustine. *Earlier Writings.* Edited and translated by J. H. S. Burleigh. Louisville: Westminster John Knox Press, 2006.

Bacon, Francis. *Advancement of Learning and New Atlantis.* Edited by Arthur Johnston. Oxford: Clarendon Press, 1980.

Bacon, Francis. *The New Organon.* Edited by Lisa Jardine and Michael Silverthorne. Cambridge: Cambridge University Press, 2000.

Baillet, Adrien. *Vie de Monsieur Des-Cartes, seconde partie.* Paris: Daniel Horthemels, 1691.

Beeckman, Isaac. *Journal tenu par Isaac Beeckman de 1603 à 1634.* 4 Vols. Edited by Cornelis de Waard. The Hague: Martinus Nijhoff, 1939–53.

Bottrigari, Hercole. *Il Desiderio, or Concerning the Playing Together of Various Musical Instruments.* Translated by Carol MacClintock. Rome: American Institute of Musicology, 1962.

Chapard, François. *Fabriciani cimeliarchii promptuarium triceps.* Aix: Roize, 1647.

Colonna, Fabio. *La sambuca lincea overo dell'istromento musico perfetto. Con annotazioni critiche manoscritte di Scipione Stella (1618–1624) (rist. anast. Napoli, 1618).* Edited by Patrizio Barbieri. Lucca: Libreria Musicale Italiana, 1991.

De Coste, Hilarion. *La vie du R.P. Marin Mersenne.* Paris: Cramoisy, 1649.

Descartes, René. *The Philosophical Writings of Descartes.* Vol. 1 of 2 vols. Translated by John Cottingham, Robert Stoothoff, and Dugald Murdach. Cambridge: Cambridge University Press, 1985.

Evelyn, John. *The Diary of John Evelyn.* Oxford: Oxford University Press, 1960.

Furetière, Antoine. *Dictionnaire universel contenant généralement tous les mots françois tant vieux que modernes* [...]. The Hague/Rotterdam: Arnout & Reinier Leers, 1690.

Gaffarel, Jacques. *Curiositez inouyes sur la sculpture talismanique des Persans, horoscope des Patriarches, et lecture des estoilles*. Paris: Masnil, 1629.

Gay, Godefroy. *Ode sur les raretés du cabinet de Pierre Trichet, advocat bourdelois*. N.p.: s.n., ca. 1635.

Hobbes, Thomas. *Leviathan*. Edited by J. C. A. Gaskin. Oxford: Oxford University Press, 2008.

Huygens, Constantijn. *Driehonderd brieven over muziek, van, aan en rond Constantijn Huygens*. Edited by Rudolf Rasch. 2 vols. Hilversum: Verloren, 2007.

Locke, John. *The Locke Reader: Selections from the Works of John Locke with a General Introduction and Commentary*. Edited by John W. Yolton. Cambridge: Cambridge University Press, 1977.

Maugars, André. *Response faite à un curieux, sur le sentiment de la musique d'Italie*. Paris: s.n., 1639–40.

Mersenne, Marin. *Cogitata physico-mathematica*. Paris: Bertier, 1644.

Mersenne, Marin. *Correspondance du P. M. Mersenne*. Edited by Paul Tannery, Cornelis de Waard, Bernard Rochot, Armand Beaulieu, and René Pintard. Paris: Centre National de la Recherche Scientifique et al., 1932–88.

Mersenne, Marin. *Harmonicorum libri XII*. Paris: Baudry, 1648. Reprint, Geneva: Minkoff, 1972.

Mersenne, Marin. *Harmonie universelle contenant la théorie et la pratique de la musique (Paris, 1636). Édition fascimilé de l'exemplaire conserve à la Bibliothèque des Arts et Métiers et annoté par l'auteur.* Introduction by François Lesure. Paris: Éditions du Centre National de la Recherche Scientifique, 1963.

Mersenne, Marin. *Quaestiones celeberrimae in Genesim, cum accurata textus explicatione: in hoc volumine athei, et deistae impugnantur & expugnantur, & vulgata edition ab hereticorum calumnijs vindicator. Graecorum, & Hebraeorum musica instauratur.* Paris: Cramoisy, 1623.

Mersenne, Marin. *Les questions theologiques, physiques, morales, et mathematiques. Où chacun trouvera du contentement ou de l'exercice*. Paris: Guenon, 1634.

Mersenne, Marin. *La verité des sciences, contre les sceptiques ou pyrrhoniens*. Paris: Du Bray, 1625.

Montaigne, Michel de. *The Complete Works of Montaigne*. Translated by Donald Frame. Stanford: Stanford University Press, 1958.

Niceron, Jean-François. *La perspective curieuse ou magie artificielle des effets merveilleux*. Paris: Billaine, 1638.

Pascal, Blaise. *Histoire de la roulette, appellée autrement la trochoïde, ou la cycloïde* [...]. N.p.: s.n., 1658.

Piccini, Alessandro. *Intavolatura di liuto e di chitiarrone libro primo*. Bologna: Heredi di Gio, 1623.

Plautus. *Stichus; Three-Dollar Day; Truculentus; The Tale of a Travelling-Bag; Fragments*. Edited and translated by Wolfgang de Melo. Cambridge: Harvard University Press, 2013.

Quiccheberg, Samuel. *Inscriptiones vel tituli theatri amplissimi* [...]. Munich: Berg, 1565.

Ripa, Cesare. *Iconologia, or, Moral Emblems*. London: Motte, 1709.

Taveau, Laurence. "Le manuscrit 2884 du P. Mersenne à la bibliothèque de l'Arsenal: Étude et édition critique." PhD diss., Université de Paris-Sorbonne, Laboratoire IREMUS, 2018.

Terzago, Paolo Maria. *Museo ò galeria adunata dal sapere, e dallo studio del Sig. Canonico Manfredo Settala nobile Milanese*. Tortona: Viola, 1666.

Trichet, Pierre. *Synopsis rerum variarum tam naturalium quam artificialium, quae in Musaeo Petri Tricheti Burdigalae reperiuntur/Denombrement de diverses et curieuses choses du Cabinet de Pierre Trichet Bourdelois*. Bordeaux: s.n., 1631.

Trichet, Pierre. *Traite des instruments de musique (vers 1640)*. Edited by François Lesure. Geneva: Minkoff, 1978.

Secondary literature

Adkins, Cecil. *A Trumpet by Any Other Name: A History of the Trumpet Marine*. Buren: Frits Knuf, 1991.

Ashworth, William B. "Catholicism and Early Modern Science." In *God and Nature: Historical Essays on the Encounter between Christianity and Science*, edited by David C. Lindberg and Ronald L. Numbers, 136–66. Berkeley: University of California Press, 1986.

Belin, Christian. "La notion de collection scientifique dans les *Questions inouïes* de Marin Mersenne." In *La Tradition rassemblée: Journées d'études de l'Université de Fribourg*, edited by Guy Bedouelle, Christian Belin, and Simone de Reyff, 169–82. Fribourg: Academic Press Fribourg, 2007.

Benedict, Barbara M. *Curiosity: A Cultural History of Early Modern Inquiry*. Chicago: University of Chicago Press, 2002.

Bianchi, Eric. "Bad Latin, Bad Manners: Giovanni Battista Doni, Marin Mersenne, and Literary Style in Seventeenth-Century Music Theory." *Music & Letters* 96, no. 2 (2015): 167–84.

Bleichmar, Daniela, and Peter C. Mancall. *Collecting across Cultures: Material Exchanges in the Early Modern Atlantic World*. Philadelphia: University of Pennsylvania Press, 2011.

Bolzoni, Lina. *The Gallery of Memory: Literary and Iconographic Models in the Age of the Printing Press*. Toronto: University of Toronto Press, 2001.

Bredekamp, Horst. *Antikensehnsucht und Maschinenglauben: Die Geschichte der Kunstkammer und die Zukunft der Kunstgeschichte.* Berlin: Wagenbach, 1993.

Buccolini, Claudio. "Mersenne et la philosophie baconienne en France à l'époque de Descartes." In *Bacon et Descartes: genèse de la modernité philosophique*, edited by Élodie Cassan, 115–34. Lyon: ENS Éditions, 2014.

Buccolini, Claudio. "Mersenne Translator of Bacon?" *Journal of Early Modern Studies* 2, no. 1 (2013): 33–59.

Cavicchi, Camilla, and Philippe Vendrix. "L'érudit et l'amateur: Collectionner la musique à la Renaissance." In *Collectionner la musique*. Vol. 1 of 3 vols.: *Histoires d'une passion*, edited by Denis Herlin, Catherine Massip, Jean Duron, and Dinko Fabris, 25–54. Turnhout: Brepols, 2010.

Cook, Harold J. *Matters of Exchange: Commerce, Medicine, and Science in the Dutch Golden Age.* New Haven/London: Yale University Press, 2007.

Cypess, Rebecca. *Curious and Modern Inventions: Instrumental Music as Discovery in Galileo's Italy.* Chicago/London: University of Chicago Press, 2016.

Daston, Lorraine, and Katharine Park. *Wonders and the Order of Nature, 1150–1750.* New York: Zone Books, 1998.

Dear, Peter. *Discipline and Experience: The Mathematical Way in the Scientific Revolution.* Chicago: University of Chicago Press, 1995.

Dennis, Flora. "When is a Room a Music Room? Sounds, Spaces, and Objects in Non-Courtly Italian Interiors." In *The Music Room in Early Modern France and Italy: Sound, Space and Object*, edited by Deborah Howard and Laura Moretti, 37–49. Oxford: Oxford University Press, 2012.

Evans, R. J. W., and Alexander Marr, eds. *Curiosity and Wonder from the Renaissance to the Enlightenment.* Aldershot: Ashgate, 2006.

Fabbri, Natacha. "Questioning Fludd, Kepler and Galileo: Mersenne's Harmonious Universe." In *Space, Imagination, and the Cosmos from Antiquity to the Early Modern Period*, edited by Frederik A. Bakker, Delphine Bellis, and Carla Rita Palmerino, 201–32. Cham: Springer, 2018.

Falletti, Franca, Renato Meucci, and Gabriele Rossi-Rognoni, eds. *Marvels of Sound and Beauty: Italian Baroque Musical Instruments.* Florence: Giunti Editore, 2007.

Findlen, Paula. "The Janus Faces of Science in the Seventeenth Century: Athanasius Kircher and Isaac Newton." In *Rethinking the Scientific Revolution*, edited by Margaret J. Osler, 221–46. Cambridge: Cambridge University Press, 2000.

Findlen, Paula. *Possessing Nature: Museums, Collecting, and Scientific Culture in Early Modern Italy.* Berkeley: University of California Press, 1996.

Frisch, Andrea. *The Invention of the Eyewitness: Witnessing and Testimony in Early Modern France.* Chapel Hill: University of North Carolina Press, 2004.

Grosslight, Justin. "Small Skills, Big Networks: Marin Mersenne as Mathematical Intelligencer." *History of Science* 51, no. 3 (2013): 337–74.

Hine, William. "Marin Mersenne: Renaissance Naturalism and Renaissance Magic." In *Occult and Scientific Mentalities in the Renaissance*, edited by Brian Vickers, 165–76. Cambridge: Cambridge University Press, 1984.

Hine, William. "Mersenne and Vanini." *Renaissance Quarterly* 29, no. 1 (1976): 52–65.

Jaffé, David. "Peiresc: Wissenschaftlicher Betrieb in einem Raritäten-Kabinett." In *Macrocosmos in Microcosmo: Die Welt in der Stube: Zur Geschichte des Sammelns 1450 bis 1800*, edited by Andreas Grote, 301–22. Opladen: Leske + Budrich, 1994.

Keller, Vera. *The Interlopers: Early Stuart Projects and the Undisciplining of Knowledge*. Baltimore: John Hopkins University Press.

Krakovitch, Odile. "La vie intellectuelle dans les trois couvents minimes de la Place Royale, de Nigeon et de Vincennes." *Bulletin de la Société de l'Histoire de Paris et de L'Ile-de-France* 109 (1983): 23–175.

Launay, Denise. "La musique à Venise vers 1645: Ismaël Boulliau, astronome français mélonáme et voyageur." *Revue de Musicologie* 77, no. 2 (1991): 269–77.

Lenoble, Robert. *Mersenne ou la naissance du mécanisme*. Paris: Vrin, 1943.

Licoppe, Christian. *La formation de la pratique scientifique: Le discourse de l'expérience en France et en Angleterre (1630–1820)*. Paris: La Découverte, 1996.

Marr, Alexander. "Introduction." In *Curiosity and Wonder from the Renaissance to the Enlightenment*, edited by R. J. W. Evans and Alexander Marr, 1–20. Aldershot: Ashgate, 2006.

Meyer, Andreas. "In Sound Similar to the Harps: Early Descriptions of African Musical Instruments." In *Instruments in Art and Science: On the Architectonics of Cultural Boundaries in the 17th Century*, edited by Helmar Schramm, Ludger Schwarte, and Jan Lazardzig, 269–79. Berlin: De Gruyter, 2008.

Miller, Peter. "Looking at the Past, Nature, and Peoples in Peiresc's Archive." In *Historia: Empiricism and Erudition in Early Modern Europe*, edited by Gianna Pomata and Nancy Siraisi, 355–97. Cambridge, MA: The MIT Press, 2005.

Miller, Peter. *Peiresc's Europe: Leaning and Virtue in the Seventeenth Century*. New Haven: Yale University Press, 2000.

Miller, Peter. *Peiresc's Mediterranean World*. Cambridge, MA/London: Harvard University Press, 2015.

Mueller, Paul R. "Marin Mersenne's 'Questions theologiques, Physiques, Morales, et Mathématiques': Agenda and Structure, with an Annotated Translation." PhD diss., University of Chicago, 2006.

Ossa-Richardson, Anthony. "Nicolas Peiresc and the Delphic Tripod in the Republic of Letters." *Journal of the Warburg and Courtauld Institutes* 74, no. 1 (2011): 263–79.

Palisca, Claude V. *G.B. Doni's Lyra Barberina: Commentary and Iconographical Study: Facsimile Edition with Critical Notes*. Bologna: A.M.I.S., 1981.

Pomata, Gianna. "Observation Rising: Birth of an Epistemic Genre, 1500–1650." In *Histories of Scientific Observation*, edited by Lorraine Daston and Elizabeth Lunbeck, 45–80. Chicago: University of Chicago Press, 2011.

Pomian, Krysztof. *Collections and Curiosities: Paris and Venice, 1500–1800*. Translated by Elizabeth Wiles-Portier. Cambridge: Polity Press, 1990.

Raphael, Renée. *Reading Galileo: Scribal Technologies and the Two New Sciences*. Baltimore: John Hopkins University Press, 2017.

Rasch, Rudolf. "*Me veux-tu voir mourir*: Joan Albert Ban versus Antoine Boësset (1640–1641)." In *Music and Science from Leonardo to Galileo*, edited by Rudolf Rasch, 211–49. Turnhout: Brepols, 2022.

Rochot, Bernard. *La correspondance scientifique du Père Mersenne*. N.p.: Conférence donnée au Palais de la Découverte le 4 juin, 1966.

Ruellet, Aurélien. "'Le grand négociant des lettres': contours, évolution, et usages du réseau épistolaires de Mersenne." In *Saint François de Paule et les Minimes en France de la fin du XVe au XVIIIe siècle*, edited by André Vauchez and Pierre Benoist, 277–89. Tours: Presses universitaire François Rabelais, 2011.

Schnapper, Antoine. *Le géant, la licorne, la tulipe: collections françaises au XVII siècle*. Paris: Flammarion, 1988.

Seipel, Wilfried, ed. *Für Aug' und Ohr: Musik in Kunst- und Wunderkammern*. Milan: Skira, 1999.

Shapiro, Barbara. *A Culture of Fact: England, 1550–1720*. Ithaca: Cornell University Press, 2000.

Smith, Pamela H., and Paula Findlen. *Merchants and Marvels: Commerce, Science, and Art in Early Modern Europe*. New York: Routledge, 2002.

Swann, Marjorie. *Curiosities and Texts: The Culture of Collecting in Early Modern England*. Philadelphia: University of Pennsylvania Press, 2001.

Taussig, Sylvie. "La réception des curiositez inouyes par Gassendi et Mersenne." In *Jacques Gaffarel Between Magic and Science*, edited by Hiro Hirai, 57–71. Pisa/Rome: Fabrizio, 2014.

Whatley, Janet. "Savage Hierarchies: French Catholic Observers of the New World." *The Sixteenth Century Journal* 17, no. 3 (1986): 319–30.

Whitmore, P. J. S. *The Order of Minims in Seventeenth-Century France*. The Hague: Nijhoff, 1967.

Wymeersch, Brigitte van. "Gabriel de la Charlonye (1588–1646): portrait d'un intellectuel de province à l'aube de la première modernité." *Albineana* 33 (2021): 181–201.

4. "The octave pleases all": The Universality of Music

Abstract: This chapter focuses on the relation between musical scholarship and transcultural contacts by investigating the moment when scientific debates on harmony and rationality in music gained new urgency in the context of mercantile networks and colonialism. I show how for several years in the early 1630s, the antiquarian Nicolas-Claude Fabri de Peiresc circulated requests to the eastern Mediterranean about various aspects of music on behalf of Marin Mersenne. I argue that it contains key insights into understanding how musical scholarship in early modern Europe turned toward music outside Europe.

Keywords: Nicolas-Claude Fabri de Peiresc, musical scholarship, cross-cultural translation, ancient music, skepticism

In his *La verité des sciences*, Mersenne stages a conversation between a skeptic, an alchemist, and a Christian philosopher, the third mirroring Mersenne's own point of view, attempting to refute the arguments of the skeptic and the alchemist. Discussing the uncertainties of the sciences in general, the skeptic laments that it is still unknown whether the octave and the fifth—the "most agreeable conjoint sounds and the principal consonances of music"—were agreeable to all countries.[1] How can one defend the rational foundation of intervals like the fifth and the octave, he writes, if it is not certain that these consonances are universally shared? This skeptic's argument introduces an anxiety that does not seem to have been uttered

Parts of this chapter have been reused with permission from: Leendert van der Miesen, "Songs at the Docks: Nicolas-Claude Fabri de Peiresc and Musical Entanglements in Seventeenth-Century Marseille," *Music & Letters* 105, no. 1 (2024): 27–51.

1 "Certainement nous ne sçavons pas même si l'octave, & la quinte, qui sont des sons tres-agreablement conjoints, & les principales consonances de la musique, sont agreable à toutes sortes de nations." Mersenne, *La verité*, 16.

Van der Miesen, Leendert. *Marin Mersenne and the Study of Harmony: From Sound to Music.* Taylor & Francis Group, 2025.
DOI 10.5117/9789048564132_CH04

before: Can one claim the universality of a musical system if one has not investigated all musical cultures? Here the factual universality of a "musica universalis" is questioned, putting pressure on the idea that the consonances as they are known might not be universal and immutable. In his *Questions harmoniques*, printed almost a decade later, Mersenne introduces the same anxiety. First giving skeptical arguments on the uncertainty of music, he again writes that the intervals are not shared across cultures and states, "The Canadians [*les Canadois*] and the peoples who are not familiar with our opinions and our musical intervals use other intervals than ours when they sing songs."[2] Following this argument, the skeptic argues, "We ultimately cannot conclude that our music is agreeable until we have consulted all those who live according the laws of nature."[3]

A remarkable feature of these texts is the epistemological anxiety uttered by the skeptic regarding the universal claims of the tonal system in light of colonial expeditions. In his "Discours sceptique sur la musique" printed in Mersenne's *Questions harmoniques*, La Mothe le Vayer comes to a similar position, which he connects to the skeptical position of suspended judgment (*epochê*) and the ten modes of skepticism mentioned by Sextus Empiricus. The first two modes concern variation among animals and variation among humans. These modes are used to argue against dogmatists who claim absolute knowledge, for example that perfume is pleasant to humans but intolerable to certain insects or that honey tastes sweet to some and bitter to others.[4] Against this view, Mersenne argues in the same book that the diatonic scale is the most natural, citing evidence from the French colonial mission in Northern America. The use of diatonic scales by musicians there, according to Mersenne, is proof that "the octave pleases all."[5] Mersenne's counterargument here is noteworthy, as colonial reports are employed to defend the universality of the tonal system in use. Two years later, in his *Harmonie universelle*, Mersenne actually prints some of the melodies he collected, namely the ones found in Jean de Lery's *Histoire d'un voyage fait*

2 "les Canadois, & les autres peuples qui ne sont point prevenus de nos opinions, & des intervalles de nostre Musique usent d'autres intervalles que les nostres quand il chantent des chansons." Mersenne, *Questions harmonique*, 29 30.

3 "nous ne pouvons pas conclure en dernier ressort que nostre Musique soit agreable jusques à ce que l'on aye consulté tous ceux qui vivent selon les loix de la nature." Mersenne, *Questions harmoniques*, 30.

4 Sextus Empiricus, *Outlines of Scepticism*, 12–38.

5 "car l'octave plaist à tous." Mersenne, *Questions harmoniques*, 75. As mentioned in chapter 2, in his *Harmonie universelle* Mersenne persisted in this view by calling musical intervals natural "when they are made by everyone" ("qui sont faits par tout le monde"), meaning the diatonic intervals. Mersenne, *Harmonie universelle*, "Livre second, des chants," 90.

en la terre de Brésil (Paris, 1578) and a melody he obtained from a French captain who was part of the colonial mission in North America.[6] Although Mersenne again cites this as possible evidence of the widespread use of the intervals of the fifth, fourth, third, second, etc., he now claims that this evidence cannot be definitive, as these cultures could have learned the intervals and they did not necessarily come out of nature.[7]

These references to non-European music within Mersenne's printed works are sporadic, but the surviving correspondence from the years 1633 through 1635 shows a remarkably high number of references to non-European music, not only from North America but also from various cultures across the Mediterranean. From these letters, it becomes clear that Mersenne actively sought out information on the tonal systems, polyphony, and instruments in use. To this end, he relied on the network of his patron Nicolas-Claude Fabri de Peiresc. In the years leading up to the publication of *Harmonicorum* and *Harmonie universelle*, he and Peiresc circulated questions on music in the Arab world, instructing travelers and missionaries to notate local songs and provide them with manuscripts and images. The increasing importance of non-European music for music theorists such as Mersenne has been scarcely addressed and is absent from most scholarly accounts of the Minim. Ultimately, Peiresc's efforts on behalf of Mersenne went well beyond the efforts of any other music theorist of the century, projecting a cross-cultural investigation of music composition throughout the Mediterranean.

This chapter examines the moment when Mersenne's questions regarding harmony were applied outside Europe by focusing on the circulation of requests and letters about music. This question of the universality or "naturalness" of the diatonic scale had been asked before but gained new urgency in an age of an increasing circulation of goods, people, and music. With the increasing number of travels, missions, and colonial expeditions, European authors encountered different instruments, musical practices, and harmonies. The immutable and universal nature of music was in increasing tension with observations on different customs and traditions. In her study on musical Stoicism in early modern France, Melinda Latour writes, "Music proved to be one of the most fascinating of human universals, although it quickly became evident that it was an example of radical local

6 De Lery only added the melodies in the Geneva edition of 1585. De Lery, *Histoire d'un voyage*, 159, 173. De Lery's transcriptions and his description have been discussed in a number of studies, for example: Greenblatt, *Marvelous Possessions*, 17–19; Tomlinson, *The Singing of the New World*, 45–48; Bloechl, *Native American Song*, 44–47.

7 Mersenne, *Harmonie universelle*, "Livre troisiesme des genres," 148.

particularity, for music was not performed, understood, or appreciated in any degree of uniformity across the globe."[8] Such experiences also percolated into the seventeenth-century projects on universal harmony. The astronomer Johannes Kepler already gave a transcription of Ottoman chant in his *Harmonice mundi*. Mersenne's Jesuit counterpart Athanasius Kircher, for example, included examples of Ethiopian, Chinese, and Ottoman music in his book to demonstrate the universality of the diatonic scale.[9] Kircher writes that he was informed by his fellow Jesuit missionaries that all people and cultures across the world use the diatonic scale, disproving any claims that the scale is a product of culture. Other examples of such musical descriptions and transcriptions come from French colonial and missionary literature, such as Marc Lescarbot's *Histoire de la Nouvelle-France* (Paris, 1609) and Gabriel Sagard's *Histoire du Canada* (Paris, 1636).[10] In 1632 Mersenne's printer Cramoisy began printing the Jesuit Relations, which are filled with musical anecdotes.[11] Such descriptions had a variety of functions, such as offering an aural souvenir to readers at home, demonstrating cultural hegemony, and exercising colonial power.

In his book *The Making of European Music*, David Irving has recently shown how the notion of "European music" was invented and cultivated during the eighteenth century, with the phrase first appearing in the second half of the seventeenth century. It was during this period that music was increasingly tied to a particular place in European thought, dependent on customs and traditions rather than cosmic harmonies. Mersenne's role in this history is ambiguous. As Irving points out, when Mersenne used the phrase "our music" (*notre musique*) it is unclear what exactly he refers to. Irving proposes that his "universalist paradigm of numerically based rationality and order" suggests that he referred to a supranational notion of "European music" rather than French music in particular.[12] Further complicating Mersenne's role in this history, in this chapter I show how an investigation of non-European music was a central aspect of his study of music, which combined humanist, textual, experimental, and ethnographic methods to question the nature of harmony. I expand the view not only to Mersenne's network and sources but also to those of his patron Peiresc in order to show how materials on music were circulated

8 Latour, *The Voice of Virtue*, 187.

9 Kepler, *The Harmony of the World*, 217 (III.13); Kircher, *Musurgia universalis*, VII, 568.

10 Lescarbot, *Histoire de la Nouvelle France*, 691–92; Sagard, *Histoire du Canada*, 310–13. Olivia Bloechl has discussed these sources most extensively in Bloechl, "The Pedagogy of Polyphony."

11 Bloechl, "Wendat Song and Carnival Noise."

12 Irving, *The Making of European Music*, 135n12.

within a wider network of letters. I look at the queries on music that were sent to Tunis, Aleppo, and Cairo and the reports that Mersenne and Peiresc received. In offering a case study of the ways in which texts and manuscripts were evaluated, I consider the treatise of the fourteenth-century music theorist Shams al-Dīn al-Ṣaydāwī, sent to Mersenne by a French salesman in Cairo. Peiresc and Mersenne engaged translators and informants to decipher the diagrams and images they received, though Mersenne ultimately chose not to include them in *Harmonie universelle*. The aim of this endeavor was manifold, namely to establish relationships between Arab and Hellenic cultures, to investigate which musical scales were used outside Europe, and to offer curiosities and rarities to the reader. As I show, this endeavor is central to understanding how musical scholarship in early modern Europe turned toward music outside Europe and how such material was adopted by polymaths like Mersenne. This chapter shows that there was an increasing awareness of regional musical differences to which Mersenne responded, influenced by both philosophical currents such as skepticism and cultural circumstances such as the growing influx of travelers and travel accounts. His approach to distant music was not necessarily proto-empirical as his approach to empirical evidence was highly selective and colored by his Catholic orthodoxy; his turn toward non-European music must therefore be critically measured and questioned.

Skepticism and Musical Difference

While Mersenne was writing his books, there was a growing interest in France in anything related to the East. Cardinal Richelieu and his political advisers sought out travel agreements with the Ottoman Empire and Safavid Iran, collected valuables and manuscripts, and promoted the study of Arabic at colleges and universities.[13] Meanwhile, in the French theaters, the so-called *Ballet des nations* staged different cultures of the world, often in grotesque manners. Although the idea of a "Turkish style" in music was especially in vogue after the premiere of Molière and Lully's *Le Bourgeois gentilhomme* (Paris, 1670), there are earlier examples of musical "exoticism" in French music theater in the early seventeenth century. The French composer Charles Tessier, for example, included two "chansons turquesques" in his collection of songs from 1610, which were adopted from actual Ottoman

13 Dew, *Orientalism in Louis XIV's France*; Bevilacqua, *The Republic of Arabic Letters*.

songs, as Kate van Orden has recently demonstrated.[14] Many of these depictions or adoptions were extremely political. In the *Ballet de Madame* (performed in 1615 in Paris), for example, the opening verses announced the French triumph over the Ottoman Empire and predicted a future conquest of the Holy Land, extending the French Empire across the globe.[15] So-called "Turkish" characters were not only staged as "exotic" and "wondrous" but were also used to incite fear in the audience, a trope closely related to nascent French imperial ambitions.

Mersenne himself had already considered forms of non-European music starting with his earliest writings on music. His 1623 *Quaestiones* demonstrates an interest in the workings and virtues of music across nations, as he writes that each nation differs "in voice and in the ways it sings." He paraphrases here the famous expression from the Italian music theorist Franchinus Gaffurius, who wrote in 1492 that "the English rejoice, the Spaniards weep, the Germans howl" and "only the French are said to sing."[16] Going beyond Gaffurius's original reference to the Genoese and "those who live by sea," Mersenne added a reference to the "barking" of "barbarians" (*barbari latrare*). He might have been inspired to include this addition by the sixteenth-century author Gioseffo Zarlino, who wrote, "An Italian sings differently from a Frenchman, and a Spaniard sings in a manner different from that of a German, and of course different from that of the barbarian nations of infidels, which is obvious."[17] Although these authors did not always offer an explanation as to why different people might appreciate different songs, at times it was explicitly connected to ideas about the role of climate and humoral constitution of inhabitants of different nations. Athanasius Kircher, for example, described musical style as the result of "the natural complexion of humans" and constitution of the region.[18] According

14 See Obelkevich, "Turkish Affect"; Locke, *Music and the Exotic*; Van Orden, "Hearing Franco-Ottoman Relations"; Psychoyou, "De la musique universelle."

15 Welch, "The Specter of the Turk," 90–91.

16 "Imprimis verò deprehendes singulas nationes in voce, atque modo canendi differre, adeout Angli concinnédo jubilare, Hispani fletus promere, ulutatus Germani, Itali, Genuenses praesertin, illisque vicini caprizare, barbari latrare, soli Galli cantare dicantur, si Franchino Gafuro in sine Theoricae sua credimus." Mersenne, *Quaestiones*, col. 1610. Gaffurius's original phrase was: "Anglici enim concinendo jubilant. Cantant Galici. Hyspani ploratus promunt. Germani ululatus. Italorum nonnullos ut genuenses et qui ad eorum littora resident capizarre ferunt." Gaffurius, *Theorica musicae*, fol. K5. On this phrase, see Bertoglio, *Reforming Music*, 143–46.

17 Zarlino, *On the Modes*, 6

18 "stylum musicae alicui loco usitatum, naturalem complexionem hominum, & particularis alicuius regionis constitutionem consequi." Kircher, *Musurgia universalis*, VII, 543.

to Kircher, the German preference for deep, calm, moderate polyphony could be explained by their complexion; the French, as one example, have a much more mobile, lively style of music making that could be understood by their (humoral) complexion. In 1623 Mersenne would also argue that inhabitants of a nation would always favor their own national style, as each listener prefers songs from their own country: "A Frenchman will be more excited by the harmonies of a French song."[19] Although Mersenne here emphasized the role of the text in a composition and its type, these considerations were for Mersenne inextricably linked to those living outside Europe, as he added that he was reminded of an unnamed Ethiopian that could not appreciate music, until a "Moorish" tune was played.[20] Later on Mersenne advised composers and performers to take into account the country of origins of their listeners, again referring to "Moorish songs" that pleased Ethiopian listeners.[21] As Bettina Varwig and David Irving have shown, humoral or environmental accounts of musical difference easily moved into ethnic and racial prejudices.[22]

There was also a humanist tradition that concerned itself with the role of music in the behavior and ethos of a country. Ancient Greek writers associated certain melodic styles with different regions and peoples: "Dorian" referred to the Dorian culture of the Greeks and "Phrygian" the kingdom

19 "Eapropter Gallus magis excitabitur harmoniâ Gallicam cantilenam referente." Mersenne, *Quaestiones*, col. 1572.

20 "quod mihi Aethiopem in memoriam revocat, quem Musicus excitare non potuit ullo cantu, donec moriscam caneret." Mersenne, *Quaestiones*, col. 1572. This stance did not mean that Mersenne had a relativist position toward music. In the same book, he derides contemporary Jews and rabbis ["recentiores Judiaei, & Rabini hoc saeculo"] and writes that one should not dwell on their possible notational systems for music for long, since it adds nothing to the understanding of music; see for example Mersenne, *Quaestiones*, col. 1700.

21 "cantilenam Aethiopibus propriam videlicet moriscam cecinit." Mersenne, *Quaestiones*, col. 1623. As Noémie Ndiaye observes, French "racial lexicons remained highly ambiguous throughout the seventeenth century." Ndiaye, *Scripts of Blackness*, 12. The word "Turk" was used loosely in early modern France; it could mean a subject of the Ottoman sultan, a Muslim, or an enslaved galley rower from Muslim lands, regardless of their religion. In writing about French slavery in the Mediterranean, Gillian Weiss has noted that "Turk" almost became synonymous with 'slave' in the mid-sixteenth century. Weiss, "Infidels at the Oar," 399. Jewish and Greek Orthodox galley rowers were occasionally described as "Turks" and "Moors" as well. The term "Moor" was used for the inhabitants of Morocco (the independent kingdom ruled by the Saadi dynasty); for those who came from the regions of Algiers, Tunis, and Tripoli; and for the people of Mauritania. Jean Nicot writes in his *Thresor de la langue françoyse* (Paris, 1606) that the French word "More" was at times used for all those of the Islamic faith, except for Muslims from Istanbul (who were often called just "Turks"). Nicot, *Thresor de la langue françoyse*, 418.

22 Varwig, *Music in the Flesh*, 52–53; Irving, *The Making of European Music*, 166–68. For such discussions outside the realm of music, see Smith, *Nature*, 58–60.

Phrygia in Anatolia.[23] Rather than adopting a universalist perspective, many authors emphasized that each country had their own music. Mersenne's sixteenth-century predecessor Pontus de Tyard, for example, stressed that music must be adjusted to the particular disposition of the listeners, citing the fact that the inhabitants of Crete listened to harp music before battles, whereas the Lacedemonians played flutes and oboes. Tyard's essay continues, "So it is that according to the climate, the nation, and the complexion, the mode of music receives its effectiveness." Tyard concludes that each listener has their own disposition for certain music: "The ploughman, with a rustic song pushes his oxen, and makes his toil easier."[24] The sixteenth-century composer Claude le Jeune would also position himself in this tradition, although he would not go as far as others who argued that region fully determines musical taste: "I will content myself with remarking that the appetites of peoples in the election of modes and measures are certain samples of the dominant affection in them."[25] Musical preference is here thus the result of some traits of the general nature of the people rather than the general nature of a culture the result of its music. Ancient authors already signaled music as one of the most important markers of civility, such as Aristides Quintilianus, who described cultures that never cultivated music as "insensitive and bovine."[26] With music's central role in education, ethics, and morality, the question of how music was cultivated by other cultures was a crucial one.

As quoted at the opening of this chapter, the topic of music and cultural difference returned two years later after Mersenne's *Quaestiones* in his *La verité*, where the skeptic argues that it is uncertain if all nations actually prefer the same harmonies.[27] In this text, the skeptic attempts to show that sensory perceptions differed widely and that knowledge, which is based on the senses, is therefore unreliable. The skeptic argues that human taste is so different that some prefer the shouting of pigs to the sound of an organ and concludes that "we see that different temperaments require

23 Barker, *Greek Musical Writings*, 14. Mersenne also referred to this tradition in Mersenne, *Quaestiones*, 1622–23.

24 "Tellemen que selon le climat, la nation, et la complexion, la Mode de Musique reçoit son efficace. Le laboureur, d'une chanson rustique pousse ses boeufs, et rend sa peine plus aisée." Tyard, *Solitaire second*, 195.

25 "Je me contenteray de remarquer, que les appetits des peuples, en l'election des Modes & mesures, sont eschantillons certains de l'affection dominante en eux." Le Jeune, *Dodecacorde*, n.p. ("A monseigneur").

26 Quintilianus, *De musica*, II.6, in Barker, *Greek Musical Writings*, 467.

27 Mersenne, *La verité*, 16.

different types of music."[28] There was a longer tradition of using cultural diversity as an argument for the instability and uncertainty of knowledge. One of the most famous authors to do so was Michel de Montaigne, who in his essay "On Cannibals" ponders the meanings of "savage" (*sauvage*) and "barbaric" and ultimately arrives at the often-quoted phrase "Each man calls barbarism whatever is not his own practice."[29] As others have shown, Montaigne's essay is deeply concerned with contemporary French topics, utilizing the example of the Brazil Tupinambá to discuss France's own civil wars.[30] Throughout the essay, Montaigne emphasizes that barbarity is not an absolute state but rather culturally dependent, concluding, "it seems we have no other test of truth and reason than the example and pattern of the opinions and customs of the country we live in."[31] Importantly, song and poetry take on a special function within Montaigne's essay, suggesting the importance of music for concepts of civility and naturalness.[32]

Mersenne would take offense at Montaigne's suggestion that there is no test of truth or reason other than our customs, as this chapter will later show. The majority of his *La verité* is devoted to refuting this position, demonstrating what could be known with the help of mathematics and the sciences in general. In the book, Mersenne does not return to this specific argument, namely that evidence from other cultures seems to suggest that the consonances are not universal. Although this claim is not discussed in his unpublished manuscript from 1628, Mersenne does write that there could be no other consonances than the ones that are already known, as "all the musicians in the world agree with this truth."[33] In his writings from 1634 onward, the question seems to have been more pressing. In *Questions harmoniques*, the problem is presented as follows: Reports from North America seemed to suggest that different tonal systems were in use on this continent, undermining the universality of the European intervals. The octave and fifth could hardly be deemed beautiful intervals if they were

28 "nous voyons que les divers temperamens requierent diverses musiques." Mersenne, *La verité*, 16. The argument regarding the diversity of sensory perception among animals also occurs on pp. 18 and 32. In his *Les preludes* this is turned into persons who appreciated the neighing of horses more than music, see Mersenne, *Les preludes*, 152.

29 Montaigne, *The Complete Works*, 152.

30 Quint, "A Reconsideration."; Reiss, "Montaigne." On the relationship between such "early ethnography" and skepticism, see Hamlin, "On Continuities between Skepticism and Early Ethnography."

31 Montaigne, *The Complete Works*, 152.

32 Tomlinson, *Singing of the New World*, 93–123.

33 "tous les musiciens du monde accordent cette verité." Paris, Bibliothèque nationale de France, MS 2884, fol. 56v. Transcription from Taveau, "Le manuscrit 2884," 481.

not accepted by all cultures. As already cited, the skeptic voice in the book argues that it is impossible to conclude "our music" is pleasant, "until we have consulted all those who live according the laws of nature."[34] Thus, the wisest option according to the skeptic seems to be to suspend judgement on such an uncertain subject like music. This point is argued even more forcefully in the short "Discourse sceptique de la musique" by La Mothe le Vayer. La Mothe le Vayer argues that it was difficult to say anything conclusive about music amid the increasing number of travel reports. He finds that the news that arrived in Paris complicates claims that the music that was heard in Europe was universal and singles out the use of silk strings in China and newly encountered instruments in the East. Drawing on a wide variety of sources, such as Fernão Mendes Pinto's *Peregrinação* (Lisbon, 1614), Nicolas Trigault's *De Christiana expeditione apud Sinas* (Augsburg, 1615), and various unnamed English travel literature, La Mothe le Vayer argues that taste is changeable, undermining the certainty of music:

> Each one follows their passion and has his particular taste here and elsewhere; the Chinese put raw silk strings on their spinets and other instruments, which they prefer over our gut or metal ones. For Father La Croix and Mendoça maintain against Trigault that they have always used harpsichords. The navigations of the English show that they saw in Java a number of musical instruments that Europe does not know. We found the New World with its own particulars that they consider the best of all.[35]

La Mothe le Vayer describes a remarkable moment; the number of travel descriptions circulating in Europe made it clear that there were many other forms of music out there, similar and different instruments with their own traditions, which were not necessarily inferior to European practices. If these cultures considered their music most beautiful, how could European authors claim that theirs was? How could anyone claim that music depended on universal and certain foundations when new instruments and songs were

34 jusques à ce que l'on aye consulté tous ceux qui vivent selon les loix de la nature." Mersenne, *Questions harmoniques*, 30.

35 "Chacun suit sa passion, & a son goust particulier icy comme ailleurs; Les Chinois mettent à leurs espinettes & autres instrumens, des cordes de soie cruë retorte, qu'ils preferent aux nostres de boyau ou de metal. (Car le Pere la Croix & Mendoça soustiennent contre Trigault qu'ils ont de tout temps l'usage des clavecins.) Les navigations des Anglois portent qu'ils virent en Java quantité d'instrumens de Musique que l'Europe ne cognoist point. Nous avons trouvé le monde nouveau avec les siens particuliers qu'il estimoit les meilleurs de tous." La Mothe le Vayer, "Discours sceptique sur la musique," 146–47.

being discovered every day? What secrets of this art were still waiting to be discovered? Interestingly, he addressed these questions in his other writings. In another essay referring to Gasparo Balbi's fantastic *Viaggio dell'Indie Orientali* (Venice, 1580), Le Mothe le Vayer mentions that monkeys are considered God's creatures first and foremost and that in Guinea monkeys are used as valets and play the lute and guitar "admirably."[36] And in his essay on doubt, he writes on the Algonquin people: "Regarding hearing, our cheerful music seems to them to be confused, while theirs is dull and heavy, and they make much more of it."[37] With these different statements that seem to counter French or European views on universality, Le Mothe le Vayer thought that it was best to suspend any judgment on music until more is known.

This skeptical position amends many earlier statements, namely those that relegated music of other cultures to the realm of noise and barbarity. In his unfinished *De inventione et usu musice*, the Renaissance theorist Johannes Tinctoris recalls encountering "Turkish" prisoners in Naples. He describes the music as "crude and absurd" and states that it simply demonstrates their "barbarity."[38] European authors often excluded all other musical forms that did not fall within their own musical system and relegated them to the realm of barbarous noise. The plethora of travel reports that were published in the sixteenth and early seventeenth centuries often dismissed the music of any country outside Europe, characterizing it as noise or music lacking any rationality. Many travelogues on the Ottoman Empire mentioned musical performances but were almost always negative. The traveler Guillaume Postel, for example, mentioned several instruments but described the music as noise "that could break the head and ears of the roughest French shepherds."[39] Jean Thevenot witnessed Ottoman musicians playing the tambur, but in "an unpleasant way," and noted that "they sing, but not according to our music."[40] Only two travel reports to the Ottoman Empire from the seventeenth century contained

36 "outre qu'ils jouënt de la fluste & de la guiterre, avec admiration." La Mothe le Vayer, "La physique du prince," 988.

37 "A l'égard de l'Ouïe, nos musiques gaies ne leur paroissent qu'un confusion, aient les leurs mornes & pesantes, dont ils font beaucoup plus de cas." La Mothe le Vayer, "Des doutes raisonnes," 312.

38 Cited and translated in Wegman, "Sense and Sensibility," 306.

39 "pour rompre la teste & les oreilles au plus gros bouviers de France." Postel, *De la republique*, 11.

40 "ou bien ils touchent une espece de lut qu'ils appellant tambour … la melodie n'en soit pas fort agreable … qu'ils chantant, non pas selon nostre Musique." Thévenot, *Relation d'un voyage*, 65, 68.

musical notation.[41] These negative descriptions were echoed in music theoretical writings; Michael Praetorius, for example, blamed the Ottomans for losing all sense and knowledge on music.[42] The Italian composer and theorist Vincenzo Galilei doubted that the "barbaric ears" of those in the Ottoman Empire would appreciate "our music," whatever that might have meant for him.[43]

Mersenne had most likely read Salomon De Caus's book *Institution harmonique* (Frankfurt, 1615). In this work, De Caus contrasts music to sciences like geometry, arithmetic, and perspective, all of which, he writes, were invented to fulfill human needs. By contrast, "the science of music is not a human invention, for everyone knows that the consonances of music are natural and not of our invention."[44] According to De Caus, music in the Americas could not be compared to that of Europe, "for the people being totally ignorant of letters have no understanding of the sciences."[45] Referring to Dutch travel literature, De Caus mentions the existence of bells and gamelan on the island of Java. According to these sources, De Caus observes, the musical system in use is similar to that already known in Europe. This should not be surprising, he continues, since the musical system in Europe is based on nature, so there could be no rational alternatives. Emphasizing that the European system is inscribed in nature, he writes, "For since we have all that nature can give, it would be wrong to think that their proportions were different from ours."[46] In other words, there were no legitimate alternatives to the proportions found in European music, whose genealogy De Caus traces back to ancient and biblical accounts. Rejecting the possibility of other sophisticated musical cultures, De Caus writes that

41 For these and other examples, see Wright, "Turning a Deaf Ear."

42 "hat sich die Music so gar verlohren | das man auch fast nichts mehr davon weis." Praetorius, *Syntagmatis musici*, II, 82–83.

43 "nemeno diletta à quelle Barbare orecchie la nostre Musica." Galilei, "Discorso di Vicentio Galilei," 164. On the wider use of "barbaric" and "barbarian" in musical discourse, see Irving, *The Making of European Music*, 201–10.

44 "mais ceste science de Musique ne vient d'aucune invention humaine, car chacun sçait, que les consonnantes de ladite Musique, sont naturelles, & non de nostre invention." De Caus, *Institution harmonique*, n.p. ("A la tres-illustre").

45 "il est tout certain qu'elle ne peut estre accomparee à la nostre, car le peuple estant du tout ignorant des lettres n'a aucune congnoissance des sciences." De Caus, *Institution harmonique*, 23r.

46 "car veu que nous avons tout ce que la nature peut donner, ce seroit erreur de penser que leurs proportions fussent autres que les nostres." De Caus, *Institution harmonique*, 23r. Although he does not give any references, De Caus might have had the travel account of Willem Lodewijcksz in mind, who included images of instruments and players; see Lodewijcksz, *Prima pars descriptionis itineris*.

"the muses have abandoned the lands of Greece, Egypt, Judea, and the rest of the earth to retire in these parts of Europe."[47]

Whereas these authors emphasized incommensurability or exclusion, others argued that taste is conditional. The Jesuit missionary Trigault, for example, wrote that although Chinese music "to the superb judgment of our ears seems to be out of tune," their musicians might consider the same about the music in France.[48] Mersenne's correspondent Peiresc frequently wrote about the remarkable music he heard from presumably Ottoman and North African singers in France and Italy, including an enslaved galley rower who sang for Peiresc in Marseille. The composer and travelogue author Pietro della Valle wrote in his "Dell musica dell'età nostra" from 1640 that he had written down "Persian, Turkish, Arabic, and Indian songs ... very curious and different from ours," suggesting a wider interest in the music of these cultures in seventeenth-century musical circles.[49] The Pole Alî Ufukî (Wojciech Bobowski) and the Moldavian Dimitrie Cantemir were only able to notate Ottoman music and participate in an Ottoman musical context through extensive study and exposure, experience that most European authors lacked.[50]

Mersenne attempted to counter authors who argued that music was relative or only sensory by pointing to the universality of music. But a rising number of published travelogues contradicted this universality. Mersenne described a similar response in his *Harmonicorum*, writing that there is no country reached by sailors that does not have a musical instrument of some kind.[51] While this fact caused La Mothe le Vayer to question the status of music as a field of knowledge, it seemed to have only strengthened his belief in the importance of his work. In 1628 Mersenne argued that there could be no other consonances than the ones that are already known and 1634 he still seemed to have been convinced that an ethnographic study of all musical traditions around the globe would confirm this. Countering statements such as La Mothe le Vayer's, he wrote that the "natural" consonances used in Europe were also used in other continents and concluded, "if we would consult all other nations, we would see that

47 "il semble que les Muses ont abandonné le pays de Grece, & d'Egypte & Judee, voire tout le reste de la terre, pour se retirer en ces quartiers de l'Europe." De Caus, *Institution harmonique*, 23v.

48 "qui au jugement suberbe de nos oreilles semble estre du tout de mauvais accord." Trigault, *Histoire de l'expédition chrestienne*, 33.

49 "alcune arie Persiane, Turchesche, Arabiche, e Indiane, e assai curiose, e diverse dalle nostre." Della Valle, "Della musica dell'età nostra," 259.

50 Haug, *Ottoman and European Music*.

51 Mersenne, *Harmonicorum*, "De campanis," 168.

they use our intervals, because they are natural."[52] Although his *Harmonie universelle* somewhat retracted the importance of these observations for this statement, Mersenne nonetheless remained convinced of this axiom throughout his life. As shown in chapters 1 and 2, in his *Harmonie universelle* he seems to remain convinced that the intervals of the diatonic scale are natural and that "no human understanding can cause a diminished fifth … to give pleasure," just as "nature has set limits to the sea that even the stormiest waves cannot surpass."[53]

Ancients and Moderns

The first surviving letter from Mersenne to his future patron Peiresc is dated May 1, 1633. The letter was highly anticipated. The philosopher Pierre Gassendi, Mersenne's friend, had already approached Peiresc on the subject of finding materials for his book on music. Peiresc had offered his services to assist the theorist as much as he could, and that he would send out letters on Mersenne's behalf to various places in the eastern Mediterranean.[54] When Mersenne eventually did write to Peiresc, he explained that he had been working on a book on music for several years now but struggled to describe music from outside Europe. He thought that Peiresc could help him with his contacts, such as with the French consul in Aleppo. Mersenne had heard that there was a Basilian monastery in Aleppo where the brothers practiced excellent singing, and he hoped to get in touch with them. Mersenne had already written to Rome, Venice, and Istanbul for this purpose but had received no answer. He was especially interested in the musical characters and tablatures used in the eastern Mediterranean, which he wrote to Peiresc while sending along a two-part composition that was to be transcribed into the music notation used in those regions.[55]

A central figure in multiple correspondence networks, Peiresc was interested in old and curious things. After living in Paris from 1616 to 1623, he

52 "Et si l'on consulte toutes les autres nations, l'on remarquera qu'ils usent de nos intervalles, parce qu'ils sont naturels." Mersenne, *Questions harmoniques*, 75.

53 "comme la nature a mis des bornes à la mer que tous les flots les plus orageux ne peuvent outrepasser, aussi nul entendement humain ne peut faire qu'une fausse Quinte, c'est à dire moindre qu'elle ne doit estre, ou qu'une fausse Octave puisse donner du plaisir." Mersenne, *Harmonie universelle*, "Livre premier des consonances," 41.

54 Peiresc to Gassendi, 21 December 1632, CM 3, 353. I abbreviate the *Correspondance du P. Marin Mersenne* with CM and the number of the volume.

55 Mersenne to Peiresc, 1 May 1633, CM 3, 392–94.

returned to Provence, where he lived until his death in 1637. From Provence, Peiresc corresponded with scholars, missionaries, and merchants all over Europe, organized scientific observations, and supplied scholars with sources. Peiresc was especially involved in the Mediterranean, maintaining a far-reaching epistolary network from Aix. From here, he obtained information on Ethiopian history, manuscripts from Cairo, reports from Jerusalem, and rarities for his cabinet. He was in close contact with the merchants and captains in Marseille and sent out his own agents, who traveled between Aix, Rome, Jerusalem, and Aleppo.[56] It was Peiresc's interest in antiquity that turned his gaze eastward. As recent studies have affirmed, the antiquarian methods of collection, observation, and comparison were also used to study materials from distant cultures. Not only were the objects of study (such as customs, religion, and material culture) closely related; the methods were similar as well, collapsing geographical and chronological distance.[57] In the field of music, this meant that images of ancient instruments depicted in marble were compared to those encountered in the Ottoman Empire, and European travelers hoped to find something of an ancient civilization when they traveled abroad. Musical instruments encountered in Tunis were compared to what was known about instruments in use in antiquity. Peiresc stimulated such research into possible common cultural practices between the ancient world and modern Mediterranean, asking his correspondents to take detailed notes of anything that might seem worthy.

Peiresc had a deep interest in ancient Greek music and hoped that Mersenne's work on music would increase the current knowledge of the ancients. His archival papers include summaries of sources on ancient instruments, and he collected numerous non-European instruments and depictions in his cabinet.[58] This was not uncommon. The collector Cassiano del Pozzo sampled musical instruments in his paper museum. One of the printed texts containing microscopic observations, Francesco Stelluti's *Persio tradotto* (Rome, 1630), includes a discussion of an Egyptian sistrum that was kept in Francesco Gualdi's cabinet in Rome.[59] In his letters to Mersenne, Peiresc repeatedly expressed the hope that the *Harmonie universelle* would uncover the secrets of Greek music, making new sources available for the humanist project of restoring the greatness of the past. Peiresc even went

56 See especially: Brentjes, "Peiresc's Interests"; Miller, *Peiresc's Mediterranean*.

57 Taylor, "Ancients and Moderns." Peiresc's archive and antiquarian approach have been meticulously researched by Peter Miller in various publications; see Miller, *Peiresc's Orient*.

58 Carpentras, Bibliothèque Inguimbertine, MS 1831, fol. 350r–v. On Peiresc's musical interests, see especially Scherpereel, "Peiresc et la musique"; Wymeersch, "Peiresc et la musique."

59 Stelluti, *Persio tradotto*, 185–88.

so far as to describe Mersenne's book as a prelude to discussing the efforts of "great men of the past."[60] As the previous chapter has made clear, Peiresc provided Mersenne with numerous manuscripts and objects, such as Greek and Byzantine manuscripts. Already in June 1633, Peiresc discussed sending Mersenne several manuscripts that included music, stressing that Mersenne would need a musician to interpret these writings.[61] Although he was afraid that the road to Paris might damage the manuscripts even further, it seems that Mersenne received several of these books, including two volumes on the rituals of the Greek Orthodox Church.[62]

Translating, commenting, and synthesizing ancient Greek sources on music was a central task of musical humanism. Since the fourteenth and fifteenth centuries, multiple Greek sources had been copied and translated, providing a more diverse picture of musical science in antiquity beyond the confines of Boethius. By the mid-sixteenth century, it was already possible for scholars to find at least one copy of the major Greek music treatises.[63] By the mid-seventeenth century, most of the ancient texts on music were printed in collections such as those of Johannes Meursius and Marcus Meibom. But was Mersenne truly discovering the secrets of antiquity, as Peiresc liked to write to his correspondents? Mersenne had in many ways aligned himself with the humanist endeavor of uncovering antiquity. His *Quaestiones* dealt extensively with ancient Greek music, inspired by the musical references of the book of Genesis.[64] The subtitle of the book *Greek and Hebrew Music Retrieved* indicates that his work must be understood within a longer line of musical humanism.[65] The belief that modern music had strayed from its original principles was widespread in early modern Europe. Humanists believed that modern music was no longer capable of inciting the ethical, moral, and medicinal effects that ancient texts described. Modern music could no longer move stones and plants, or cure diseases.

60 "serviroit de prelude de preparatif à l'intelligence des conceptions de ces grandz hommes du vieux temps." Peiresc to Mersenne, 13 August 1634, CM 4, 293.

61 Peiresc to Gassendi, 11 June 1633, CM 3, 429.

62 Paris, Bibliothèque nationale de France, MS fr. 5174, fol. 21r. Mersenne had at least a manuscript copy of Aristides Quintilianus's work on music and was helped by several professors of Greek in Paris, for example Pierre de Montmaur; see Bredeau to Mersenne, 21 October 1621, CM 1, 71; Titelouze to Mersenne, 2 March 1622, CM 1, 78. In his *Quaestiones*, he expresses the hope that the codices from the Vatican library or in Venice would be restored and printed, for the glory of God and benefit of all. Mersenne, *Quaestiones*, 1882. See also Doni to Mersenne, 15 October 1533, CM 3, 509–11; Naudé to Mersenne, 12 November 1633, CM 3, 535.

63 Palisca, *Humanism*, 35.

64 Mersenne, *Quaestiones*, col. 1529–40.

65 "Graecorum, & Hebraeorum Musica instauratur." Mersenne, *Quaestiones*, title page.

To remedy this, Mersenne, in his *Quaestiones*, urged his contemporaries to return to ancient techniques of composition by means of the ancient Greek musical modes, tunings, and microtonal intervals.[66] If this was successful, a true spiritual and moral understanding could be reached, whereas the path of most modern composers had almost extinguished the beauty and excellence of music.

In the introduction of *Quaestiones*, Mersenne connects this restoration to his project seeking to "restore or find anew" ("restituant vel de novo reperiant") ancient musical practices so that musicians could find a deeper and more divine connection to music, performing songs with the joy of heart.[67] He was interested in restoring "those noble ancient musical effects" for the French in particular, making clear that his project, like those of his sixteenth-century predecessors, was aimed at the nation.[68] In their own aspirations for their Académie de poésie et de musique, Jean-Antoine de Baïf and the musician Joachim Thibault de Courville hoped that the revival of ancient poetic practices would lead to a restructuring and revitalization of society in general.[69] The introduction of classical meters of long and short syllables for the vernacular and the development of a *musique mesuree* was thus a highly moral and political event. Although the academy had already disbanded before Mersenne came to Paris, he knew the academy composer Jacques Mauduit personally and learned from him about the academy and its principles.[70] And in his earliest works, Mersenne applauded the efforts of Jean-Antoine Baif and Joachim Thibault de Courville to remove "barbarism" in France and exhorted the French to adopt the *vers mesurés* of Baif.[71]

66 Mersenne, *Quaestiones*, col. 1683–84, 1687; Yates, *The French Academies*, 284–90; Duncan, "Mersenne and Modern Learning."

67 "quae praesertim Dei gloriae studeat, & habeat peritissimos viros, qui Metricam, Rhythmicam, & alia restituant, vel de novo reperiant, quae à nobis discussa, perspexerint." Mersenne, *Quaestiones*, n.p. ("Per illustri").

68 "nobiles illos antiquae musicae effectus." Mersenne, *Quaestiones*, col. 1513.

69 These sixteenth-century scholars were in turn inspired by Marsilio Ficino's academy in Florence in the fifteenth century, see especially Prins, *Echoes of an Invisible World*.

70 Yates, *The French Academies*, 64–69. On Mersenne's subsequent critique of Ficino, see Prins, "Patrizi's and Mersenne's Critiques."

71 "ut barbariem è Gallia pellerent." Mersenne, *Quaestiones*, col. 1683. Mersenne frequently referred to Mauduit's help and observations, for example recounting his experiences concerning the moving of certain stones when a certain pipe of a church organ was heard, see Mersenne, *Quaestiones*, col. 1699. In his *Traité* he wrote that the book had been approved Mauduit before his passing, see Mersenne, *Traité*, n.p. ("Advertissement de l'autheur"). He included a eulogy for Mauduit toward the end of his *Harmonie universelle*, praising the composer as the "father of music" and his knowledge of a wide variety of disciplines. Mersenne, *Harmonie universelle*, "Livre septiesme des instrumens de percussion," 63–69.

Like the humanists of the academy, Mersenne was drawn to the idea that music achieved its perfection in antiquity. In fields such as mathematics, optics, or mechanics, he found that the moderns could match the achievements of the ancients, or even eclipse them. But not in the field of music. Music, Mersenne writes in his 1627 *Traité*, "has been so abandoned that it seems that it is no longer practiced except for profit, or sensual delight."[72] In the same book, he includes works by Euclid and Bacchius to compare the theory and practice of French music with that of the ancients, making the case for the perfection of music by imitation of its ancient principles.[73] But already in his earliest works, he also expressed some skepticism. In his *Quaestiones*, Mersenne had already acknowledged that the ancient authors likely exaggerated some of their claims (calling Greece "a fountain or ocean of fables") but that there might have been many excellent musicians.[74] Mersenne also did not accept all claims of ancient philosophers, such as the idea that the human soul was made up of musical harmonies or harmonious relationships, found in Plato and Ptolemy. Mersenne found these theories futile.[75] Following his 1623 publication, the question of whether ancient Greek music should be imitated by modern composers was a recurring topic in Mersenne's correspondence in the 1620s and 1630s. The composer Jean Titelouze answered questions on the availability of musical examples and their possible use for contemporary composers. Regarding the possibility of recreating the effects described in ancient sources, Titelouze believed that modern taste had ventured too far from the ancient principles. Pointing to the change in musical styles observed during his life, he asked Mersenne to consider how much musical style has changed over thousands of years.[76] Other correspondents were asked similar questions. René Descartes, for example, wrote that the ancients were more impressionable than the moderns. In antiquity, Descartes assumed, listeners were much easier to surprise, as they were not accustomed to fine music.[77] Christoph de Villiers did not believe the myths about music and wondered if the ancients really were excellent in music.[78]

72 "la Musique a tellement esté abandonnée, qu'il semble qu'elle ne soit plus maniée que pour le lucre, ou pour la volupté." Mersenne, *Traité*, n.p. ("Preface au lecteur").

73 Mersenne, *Traité*, 106–146.

74 "Graeciam esse fabularum fontem, ac velut oceanum." Mersenne, *Quaestiones*, col. 1534.

75 Mersenne, *Quaestiones*, col. 1533.

76 Titelouze to Mersenne, 2 March 1622, CM 1, 74.

77 Descartes to Mersenne, 18 December 1629, CM 2, 351.

78 Villiers to Mersenne, 15 October 1633, CM 3, 523–24.

In his works from 1634 and onward, Mersenne limited the role of ancient music and its virtues in his books, emphasizing that his own musical system should suffice for modern composers to achieve perfection. Whereas he previously described antiquity as a golden age of science, his writings from 1634 showed more faith in progress, as he observed in his *Questions harmoniques*: "our century has men who are more knowledgeable in music theory than Ptolemy."[79] Straying further from his previous beliefs, he did not want musicians to imitate the ancients but rather argued that they should learn about the different musical systems in use.[80] Attempting to recreate the effects described in ancient texts was a vain endeavor that could never succeed. And Mersenne would distinguish his own work from "those who work on the restoration or restitution of Greek melody," possibly hinting at the work of Girolamo Mei or his colleague Giovanni Battista Doni.[81] Upon receiving *Harmonie universelle*, Doni had to acknowledge the beautiful inventions of the last century but still argued to Mersenne that there has been "more corruption than improvement" compared to the ancients.[82]

The search for unknown sources in the eastern Mediterranean was part of the humanist movement that started in the fifteenth century. Humanists also weighed the possibility of relating ancient texts to what is known about music in the territory of Greece, the Ottoman Empire, and Safavid Iran. As early as the mid-sixteenth century, the traveler and naturalist Pierre Belon urged humanist scholars to put away their books on ancient music and become ethnographers. He writes, "Those who desired to understand something of the music of the ancient instruments would learn more from knowing the instruments to be seen in Greece and Turkey than what is said about them in writing."[83] Instead of investigating ancient manuscripts as many scholars had done since the fifteenth century, Belon argued that it would be wiser to look at the music of the present-day Greek peninsulas and the wider Ottoman Empire. Mersenne seemed to share this belief when he

79 "nostre siecle a des hommes qui sont aussi sçavans que Ptolomée dans la Theorie de la Musique." Mersenne, *Questions harmoniques*, 260.

80 Mersenne, *Harmonie universelle*, "Livre quatriesme de la composition," 282.

81 "dont je laisse la recherche à ceux qui travaillent à la restauration, ou restitution de la melodie des Grecs." Mersenne, *Harmonie universelle*, "Livre cinquiesme de la composition de musique," 326v. The role of ancient music was subject for disagreement between Mersenne and Doni, see Bianchi, "Bad Latin, Bad Manners."

82 "bien plus de corruption que d'amandement." Doni to Mersenne, 26 July 1638, CM 7, 393–94.

83 "Qui voudrait esclarcir quelque chose de la Musique des instruments anciens auroit meilleur argument de l'experience de ceux qu'on voit en Grece & Turquie, que de ce que nous en trouvons par escrit." Belon, *Les observations de plusieurs singularitez*, 367.

wrote in *Harmonie universelle* that musicians in the eastern Mediterranean might have preserved elements of ancient music, since they appeared to sing monophonically.[84]

This idea recurs in several letters from 1634. Giovanni Battista Doni, for example, considered the possibility that contemporary Greeks and Ottomans preserved some musical elements from antiquity, but he saw only degeneration. Doni had a Greek singer perform before him in Rome and described the style as the "Turkish way." Dismissing what he heard, Doni wrote to Mersenne, "They say that they have lost of the old music, I think that of a hundred parts they have only one part of left."[85] In another letter, he connected his own project of restoring music to "its former glory" after what he perceived as barbary after "the invasions of the German and Arab nations."[86] Writing from Rome, the French Jean-Jacques Bouchard also derided the knowledge of music among contemporary Greeks.[87] Both men were working toward restoring ancient poetic and musical practices at the court of Cardinal Francesco Barberini and vehemently dismissed contemporary Greek culture. In a similar way, Mersenne had dismissed any contemporary "common" Jewish music, as he wrote that has nothing to please "learned men" and saw only a decline in the power of music as it was described in the Old Testament.[88] Such statements were commonplace in early modern musical discourse, as David Irving has recently shown. Irving describes how contemporary Greeks were "excluded from the circumscribed cultural geography of modern Western art music." Early modern scholars distinguished themselves from both "ancients" and "others" by projecting themselves as "modern." Irving writes, "Analogies such as these were used to reinforce the alterity of 'modern' Western European culture, acknowledging Western Europe's descent from Ancient Greece but setting it explicitly on a different plane: that of modernity."[89] By distinguishing themselves from both the ancient Greeks and the rest of the world, early modern European

84 Mersenne, *Harmonie universelle*, "Livre quatriesme de la composition de musique," 197.

85 "Ils disent d'havoir perdu la moitié de l'ancienne musique, mais je crois que de cent parties il ne leur en reste une entier"; "la maniere Turquesque." Doni to Mersenne, 8 April 1634, CM 4, 89.

86 "depuis les inondations des nations Germaniques et Arabiques, en son ancienne splendeur." Doni to Mersenne, 15 October 1633, CM 3, 509. See also Irving, *The Making of European Music*, 204–5.

87 Bouchard to Mersenne, 14 January 1634, CM 4, 4.

88 "Quemadmodum verò vulgaris ista Judaeorum musica, quae à nostris recentioribus mutuò sumpta ... viris doctis." Mersenne, *Quaestiones*, col. 1700.

89 Irving, "Ancient Greeks, World Music," 36. This is expanded in Irving, *The Making of European Music*.

writers asserted their own modernity, situating themselves between ancient manuscripts and ethnographic observations.

This interplay between ancients, moderns, and others was central to Mersenne's and Peiresc's interests. Although Peiresc did not publish on music and the topic only sparsely appears in his many notes, he would become a fundamental node in Mersenne's network, providing him with manuscripts, contacts, and translators. He repeatedly stressed the hope that Mersenne's work would uncover the secrets of the ancients. For many humanists like Peiresc, the search for sources in contemporary Arab and Persian cultures was a logical next step. With the help of Peiresc, Mersenne turned his eye and ear to the eastern parts of the Mediterranean.

A Request on Music to Tunis

Music's ephemeral nature has been characterized as both a virtue and a vice in transcultural encounters. Whereas some argue that sound has the ability to transgress cultural differences and boundaries, others have pointed to the difficulties. As Owen Wright notes in his study of European-Ottoman musical contacts in the Renaissance, "music does not generate objects."[90] As a cultural and social activity, it is a transient experience. Musical objects such as scores have very specific applicability. Although staff notation can be used to export some idea of European music, as a technology to capture something of other musical traditions, it is less usable. In contrast to plants, decorative objects, and other exotic objects, music traveled less well. Although there was a deep desire to communicate and share in Mersenne's correspondence, this project was hindered by numerous difficulties. How might one communicate knowledge on music across large distances, with cultures that relied on entirely different musical principles and instruments?

In the letter from Mersenne to Peiresc mentioned above, Mersenne specifically asked for help with the French consul in Aleppo. But through the mediation of Gassendi and the brothers Pierre and Jacques Dupuy, Peiresc was aware of Mersenne's interests and had already worked on formulating questions. In February 1633, Peiresc sent the first request on music on behalf of Mersenne to his local agent in Tunis, named Osman

90 Wright, "Turning a Deaf Ear," 144. Musical objects did of course travel between cultures. Focusing on the lute, Charles Burnett has shown "curious disjunction between the Western reception of Arabic music and musical instruments and its lack of reception of Arabic ideas of the symbolic and cosmic aspects of music." Burnett, "Musical Instruments," 164.

d'Arcos, born in Rouen.[91] Several years earlier, Peiresc had helped with negotiating d'Arcos out of captivity after he was captured by corsairs. After his release, d'Arcos converted to Islam and served a Muslim patron, which caused several tensions back in France. In Tunis he was part of a diverse group of converts and Muslims. In his letters to Peiresc, he continued to identify as Catholic, which allowed him to serve French patrons. And in this capacity, he performed a wide variety of services for the antiquarian, such as obtaining a giant's teeth, ostrich eggs, and clothing. He also reported on natural phenomena and was asked to make astronomical observations. Peiresc described d'Arcos as an excellent lute player and organist, making him especially fit to report on music.[92] A copy of the request is still among Peiresc's papers at the Bibliothèque nationale and is worth quoting in full (fig. 4.1):

> For an excellent work on music that will be brought to the press, we would like to have learned of the music and the ways of singing in use today of the vulgar Greeks, the Turks, the Persians, the Egyptians, the Moors, and other peoples of these countries. And for this, it would be necessary to address those who have some particular understanding of their songs, because we would like to know what notes they use, and what keys, and which voices; how do they express, for example, *ut, re, mi, fa, sol, la*, either a B-flat or B-natural; how much do they go up or down, and by what degrees, and in what way do they practice their harmonies. And to see if they sing in two, three, four, or more parts, and by which method they express, for example, this song in four parts:

> And other similar examples. You will advise, Monsieur, what is possible to do on this subject without a noticeable inconvenience. And if you are

91 Peiresc to d'Arcos, Late January/Early February 1633, CM 3, 373–74. D'Arcos adopted the name Osman after his conversion to Islam. For a closer investigation into the relationship between d'Arcos and Peiresc, see the work by Jane Tolbert and Jocelyne Dakhlia: Tolbert, "Ambiguity and Conversion"; Dakhlia, "Une archéologie du même."

92 Peiresc to Pierre Dupuy 16 May 1633, CM 3, 399; Peiresc to Aycard 9 April 1633, CM 3, 383.

able to help this good character, entirely or partly, I assure you that he will not be ungrateful and that he is a man well worth it.

It would still be necessary to add in Greek characters or other notes, which can serve these eastern peoples, the value of the following notes:[93]

The request is remarkable in that it asks about the ways of singing, musical notation, and modes of a wide range of cultures. It does not ask about music in Tunis specifically, but rather about Greek, Persian, Egyptian, and Ottoman cultures in general. D'Arcos was asked not only to rely on his own judgment but to find an expert, someone with "particular knowledge" on their music. This shows that the instructions were not to observe music from a distance but rather to engage in a conversation with local musicians. D'Arcos was asked about their systems of keys, clefs, and notes and how musicians in Tunis would represent European music. The reference to the B-flat and B-natural suggests an interest in modal systems, perhaps stemming from the assumption that there was a relationship with ancient Greek music. Peiresc added the different values and pitches used in Europe, proposing that they could be transcribed by d'Arcos to collect examples of notation in use in Tunis. Peiresc already had in mind how this could work. He had conducted his own musical research at the docks of Marseille, where he notated the song of an enslaved galley worker, relying on a local musician and a table of the Arabic syllables.[94]

93 "Pour un ouvrage excellent de la musique qui se va mettre soubz la presse, on desireroit d'avoir quelque cognoissance de la musique et façon de chantér dont se servent aujourd'huy les Grecs vulgaires, les Turcs, les Persans, les Aegyptiens, les Mores et aultres peuples de ces paÿs-la. Et ce cas la, il faudroit qu'on s'adressat à ceux qui auroient quelque particuliere intelligence de leurs chants, car on voudroit sçavoir de quelles nottes ils se servent, et quelles clefs, et quelles voix; comment est ce qu'ilz expriment, par Exemple, ut, re, mi, fa, sol, la soit par b mol, soit par b quarré; combien est ce qu'ilz montent hault ou dessendent bas, et par quelz degretz; de quelle sorte ilz pratiquent leurs accords. Et voir s'ilz chantent à deux, trois, quatre ou plusieurs parties, avec quelle methode ilz expriment par example ce verset chanté à quatre parties .... Et autres semblables. Vous adviserez Monsieur, qu'est ce qu'il sera possible de faire sur ce sujet sans une notable importunité. Et sy en tout, ou en partie, vous pouvez obligér ce bon personnage, je vous asseure qu'il n'en sera pas ingrat, et que c'est un homme qui le vaut bien. Il faudroit encore adjouster en caracteres Grecz ou aultres notes, dont se peuvent servir ces peuples orientaux la valeur des notes suivantes." Paris, Bibliothèque nationale de France, MS fr. 9531, fol. 268r. The letter is also transcribed (with some differences) in Peiresc to d'Arcos, end of January/early February 1633, CM 3, 373–74.

94 Peiresc to Mersenne, 18 June 1634, CM 4, 179–80. For a more extensive investigation of this anecdote, see Van der Miesen, "Songs at the Docks."

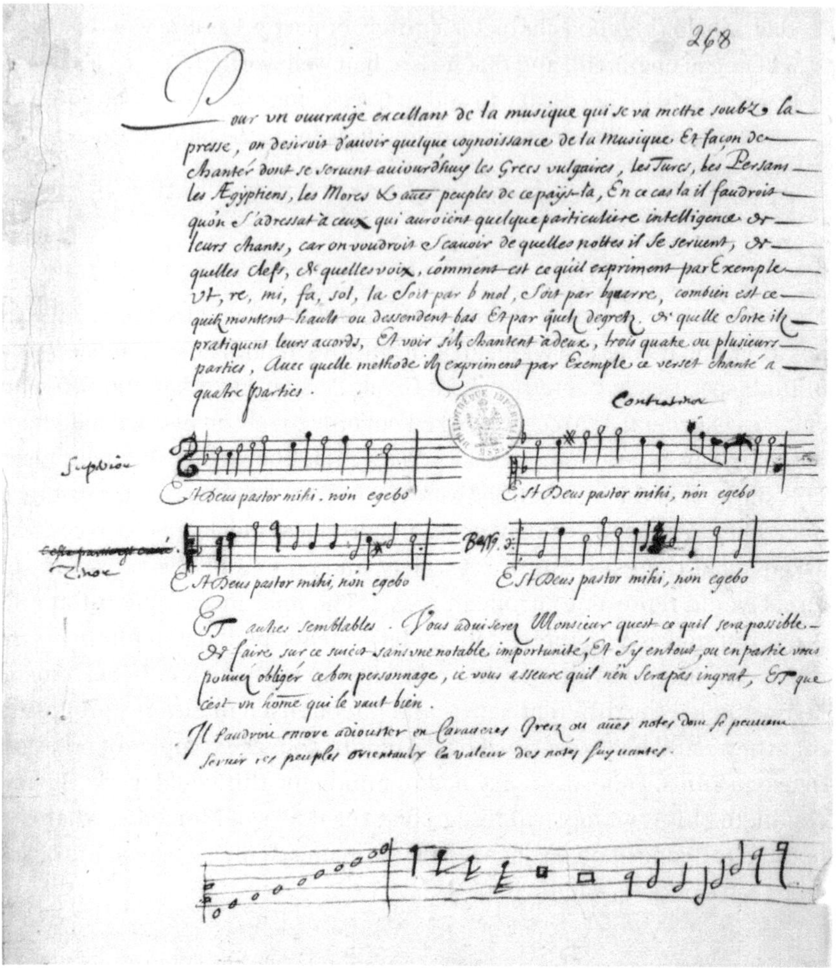
268

Pour un ouvraige excellant de la musique qui se va mettre soubz la presse, on desiroit d'avoir quelque cognoissance de la Musique et façon de chanter dont se servent aujourd'huy les Grecs vulgaires, les Turcs, les Persans les Ægyptiens, les Mores & autres peuples de ce pays-la, En ce cas la il faudroit quon s'adressat à ceux qui auroient quelque particuliere intelligence de leurs chants, car on voudroit scavoir de quelles nottes il se servent, de quelles clefs, et quelles voix, comment est ce quil expriment par Exemple ut, re, mi, fa, sol, la soit par b mol, soit par bquarre, combien est ce quils montent hault ou descendent bas Et par quels degrez, et quelle sorte ils pratiquent leurs accords, Et voir s'ils chantent a deux, trois quatre ou plusieurs parties, Avec quelle methode ils expriment par Exemple ce verset chanté a quatre parties.

Et autres semblables. Vous adviserez Monsieur quest ce quil sera possible de faire sur ce suject sans une notable importunité, Et s'y en tout ou en partie vous pouvez obliger ce bon personnage, je vous asseure quil n'en sera pas ingrat, Et que c'est un home qui le vaut bien.

Il faudroit encore adiouster en Caracteres Grecz ou autres notes dont se peuvent servir ces peuples orientaulx la valeur des notes suyvantes.

Figure 4.1: A request on music. Paris, Bibliothèque nationale de France, MS fr. 9531, fol. 268r.

The musical example included in the request is a four-part "Est Deus pastor mihi, non egebo." The text is by the sixteenth-century poet Jean-Antoine de Baïf, set to music by Jacques Mauduit (example 4.1), which makes for a curious conflation of the interest in the revival of ancient musical practices found in the context of the Académie de Poésie et de Musique and a desire to obtain musical information from non-European cultures. The composition has only survived because Mersenne printed the music in his *Quaestiones* from 1623. The compositional style of Mauduit (the so-called "musique mesurée") was itself an attempt to come to a closer understanding of the music of the ancient Greeks. By experimenting with stress, rhythm, and homophony, Mauduit tried to arrive at a more declamatory style of music, inspired by descriptions of ancient music. Mersenne would later admit that Mauduit's setting of "Est Deus pastor mihi" was one of his favorite compositions, sending it to the Dutch Joan Albert Ban as well.[95] Here Mauduit's music was used in this request to the Tunisian coast in search of music that might sound similar.

Example 4.1: Jacques Mauduit, "Est Deus pastor mihi, non egebo." From Mersenne, *Quaestiones celeberrimae in Genesim*, col. 1633–34.

D'Arcos responded to the request in June of 1633 with a brief report on music, sent together with a treatise on vases and measurements used in Tunis.[96]

95 Ban to Mersenne, 12 April 1639, CM 8, 374–78.

96 D'Arcos to Peiresc, June 1633, in Peiresc, *Les correspondants de Peiresc*, 27.

Peiresc described the response as "what the good father [Mersenne] wanted to know from the music of the Turks."[97] A copy of this report is still found at the Bibliothèque nationale.[98] D'Arcos's report does not contain anything on the musical systems used in Northern Africa, nor how musicians would express the musical example. Rather, it repeats the same tropes of the many travel accounts of the period. In it d'Arcos states that he did not discern any rules of harmony and writes, "It is true that the Moors sing, but savagely, without any rules of the art."[99] Alluding to the fact that Peiresc was mostly interested in the music of the ancients, he continues that he could not discover the diatonic, enharmonic, or chromatic genera but that the Tunisian musicians seemed to intersperse whole, half, and quarter tones at random. Nor could d'Arcos observe metrical units in the songs. Instead, he laments that "everything goes uncontrollably in confusion."[100] D'Arcos found that the locals showed little interest in music from the European continent. He had a harpsichord from Italy shipped to Tunis but wrote that the people in his surroundings did not care much for the instrument. He did mention several instruments in use, such as lutes, tambours, zithers (the qanun), and whistles, but included no further information on their use. The entire essay is written in a derogatory and dismissive tone, as he disparaged the barbarity he encountered and found no "musical art, ordering of tones, or harmony."[101]

The questions asked by Peiresc and Mersenne relied on close cooperation with musicians, knowledge of language, and experience with different musical systems. Although the request to d'Arcos asked for "someone with particular knowledge" on music in Tunis, native informers were often shunned. In other cases, Peiresc often asked for a French or Morisco informer.[102] Mersenne would later write that any informer must know music to be reliable source.[103] But information on notational systems and modes could only be obtained in close conversation with local musicians. D'Arcos's musical training and access to instruments like his harpsichord

97 "J'ay receu de Tunis un memoire sur ce que le bon Pere desiroit sçavoir de la musique des Turcs." Peiresc to Pierre Dupuy, 6 September 1633, CM 3, 472.

98 Paris, Bibliothèque nationale de France, MS Dupuy 668, fol. 62r–v. A transcription is found in Scherpereel, "Peiresc et la musique," 159–61.

99 "il est vray que le Mores chantent, mais sauvagement sans aucune regle de l'art." Paris, Bibliothèque nationale de France, MS Dupuy 668, fol. 62r.

100 "tout va pesle-mesle en confusion." Paris, Bibliothèque nationale de France, MS Dupuy 668, fol. 62r.

101 "Ils n'ont aucune ... de l'art de la musique, de l'ordre des tons, ni d'aucune concordance." Paris, Bibliothèque nationale de France, MS Dupuy 668, fol. 62v.

102 Tolbert, "Ambiguity and Conversion," 23

103 Mersenne to Peiresc, 26 July 1634, CM 4, 257.

suggests that he would have had opportunity to investigate to the music in Tunis more in depth, but it seems that he did not attempt this. Rather, he repeated the musical stereotypes about Ottomans and Africans that were already circulating in Europe, perhaps thinking that Peiresc wanted to hear these stereotypes reaffirmed.[104]

When Peiresc received the report from d'Arcos, he could not hide his disappointment.[105] As he writes in response, he thought that the people in Tunis must be fond of music, "whatever barbarity there may be."[106] He began this endeavor with the conviction that the Ottoman world surely must have many capable musicians and a sophisticated musical system. He had encountered several "Moorish musicians" in southern France and Persian musicians in Rome. These examples demonstrate that there were opportunities to hear musicians from the Ottoman Empire in early modern Europe. For Peiresc, these experiences affirmed his belief that published travelogues that dismissed their music could not be trusted.[107] D'Arcos, on the other hand, seemed to be convinced that it was his task to emphasize the barbarity of the land he lived in. Although the report of d'Arcos might have been a disappointment, it did reignite Peiresc's desire to find more materials. Shortly after they received the response, Peiresc and Mersenne would again focus on the endeavor with new methods and new locations, for example by sending the engraved images from Mersenne's forthcoming *Harmonie universelle* and *Harmonicorum* (figs. 3.2 and 3.3). Perhaps inspired by the response from d'Arcos, who described instruments like the lute and the cymbal, the two scholars now turned to images to define more clearly which instruments they were interested in.

A month later, Peiresc wrote several requests on music to his agents in Aleppo, Cairo, and Jerusalem, attaching Mersenne's images and several requests on music.[108] One of these correspondents was the Minim friar Théophile Minuti, who was traveling in the area of Aleppo at the time. Minuti had been approached by Peiresc some years earlier to undertake a journey to the eastern Mediterranean and eventually undertook this travel

104 This has been pointed out in Jocelyne Dakhlia's comprehensive study of d'Arcos; see Dakhlia, "Une archéologie du même."

105 Peiresc to Mersenne, 6 September 1633, CM 3, 474.

106 "quelque barbarie qu'il y aye." Peiresc to d'Arcos, 5 January 1634, in Peiresc, *Lettres de Peiresc: Tome Septième*, 118.

107 This conviction was already expressed in a previous letter, see Peiresc to Gassendi, 21 December 1632, CM 3, 352.

108 Carpentras, Bibliothèque Inguimbertine, MS 1874, fols. 314r–318v. For some of the letters, see Miller, *Peiresc's Mediterranean*, 106–17.

in 1629, sending books and an entire mummy back to Peiresc.[109] After a short return to France, Minuti traveled back to the eastern Mediterranean, in time to receive requests to gather materials for Mersenne's project. In the first request on music, Peiresc asked for a report on the music of dervishes, any Ottoman musicians, and the musical practices within the Greek Orthodox Church. He asked for a complete piece of music, hymn, or psalm in their musical notation and perhaps an interpretation by "someone who knows a bit of our music."[110] Minuti was asked to search in Aleppo and Jerusalem. In his second letter, Peiresc sent another request on music together with Mersenne's images of instruments quoted in chapter 3.[111] Minute was asked to pay special attention to the use of musical notes and methods for singing and playing. Peiresc thought Minuti might find capable musicians at the court of Fakhr-Al-Din II in Deir el-Qamar (Lebanon), who had lived in exile in Italy two decades earlier.[112]

As Peter Miller has discovered in Peiresc's extensive archive, the antiquarian sent other correspondents similar requests. The French merchant in Cairo Jean Magy was asked for a treatise on music and the singing of Greeks, Copts, Turks, and Arabs.[113] The Franciscan Guardian of Jerusalem Jacques de Vendôme was asked about the music of different Christian sects and the names they give to different musical instruments. This comparative project returned in their correspondence in the following year.[114] After Peiresc sent several requests to his contacts in October 1633, he summarized the whole endeavor again to Mersenne. As mentioned in the previous chapter, in this correspondence Peiresc also emphasized the role that the French ambassador in Istanbul, Henri de Gournay de Marcheville, could play, who according to Peiresc had brought several French musicians with him.[115] In March 1634, Peiresc tried again to encourage the ambassador to help with Mersenne's project, asking him about "the musical instruments and ways of

109 Miller, *Peiresc's Mediterranean World*, 74–75, 79, 88. On Minims traveling to colonial contexts, see Kellman, "Mendicants, Minimalism, and Method."

110 "quelqu'un qui entendoit un peu de nostre musique." Peiresc to Minuti, 11 October 1633. Quoted and translated in Miller, *Peiresc's Mediterranean World*, 107.

111 Peiresc to Minuti, 12 October 1633, CM 3, 494.

112 Peiresc to Mersenne, 13–15 October 1633, CM 3, 498.

113 Peiresc to De Vendôme, 12 October 1633. Quoted and translated in Miller, *Peiresc's Mediterranean World*, 107.

114 Miller, *Peiresc's Mediterranean World*, 108. For example, in the letters to the Capuchin friar Gilles de Loches, mentioned in the previous chapter.

115 "de la cotter sur le papier." Peiresc to Mersenne, 13–15 October 1633, CM 3, 499. For the diplomatic mission of De Gournay, see Hamilton, "'To Divest.'"

singing of the different nations who practice their religion in this land."[116] He emphasized its use not only for Mersenne but also for Cardinal Francesco Barberini and his secretary Doni, raising the esteem of the operation.

It is not known if De Vendôme, Minuti, or Magy sent responses on music to Peiresc, but the lack of sources seems to indicate that they did not. But they did not necessarily have to put their observations in writing. They could also relay their experiences to Peiresc and Mersenne in person, as Théophile Minuti and the Capuchin friar Gilles de Loches would do. Questioning travelers was an important way of obtaining information. Peiresc would often be visited by travelers, and he would question them about their journey to see if they were telling the truth and had a good memory.[117] Minuti, for example, testified upon his return to Peiresc that he saw dervishes whirling beautifully in Damascus, which proved that they were "great lovers of harmony."[118] Minuti also talked to Mersenne when he visited Paris.[119] As became clear early in the endeavor, obtaining information was more difficult than both men expected. There were many contradicting reports, letters got lost, and agents were not always trained in music. Mersenne himself wrote in July 1634 to Peiresc,

> I did not ask him [Théophile Minuti] about the music of the Turks, as I was sure of their savagery [*brutalité*] on the basis of your report, which you no longer approve of, this is why I am very happy not to have used him. But I will not allow the father [Minuti] to make a report on it [music] that could be of use, on the grounds that it would be necessary to know the music for this subject. And we make a thousand songs and cadences similar to those of your galley rower. We have here a knight of Malta, Monsieur de Rady, who made me hear some *airs* of the Turks and the Maltese on his spinet at the first president's house. He has promised [to send] them, but he is not sending me anything.[120]

116 "des instrumens de musique et façons de chanter des differentes nations qui ont quelque exercise de leur religion en ce pais là." Peiresc to De Gournay, 19 March 1634, CM 4, 80.

117 As Gassendi writes in his biography of Peiresc: "he would ask them, what Rarities they had seen, either in their Countrey, or upon their journey." Gassendi, *The Mirrour of True Nobility*, 176.

118 "qu'ilz sont grandement friands de l'harmonie." Peiresc to Mersenne, 18 June 1634, CM 4, 179.

119 Mersenne to Peiresc, 2 July 1634, CM 4, 226.

120 "je n'ay gardé de l'interroger de la musique des Turcs, puisque je m'assurois de leur brutalité sur vostre relation, que vous n'approuvez plus maintenant; c'est pourquoy je suis bien ayse de ne l'avoir pas employee. Mais je ne permettray pas que le Pere en puisse faire une relation qui puisse servir, à raison qu'il faudroit sçavoir la musique pour ce sujet. Et nous faisons mille chants et cadences semblables à celles de vostre galerien. Nous avons icy un chevalier de Malthe, Mr

Minuti had visited the convent in Paris but apparently Mersenne had not asked anything about music, raising the thought that the whole endeavor was not so important to the music theorist. Mersenne wrote that he did not ask anything, as he was convinced of the brutality of the Ottomans on the basis of d'Arcos's report. Moreover, Minuti lacked musical knowledge. He also once more dismissed the song that Peiresc had sent to him, writing that similar songs could be heard in Paris.

The requests to transcribe music encountered in the churches of Jerusalem point to multiple difficulties in the transcription and circulation of music. How could one transmit a musical experience across seas? Transcriptions of non-European melodies circulated widely and gave the reader an idea of what music could be found in other lands, turning an aural experience into a written format. But transcribing music required extensive training and familiarity, which not every traveler or informer possessed. It took time and considerable effort, as the examples of Alî Ufukî and Dimitrie Cantemir demonstrate.[121] There was no systematic form of collecting musical examples, and European travelers were generally not equipped to notate the complexities of, for example, Ottoman music.[122] Although staff notation works well as a system to transmit pitches and rhythms of music that is familiar, it works less well as a method to transmit something about completely unknown music. Even if a notated music example could be obtained, its aural nature resisted comparison. One example is the song Peiresc ordered to be transcribed at the docks of Marseille, mentioned in the excerpt by Mersenne cited above. To Peiresc, the song sounded completely different than French music but was still similar enough to be captured in French notation. When he received the notated examples, Mersenne admired the song, but he doubted that it was representative of music in the Ottoman Empire. Circulating it among his friends, Mersenne described it as similar to "our music" and did not think it was particularly remarkable, similar to the "thousand songs and cadences" currently in Paris.[123] Meanwhile,

de Rady, qui m'a fait entendre chez Mr le premier President des airs des Turcs et des Maltez sur son epinette. Il les a promis, mais il ne m'envoye rien." Mersenne to Peiresc, 26 July 1634, CM 4, 257. I have found no further reference to these "Maltese songs" in the correspondence.

121 Behar, "The Ottoman Musical Tradition." Judith Haug has extensively considered the complexities of cultural transfer and notation in her work on Alî Ufukî, see Haug, *Ottoman and European Music*.

122 Goodman, "Sounds Heard, Meaning Deferred."

123 "nostre musique." Mersenne to Peiresc, 2 July 1634, CM 4, 230; "mille chants et cadences semblables à celles de vostre galerien." Mersenne to Peiresc, 26 July 1634, CM 4, 257; Van der Miesen, "Songs at the Docks."

Peiresc assured that the singer had learned it "in barbaric lands" and that it consisted of an actual example of Ottoman music.[124] A single melody could mean many different things, depending on what the listener expected to hear from it.

A Treatise on Music

Throughout this exchange with Peiresc, Mersenne grew increasingly impatient. Peiresc needed to explain multiple times why they had to wait so long for any answers from his informers. But in May 1634 (one and a half years after their first letters), Peiresc could report his first success. He had been able to obtain a book on music sent from Cairo: the manuscript by Shams al-Dīn al-Ṣaydāwī, who came from the coastal city of Sidon and died in Damascus in 1506.[125] The treatise is entitled *Kitāb al-inʿām bi-maʿrifat al-anghām* (Book of generosity on the understanding of melodies) and is striking in its use of colored diagrams. This treatise was bought in Cairo by the French merchant Jean Magy and sent to southern France. The manuscript was part of the wave of materials from the Ottoman Empire that were brought to Europe for private and public collections.[126] When Peiresc saw the manuscript in the spring of 1634, he was immediately drawn to the figures and colors, describing it to Claude Saumaise as "an Arabic music manuscript that includes geometrical figures for the demonstration of the notes and their proportions, distinguished by different colored lines."[127] He thought that the treatise could illuminate the secrets of music in the Ottoman Empire, of which most scholars in Europe knew nothing. Peiresc continued his letter by writing that "these barbaric people" produced many excellent minds, more than he and his contemporaries previously believed.[128] He wrote to Mersenne that the manuscript contained "many artful things

124 "en Barbarie." Peiresc to Mersenne, 16 July 1634, CM 4, 242.

125 Peiresc to Mersenne, 1 May 1634, CM 4, 107. For the manuscript and its history in Europe, see Shiloah and Berthier, "A propos."

126 Mersenne's colleague, the Dutch mathematician and orientalist Jacobus Golius, for example, brought more than 300 manuscripts from his travels through the eastern Mediterranean to Leiden; see Bevilacqua, *The Republic of Arabic Letters*, 37.

127 "Je j'ay eu qu'un ms. arabe de la musique enrich de figures geometriques pour la demonstration des tons et de leurs proportions, distinguées par des lignes de differentes couleurs." Peiresc to Saumaise, 22 May 1634, in Peiresc, *Lettres a Claude Saumaise*, 84.

128 "et qu'il ne faille advouer que cez peuples barbares ne laissent pas de produire des espritz bien curieux et bien plus punctuelz que nous ne croyons." Peiresc to Saumaise, 22 May 1634, in Peiresc, *Lettres a Claude Saumaise*, 85.

that you would not have expected from these barbaric people," applauding the use of geometrical figures.[129] To his correspondent and orientalist Gilles de Loches, Peiresc again shared his excitement about "the geometrical figures representing the proportions and ratios of the tones," concluding that he had never seen such a thing "in all the books and manuscripts that had passed through my hands."[130]

Little is known of the author al-Ṣaydāwī, who came from coastal city of Sidon and died in Damascus in 1506. In contrast to what Mersenne and Peiresc believed, his treatise does not include any mathematical proportions or tunings. Rather, the didactic poem distinguishes the four fundamental modes used in music: *Rāst*, *'Irāq*, *Zīrāfkand*, and *Iṣfahān*. All modes are constructed from a set of tetrachords that relate to each other. Al-Ṣaydāwī also distinguishes six secondary modes apart from these fundamental modes; these are again divided in two pairs of two, coming to a total of twenty modes.[131] The circular diagrams of al-Ṣaydāwī, each with eight horizontal lines, represent the degrees of the scales. Using different colors and symbols, the possible pitches of each different melody types are marked. This was not intended to be prescriptive but rather to illustrate possibilities for singers and instrumentalists. In the Ottoman tradition, the teaching and transmission of music was largely oral, based on a face-to-face relationship between master and pupil.[132] Al-Ṣaydāwī developed a representational system to indicate the musical material from which performers could choose. Although the text survived in multiple copies, no other music theorist seemed to have adopted this system. The fact that both scholars were ignorant of Arabic and Persian did not stop Mersenne and Peiresc from believing that this manuscript could be the key to understanding the music of Arab cultures and subsequently the music of the ancient Greeks.

Peiresc sent the manuscript to Mersenne in Paris, who described it as "a great ornament" and thanked Peiresc for all his help.[133] Although Mersenne had already given up on the endeavor, he now became interested again in adding some observations on Ottoman music in his *Harmonie universelle*.

129 "beaucoup plus d'art que vous ne vous en promettez de ces peoples barbares." Peiresc to Mersenne, 1 May 1634, CM 4, 107.

130 "des figures geometriquement representatives des proportions et rapportz des tons … je n'avois jamais veu en tous les livres imprimés et mss qui m'estoient passez par les mains." Peiresc to De Loches, 20 May 1634, CM 4, 159.

131 Shiloah and Berthier, "A propos"; Ghrab, "Livre de la générosite"; Norton-Wright, "Innovations in Music Notation."

132 Behar, "The Ottoman Musical Tradition."

133 "un grand ornement." Mersenne to Peiresc, 14 May 1634, CM 4, 132.

The last parts of the book remained to be printed, and Mersenne hoped to be able to translate and print the treatise at the end of his book.[134] If successful, this would be the first time that a work of music theory from the Ottoman Empire would be printed in Europe. But finding a capable translator was difficult. In the following year, the manuscript would go through the hands of many translators and scholars in Paris, Rome, Aix, and the Low Countries in search of somebody to make sense of it.[135] One of the most likely translators was the Maronite scholar Gabriel Sionita, born in Ehden (now in Lebanon), who had come to Paris with the French ambassador François Savary de Breves. Sionita would be essential in publishing the Paris Polyglot Bible, which featured the Syriac Old Testament together with various Arabic versions of the text. Upon receiving the manuscript from Peiresc, Mersenne tried to persuade Sionita to work on the music manuscript, but the translator had no time.[136] After a translation attempt by the mathematician and lawyer Claude Hardy, Mersenne brought the manuscript to the orientalist Gilbert Gaulmin in 1635. Peiresc was visibly displeased with Mersenne's decisions to give the treatise to men seemingly incapable of translating it.[137] Others, Peiresc wrote, were eager to see the manuscript, for example the Dutch mathematician and orientalist Jacobus Golius.[138] Around March 1635, Mersenne informed Peiresc that the translation was almost complete but that certain parts were left blank, as they were in Persian.[139] Mersenne returned the manuscript to Peiresc in May 1635.[140] He had been able to study the translation by Claude Hardy but wrote to Peiresc that Hardy had been struggling. Hardy also insisted to Mersenne that "one cannot print the work if one does not use the same colors that are used in it [the manuscript], as understanding it depends on it greatly."[141] Peiresc agreed that it was better not to print the treatise and hoped that one day more capable people might look at it: "I am glad it is no longer worthy of your curiosity," he wrote to Mersenne, "since you find

134 Peiresc to Mersenne, 20 March 1635, CM 5, 108.

135 Peiresc to Mersenne, 18 June 1634, CM 4, 177.

136 Mersenne to Peiresc, 26 July 1634, CM 4, 256–57; Mersenne to Peiresc, 2 August 1634, CM 4, 280. Poylglot Bibles contained versions of the sacred text in nine versions (Hebrew, Greek, Latin, Aramaic, Arabic, Samaritan, Syriac, Ethiopic, and Persian).

137 Peiresc to Mersenne, 13 August 1634, CM 4, 289; Mersenne to Peiresc, 24 August 1634, CM 4, 329.

138 Saumaise to Peiresc, 2 September 1634, CM 4, 343.

139 Peiresc to Dupuy, 20 March 1635, in Peiresc, *Lettres de Peiresc aux frères Dupuy*, 285.

140 Mersenne to Peiresc 25 May 1635, CM 5, 214.

141 "qu'on ne peut faire imprimer le livre si l'on n'use des mesmes couleurs qui y sont, d'autant que l'intelligence en depend particulierement." Mersenne to Peiresc, 17 May 1635, CM 5, 205.

nothing in it but simple terms of art."[142] Mersenne did suggest that there might be "some knowledgeable Persian or Arab in your area that will be explain [the work], and in that case, I will be happy to add a few leaves to the book [*Harmonie universelle*], even when it is finished."[143]

When Peiresc received the manuscript again in 1635, he did exactly this. He engaged a translator that previous research has identified as Mattouk or Mattou Chiassan, who was described by Peiresc as a "native Turk" (*turc natural*) from Aleppo.[144] Peiresc came into contact with Chiassan through a Marseille merchant and described him as very intelligent and from a good family.[145] In the letters to Mersenne, Peiresc simply called him "our Turk," but in a letter to Gassendi he revealed his name to be "Matton Sassan," most likely a Frenchified version of the name "Chiassan" found in the archival papers.[146] Throughout the early 1630s, Chiassan helped Peiresc with information and translations over the course of several years, for example by translating Arabic inscriptions on artifacts. Before the French actively began training young men in eastern languages (the *École des Jeunes de langues* was founded in 1669), translators like Chiassan were essential for French merchants in trading their goods, and subsequently also for Peiresc's musical endeavors.

In the summer of 1635, he wrote to one of his Parisian correspondents, "I have put the one [manuscript] you have sent me in the hands of our Turk, who made great exclamations of joy, understanding it very well."[147] In this process Chiassan was aided by a French helper. Traces of this translation process can be found among Peiresc's papers at the Bibliothèque nationale de France in Paris, where the Arabic titles of the figures and tables are compiled together with transcriptions and French translations of the title,

142 "je suis marry qu'il ne soit plus digne de vostre curiosité ... vous n'y trouvez rien que les simples termes de l'art." Peiresc to Mersenne, 5 May 1635, CM 5, 164; Peiresc to Aubery, 8 May 1635, CM 5, 173–74

143 "qu'il se rencontrera encore quelque Persan ou Arabe sçavant vers vos quartiers qui pourra expliquer, et en ce cas il me sera aysé de fere ajouter quelques feuilles au livre, encore qu'il seroyt achevé." Mersenne to Peiresc, 17 May 1635, CM 5, 205.

144 Peiresc to Mersenne, 5 May 1635, CM 5, 168; Peiresc to Jacques Dupuy, 5 June 1635, CM 5, 226; Miller, *Peiresc's Mediterranean World*, 342. Sonja Brentjes presents another possible identification, namely an Armenian merchant, see Brentjes, *Peiresc's Interests in the Middle East*, 7.

145 Peiresc to Mersenne, 5 May 1635, CM 5, 168; Miller, *Peiresc's Mediterranean World*, 277, 322, 342, 345.

146 "le Turc que vous cognoissez Matton Sassan." Peiresc to Gassendi, 17 June 1635, in Peiresc, *Lettres de Peiresc*, 4: 515; Paris, Bibliothèque nationale de France, MS fr. 9532, fol. 43r.

147 "J'ay mis celuy que vous m'avez renvoyé, ez mains de nostre Turc, qui en fit de grandes exclamations de joye, l'entendant fort bien." Peiresc to Jacques Dupuy, 26 June 1635, CM 5, 255. See also Peiresc to Mersenne, 3 July 1635, CM 5, 275.

author, and prayer that opens the book.[148] The translated headings further list the different modal branches and scale degrees, with French terms for the Arabic modal terminology. The name for the six secondary modes (*awāzāt*), for example, is translated as "songs" (*airs*).[149] Al-Ṣaydāwī also used some terms in multiple ways; the term *buhūr* is used to denote the seven degrees of the scale but is also used as a modal category in itself. The term is literally translated as "sea" (*mer*).[150] Such literal translations do not mean Chiassan did not understand the treatise but rather point to the complex task of translating musical and poetic terminology. If a full translation was ever completed, it is unfortunately no longer among Peiresc's materials.

The treatise of al-Ṣaydāwī was not the only musical manuscript Peiresc received. In June 1635, Peiresc received another collection of texts from Cairo, consisting of three separate treatises on music, including Ṣafī al-Dīn al-Urmawī's *Kitāb al-Adwār* (Book of the circles, i.e., modes) from the thirteenth century.[151] The court musician al-Urmawī is generally regarded as the founder of the theoretical tradition called the Systematist school. Again, these treatises to Peiresc seemed to confirm that the Arabic music treatises contained valuable information on the music of the ancients. He wrote to Peiresc that these manuscripts were "much more extensive than the one [manuscript] you saw, and adorned with a good number of figures very much in line with those of the Greek authors on this subject."[152] Although he could not read the words, the diagrams suggested the close connection to ancient Greek music Peiresc was looking for. In the same letter, Peiresc once more mentioned the great influence that studying these materials had on Chiassan when he showed them to him: "He shuddered with joy when he saw the books, and although he has not studied music professionally, he never fails to sing quite well, and he has no trouble in agreeing with his interpreter [*truchement*] in order to convey the meaning of the more important words."[153]

148 Paris, Bibliothèque nationale de France, MS fr. 9531, fols. 233r–237v.

149 Paris, Bibliothèque nationale de France, MS fr. 9531, fols. 236v–237v. For the use of the term, see Ghrab, "Livre de la générosite," 56.

150 Paris, Bibliothèque nationale de France, MS fr. 9531, fol. 236r; Shiloah and Berthier, "A propos," 172; Ghrab, "Livre de la générosite," 84–86.

151 The treatise of Ṣafī al-Dīn al-Urmawī' was one of the most popular Ottoman treatises and a significant number of copies have survived. On al-Urmawī's position in the history of Arab music, see Shiloah, *Music in the World of Islam*, 55–58.

152 "bien plus amples que celuy que vous avez veu et enrichy de bon nombre de figures fort conformes à celles des autheurs Grecs en cette matiere." Peiresc to Mersenne, 3 July 1635, CM 5, 275.

153 "Il tressaillit de joye quand il veid les livres et bien qu'il n'ayt pas faict d'estude ex professo de la Musique, il ne laisse pas de chanter assez passablement." Peiresc to Mersenne, 3 July 1635, CM 5, 275.

The manuscript Peiresc received included two images of the oud and zither, which he described as a lute and spinet. When Peiresc showed these images to Chiassan, he detailed that Harun ar-Rashid (ca. 763–809) sent an organ to King Charlemagne in Europe.[154] These images copied from al-Urmawī's treatise are still found among Peiresc's papers in Paris.[155] The five-string oud is tuned in fourths with seven fixed frets and was the primary demonstration instrument for al-Urmawī's explanation of the tonal system (fig. 4.2). Mersenne was not yet convinced by Peiresc's account of Chiassan and wanted more information on him and the manuscripts, particularly about notation: "to see if they use their letters or other characters or ciphers to signify the notes as the ancient Greeks did."[156] Mersenne suggested that Chiassan be taught the harmonic hand or *gamme*, writing that the philosopher Pierre Gassendi could do this in an hour. If this was successful, they might learn how Ottomans utilized musical scales and if their system was indeed similar to the system of the ancients.[157] Previously a translator, Chiassan was now becoming a possible source himself. Peiresc answered Mersenne, saying that Gassendi was not around and that Chiassan was too old to learn "our music."[158] He would send him to the town of Angoulême to meet some church musicians.

Whether Chiassan was actually sent to Angoulême is unclear, as there are no more references to him among Peiresc's correspondence. He was part of a longer line of scholars like Hardy and Golius who worked on the treatise, but not much is known about him or his work for the merchants of Marseille. Peiresc's descriptions of Chiassan's excitement and musical outbursts point to the presence of Arab and African musicians and translators who performed and worked across Europe in the early modern period. Although there is a danger of reading too much into Peiresc's description, working on the translation and seeing images of familiar musical instruments was perhaps a profound experience for Chiassan, who was already of advanced age. Possibly, he was singing one of the strophic songs dealing with love and war that were so popular in cities like Istanbul.[159] Peiresc's insistence on the joy of the translator raises the question of whether these words may have masked the use of force. Peiresc's letters also show no awareness of the contingent nature of the project: a translator

154 Peiresc to Mersenne, 3 July 1635, CM 5, 275.

155 Paris, Bibliothèque nationale de France, MS fr. 9531, fols. 225r and 231r.

156 "de voir s'ils usent de leurs lettres ou autres caracteres ou ciphres pour signifier les notes comme faisoient les Grecs anciens." Mersenne to Peiresc, 15 July 1635, CM 5, 300.

157 Mersenne to Peiresc, 15 July 1635, CM 5, 301.

158 "Nostre pauvre Turc n'est pas assez jeune pour pouvoir apprendre nostre Musique." Peiresc to Mersenne, 20 August 1635, CM 5, 358.

159 As described by Alî Ufukî in Haug, *Ottoman and European Music*, 327.

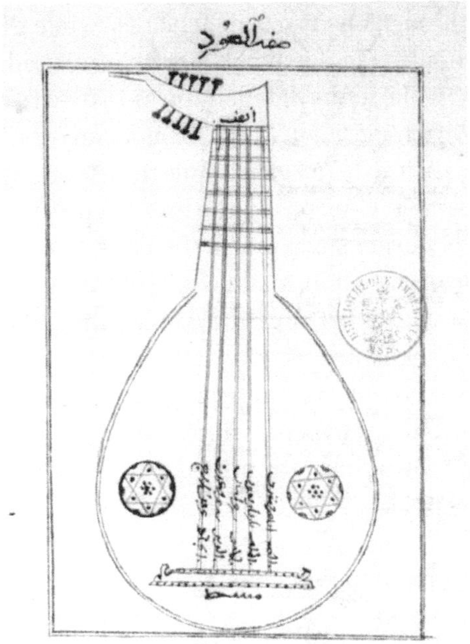

Figure 4.2: Oud from Ṣafī al-Dīn al-Urmawī's *Kitāb al-Adwār*. Paris, Bibliothèque nationale de France, MS fr. 9531, fol. 225r.

from Aleppo was asked to work on music treatises from the thirteenth and late fifteenth centuries whose connections to contemporary musical practice were far from clear, let alone representative of the huge cultural differences within the Ottoman Empire or the eastern Mediterranean at large. Finally, it demonstrates how Mersenne's project of studying universal harmonies in Paris took part in the globalizing world of the seventeenth century. The suggestion that Chiassan learn the Guidonian hand in order to compare his testimony with what was known on ancient music was one of the many ways in which Mersenne sought materials for his *Harmonie universelle* and hoped to gather evidence on the nature of harmony. Peiresc's network supplied him with images and manuscripts, but these needed to be translated and explained, demonstrating the essential role of translators and informers like Chiassan.

Mersenne ultimately did not use the images of the oud and zither in his publications, but he did include an image of an Indian vina and an image of a Chinese sheng.[160] He had not seen these instruments in person but rather

160 Mersenne, *Harmonie universelle*, "Livre quatriesme des instrumens a chordes," 227–28; Mersenne, *Harmonie universelle*, "Livre cinquiesme des instrumens a vent," 308; Mersenne, *Harmonicorum*, "Liber secundus de instrumentis," 111–12.

received images through his correspondence. He wrote that he had received more images but omitted those, since "those that are used nowadays surpass them in all ways."[161] Mersenne did mention al-Urmawī's instruments at the end of the Latin *Harmonicorum*, which includes an extra page of text at the end of the book on bells where he mentions the use of gamelan and flutes in Java and an instrument depicted on the bronze cup sent by Peiresc. Before the final words "Let every spirit praise the Lord" ("Omnius spiritus laudet dominum"), some of the penultimate words of the book are devoted to the image of a five-course oud of al-Urmawī. According to Mersenne, the lute was tuned in four fourths, and he writes that it is easy to imagine the tuning, dismissing the instrument as an afterthought.[162]

With the printing of *Harmonicorum* and the first books of *Harmonie universelle*, the story was not yet finished. Now that Peiresc had the name of a seller of music books in Cairo through which he obtained al-Ṣaydāwī's treatise, he figured that it made sense to send something back to the previous owner, a music master in Cairo named "Jann Lazaravam." Although Peiresc did not know who he was—he had only heard the name from Magy—he thought that this person might be persuaded in communicating what he knew of music if he received some books on European music.[163] Mersenne was put in charge of deciding what materials would be sent and the brothers Pierre and Jacques Dupuy could take care of the expenses.[164] When he heard about Peiresc's plans, Mersenne reacted skeptically to the endeavor, reiterating his disbelief in any musical abilities of the musicians in the eastern Mediterranean and mocking the idea itself:

> For the music master of the country of the Moors, if we send him some music book, Latin or French, how can he possibly understand it! Would you like me to send you some of Boësset's songs, or some of the newest ones printed by Ballard, so that he can hear them? But how will he understand our notes? If someone from the Levant who understands French, Latin, Italian or Greek could explain this book or the practice of their music to us, he would be most indebted to us. When you have informed me of what will be appropriate for this man, I will prepare this and send it to you. I am already sending you, at your own risk, the first book of my

161 "dont on use maintenant les surpassent en toutes façons." Mersenne, *Harmonie universelle*, "Livre quatriesme des instrumens a chordes," 227–28.

162 Mersenne, *Harmonicorum*, "De campanis," 168.

163 Peiresc to Mersenne, 1 May 1634, CM 4, 107; Peiresc to Dupuy, 1 May 1634, CM 4, 113.

164 Peiresc to Mersenne, 1 May 1634, CM 4, 107–8.

> music [*Harmonie universelle*] that you have already seen, if you think it appropriate to inform them of this; if not, I will send whatever you think is appropriate.[165]

At the end of the letter, Mersenne suggested that he could send the parts from his *Quaestiones* to Cairo, since it contained a lot of Greek and Hebrew. If this were to Peiresc's liking, he could send it with the next batch of letters.[166] Peiresc responded that the *Quaestiones* served the purpose well, but he also wanted to receive "all the books of *airs* of Boësset or Ballard, and others who are better and newer, or who are more popular, without troubling ourselves if these people can understand Latin or French."[167] The question of language and music notation should not concern them at the moment, Peiresc argued, and he asked Mersenne to prepare a small collection of music to send to "any friends curious of our music," with some new songs and music for lute or theorbo.[168] Mersenne eventually sent Peiresc the parts on music from the *Quaestiones* and a musical tablature, now lost.[169] Whether this material was really sent to Cairo is impossible to determine. In January 1635, Mersenne asked if Peiresc had already received "the question on music" and sent it to Cairo, but sources are missing on what happened after this.[170] By the summer of 1635, Mersenne and Peiresc seemed to have given up on the endeavor, and Mersenne brought his book to the printer. It would be no surprise if some *air de cour* by Boësset and the *Quaestiones* eventually had ended up in Cairo. Early modern books on music traveled far.[171]

165 "Pour le maistre de musique du pays des Mores, si nous luy envoyons quelque livre de musique, latin ou françois, le moyen qu'il le puisse entendre! Voulez-vouz que je vous envoye des airs de Boisset, ou des plus nouveaux imprimez chez Balard, pour les luy faire-tenir? Mais comment entendra-il nos notes? Si quelqu'un du Levant qui entende le françois, le latin, l'italien, ou le grec, nous pouvoit expliquer ce livre ou la pratique de leur musique, il nous obligeroit bien fort. Quand vous m'aurez averti de ce qui sera propre pour cet homme, je m'efforceray de le treuver pour le luy envoyer. Je vous envoye desja à tout hazard le 1er cayer de ma musique que vous avez desja veu, si vous jugez à propos de la prevenir de cela; si non, je luy envoyeray tout que vous aurez autrement jugé." Mersenne to Peiresc, 14 May 1634, CM 4, 133–34.

166 Mersenne to Peiresc, 14 May 1634, CM 4, 136.

167 "et non seulement cela, mais aussy tous cez libvres d'airs de Boisset ou Ballard et aultres des meilleurs plus nouveaux, ou qui sont plus en vogue, sans nous mettre en peine si cez gens-là entendent le latin ou le françoys." Peiresc to Mersenne, 18 June 1634, CM 4, 178.

168 "un petit recueil de ce que vous voudriez envoyer à quelque amy curieux de nostre musique." Peiresc to Mersenne, 18 June 1634, CM 4, 178.

169 Mersenne to Peiresc, 2 July 1634, CM 4, 230.

170 "la question de musique." Mersenne to Peiresc, 15 January 1635, CM 5, 28.

171 Irving, "The Dissemination."

It is difficult to say what Ottoman musicians might have found useful in Mersenne's books or beautiful in Boësset's *airs de cour.* As Owen Wright summarizes, there was a "general Ottoman indifference to, or disdain for, European music."[172] Jeffrey Levenberg has recently stressed that "much Arabic-Persian music history between the fourteenth and seventeenth centuries, whether in and of itself or in its relationship to that of the West, is still uncharted territory, on both sides of the Mediterranean."[173] Whereas historians of astronomy, medicine, and mathematics have uncovered a plethora of sources and moments of exchange, in the field of music much is yet to be discovered.[174] In Italy Zarlino, for example, referred to "Barbarian writers" (*Barbaro scrittore*) in his *Dimostrazioni harmoniche* (Venice, 1571), suggesting that he had come across Arab authors who he could not read.[175] Peiresc's efforts to obtain translations of music theoretical works from across the Mediterranean beg the question of whether other examples might still be uncovered.

Whereas Peiresc was interested most of all in antiquity, Mersenne increasingly stressed that the ancient musical system did not add anything to contemporary musical practice. He asked his contemporary musicians to follow his own writings, not those of the ancients. As he wrote in *Harmonie universelle,* he was interested in making a foundation for musical science, not restoring Greek melodies. Already in March 1634, he announced to Peiresc that he did not place any further hope in antiquity or the East. Hearing about al-Ṣaydāwī's treatise, Mersenne answered that he would save some space in his books on instruments or print some extra sheets at the end.[176] When Peiresc promised more materials, Mersenne was even willing to prolong the printing process.[177] As he writes to Peiresc in May 1634, "I am slowly prolonging the printing until you have received all that you are hoping for."[178] But as he wrote to Peiresc, he did not expect Ottoman music

172 Wright, "Turning a Deaf Ear," 153.

173 Levenberg, "A Diagram," 393.

174 Historian of science Avner Ben-Zaken, for example, suggests looking at overlapping "cultural margins" that create "a stimulating, mutually embraced zone where intense cross-cultural exchanges transpire," putting pressure on the idea that Ottoman and European cultures were scientifically "incommensurable." Ben-Zaken, *Cross-Cultural Scientific Exchanges,* 168.

175 Zarlino, *Dimostrazioni harmoniche,* 265; Irving, *The Making of European Music,* 204. For the presence of Arabic sources in earlier periods, see: Randel, "Al- Fārābī"; Burnett, "European Knowledge."

176 Mersenne to Peiresc, 20 March 1634, CM 4, 82.

177 Mersenne to Peiresc, 14 May 1634, CM 4, 134.

178 "Je prolongeray tout doucement l'impression jusques à ce que vous ayez recouvré tout ce que vous esperez." Mersenne to Peiresc, 14 May 1634, CM 4, 134.

to approach "the perfection of our music, nor our speculations." Instead, printing the treatise would be "an ornament" and demonstrate the "diligence of those who love the good letters."[179] If the treatise of al-Ṣaydāwī would be printed, he seemed to say, it would be a testimony to Peiresc's enduring efforts to obtain materials.

The antiquarian Peiresc continued to be puzzled by the treatise, describing "the rules of the art" in Ottoman music as "difficult" in a letter from 1636.[180] He would continue to write to Mersenne that harmony did exist outside the confines of Europe. As he wrote to Mersenne in July 1634, "harmony was cultivated by these barbarians, a bit more than you were made to believe."[181] The following year, he expressed the same opinion and wrote, "everywhere birds have their beak [*que partoutes oysons ont leur bec*], and there are hardly any nations so barbaric and unrefined, whatever opinion we have to the contrary, that do not cultivate harmony in some way."[182] Even if barbarous peoples exists, Peiresc argued again and again, that does not necessarily mean that they have not developed musical harmony in some way. Mersenne, on the other hand, turned out to be difficult to convince.

The Limits of Harmony

The small sections in *Harmonie universelle* and *Harmonicorum* that hint at Mersenne's interest in non-European music represent in many ways the end of his interest in finding more materials. When Peiresc died in June 1637, the latest additions to the *Harmonie universelle* still remained to be printed. With Peiresc's death, Mersenne not only lost his patron who supplied him with money for printing his books but also Peiresc's elaborate network of merchants, missionaries, translators, and artists. But the topic of music outside Europe continued to reappear in the correspondence, albeit marginally. In 1642, for example, Mersenne received a letter from a man named

179 "Je n'espere pas qu'ils approchent en aucune façon de la perfection de nostre musique, non plus que de nos spéculations, mais cela servira du moins d'ornement et fera paroistre la diligence de ceux qui affectionnent les bonnes lettres." Mersenne to Peiresc, 14 May 1634, CM 4, 134.

180 "les regles de l'art assez difficile." Peiresc to Saumaise, 2 June 1636, in Peiresc, *Lettres a Claude Saumaise*, 282.

181 "que l'harmonie estoit cultivee parmy ces barbares, un peu plus que l'on ne vous l'avoit faict à croire." Peiresc to Mersenne, 16 July 1634, CM 4, 241.

182 "afin de faire voir ce que porte le commun proverbe, que partoutes oysons ont leur bec, et qu'il n'y a guieres de nations si barbares et si mal polies, quelque opinion que nous ayons au contraire, qui ne cultivent en quelque façon l'harmonie." Peiresc to Mersenne, 10 and 15 May 1635, CM 5, 185–86.

Bomot regarding the musical system in use in China and the naturalness of the solmization syllables.[183] In 1640 the English mathematician and diplomat John Pell wrote to him, complimenting him on his *Harmonie universelle*. Music was a worthy topic of investigation, Pell wrote, as it extends to "even the most remote and barbarous nations."[184] He also offered his summary of a musical encounter several decades earlier by the East Indiana Company Captain John Saris on the Hirado Island in Japan, where the English captain John Saris observed the use of music notation at a banquet.[185] In his next letter, Mersenne did not respond to this information, but it seems that he did not change his opinion regarding the naturalness of the intervals commonly in use. In same year, he wrote a letter to Jean-Baptiste Geoffroy regarding the strings of the harp that David used to cure King Saul, where Mersenne stated that it must have used the same intervals as the modern harp, because they "alone are natural," an answer that did not seem to satisfy Geoffroy.[186]

But Mersenne was not solely interested in gathering knowledge on the use of musical intervals. He also considered how to improve its transmission. In his *Harmonicorum*, he wondered why there has not yet been a universal method to transmit diatonic music, "which we have proved to be natural to all humans."[187] He gave his own answer to this question in a peculiar table in both versions of his book. Instead of incorporating the observations on Ottoman music by Minuti or the treatise by al-Ṣaydāwī, Mersenne included a suggestion for all nations of the earth to conform to European music, printing a "universal tablature with the letters of the alphabet," listing musical notes and letters in different language. Its use is described as showing "how the Jews, Arabs, and Greeks, and all the other nations can conform [*se conformer*] to our manner of singing and writing all kinds of songs."[188] In this table, Mersenne collects the major eastern alphabets of Armenian, Greek, Arabic, Samaritan, Syriac, Rashi script, and Hebrew. Each of the first fifteen letters of these scripts are attached to a note from the diatonic scale (fig. 4.3). Interestingly, in Ottoman music theory, among various other graphic systems, the Abjad

183 Bomot to Mersenne, 29 June 1642, CM 11, 187.

184 "quae ad ipsas remotissimas et barbaras nationes si hinc inde extendit." Pell to Mersenne, 29 March 1640, CM 9, 232.

185 Pell to Mersenne, 29 March 1640, CM 9, 232–33.

186 "qui seulz sont naturalz." Geoffroy to Mersenne, 20 May 1640, CM 9, 326.

187 "Quî fieri potest ut Cantus Diatonicus, quem omnibus hominibus naturalem esse alibi probavimus, nondum invenerit characteres generales quibus incolae terrae suos sibi cantus communicent?" Mersenne, *Harmonicorum*, "De compositione musica," 164.

188 "monstrer comme les Juifs, les Arabes, les Grecs, & toutes les autres nations peuvent se conformer à nostre maniere de chanter, & d'escrire toutes sortes de chants." Mersenne, *Harmonie universelle*, "Livre quatriesme de la composition de musique," 251.

Figure 4.3: Table with letters of the alphabet. Mersenne, *Harmonicorum libri*, "De compositione musica," 165. Paris, Bibliothèque nationale de France, département Réserve des livres rares, RES-V-580.

numerals of the Arabic alphabet were frequently used to denote intervals.[189] This was exactly what Mersenne asked about in his letter to Peiresc from July 1635, when he suggested that Chiassan would be taught the harmonic hand. The method is also found in al-Ṣaydāwī's text that he himself had translated, but it is uncertain whether Mersenne was aware of this.

As Mersenne confessed, he still had little idea about the notational forms of these cultures but thought it was better if they adopted this manner: "nevertheless, we have so little knowledge of it that it is better to advise them to use the first eight or fifteenth letters of their alphabet, as we do."[190] Though he knew presumably hardly anything on the subject, Mersenne did not hesitate to suggest what he deemed a better method. Such efforts were

189 See Wright, "The Sight of Sound," 360–61. I would like to thank Mona Mirkamali for pointing this out to me.

190 "l'on en a si peu de connoissance, qu'il vaut mieux leur conseiller d'user des 8 ou 15 premieres lettres de leur Alphabet, comme nous faisons, pour chanter tout ce qu'ils voudront." Mersenne, *Harmonie universelle*, "Livre quatriesme de la composition de musique," 252.

further operationalized by Athanasius Kircher, who designed a musical composition machine, the "Arca Musarithmica," suggesting that it could be used by missionaries in their efforts to evangelize other cultures, allow non-musicians to compose simple music, and allow melodies to be made easily for other languages.[191] Mersenne did not propose his table explicitly for such evangelical goals, but he did connect it to the search for a universal musical language. In his *Harmonicorum libri*, he suggested that by adoption of the table, songs and musical concepts could be communicated to the "Persians, Tartars, Chinese and other inhabitants of both Indies."[192] Diatonic music could now be easily communicated across the globe. Although Mersenne described the table as a communication device, such method of notation would necessarily eliminate musical aspects that cannot be captured in such a modified solmization system. As Olivia Bloechl has shown in her work on the colonial logic of French early modern music, depictions of musical sameness and universality were ambivalent, as they stressed sameness as a form of the "symbolic integration of colonial peoples as quasi-French subjects" while at the same time emphasizing their alterity.[193] Mersenne's table would aid such further integration, namely by providing an alternative method to introduce or teach the diatonic scale to various cultures outside Europe. Within this system, there is no room for microtonal intervals or inflections, and thus it would necessarily have an impact on what music would be transmitted by this system. In doing so, he was explicit about the wider cultural implications of his work, much as Kepler connected his investigation of world harmonies with the current political situation in Europe in his *Harmonice mundi* or Athanasius Kircher's *Musurgia universalis* addressed the global aspirations of the Jesuit Society.

Interestingly, the characters for this table were made available to Mersenne by the Royal Printers Antoine Vitré and Guy Michel Lejay, who worked on the Paris Polyglot Bible, which was produced with the help of Gabriel Sionita, the same scholar Mersenne tried to engage for his translation.[194] The Paris edition consists of ten volumes and was printed over seventeen

191 Kircher, *Musurgia universalis*, VIII, 3, 76. On Kirchers "Arca" in the context of Roman musical life, see Klotz, *Kombinatorik* 15–48; Stoessel and Collins, "La grandezza."

192 "Superest igitur ut aliquâ ratione characteres harmonicos, quibus utimur, ad cantilenas exprimendas ita litteris Orientalium accommodemus, ut illos concipere, suisque characteribus describere, & conceptos vicinis Persis, Tartaris, Sinensibus, & alijs utriúsque Indinae incolis communicare possint." Mersenne, *Harmonicorum*, "De compositione musica," 164.

193 Bloechl, *Native American Song*, 143.

194 Mersenne was especially interested in the Biblical scholarship of Jean Morin and Jacobus Golius, see Malcolm, *Aspects of Hobbes*, 419–20.

years (1628–45), exactly the period in which Mersenne was working on his *Harmonie universelle.* The goal of such Bibles was to observe textual variants, and they stimulated research into Arabic and other eastern languages.[195] The project of the Parisian Bible was closely tied to the French interest in the Ottoman Empire. But Mersenne's table would have an opposite effect. Instead of identifying textual or musical differences, its goal was to create a uniform system, eliminating differences rather than studying them.

Cultural differences in music making were not viewed as different forms of taste or customs, changeable over time and place. Rather, Mersenne continued to believe that music was a part of the fabric of the world, a natural law that should be followed. At the same time, his position was remarkably flexible. In his *Questions harmoniques* of 1634 he still maintained that if one would consult all nations in the world, one would find that they all use the same intervals, but only two years later he seemed less certain. It was not possible to exclude any element of learning or custom and assume that all nations share the same musical materials. This difficulty could not be resolved until one encountered "someone who sings without having heard singing," an idea that was expressed by Jean-Philippe Rameau a century later.[196] If the endeavor with Peiresc achieved anything, it seems to have made Mersenne less certain of his convictions that French music was truly universal, but he would not let go of his emphasis on its divine and universal nature. When Mersenne did acknowledge cultural differences, it did not lead to a form of cultural relativism that some of the skeptics propagated. In his early *La verité*, the figure of the philosopher goes as far as to say

> if we are thought foolish by the barbarians, it matters little to us, for besides the fact that we are ready to defend our customs of living, we have natural and divine laws, that guide us in everything that pertains to our salvation.[197]

Here it becomes clear that for Mersenne such an encounter could never happen in reverse or that cultural differences could be discussed on an equal playing field. As discussed in chapter 2, Mersenne remained convinced of

195 The publication of the Paris Polyglot Bible stimulated research into several languages and was closely associated to the efforts of Peiresc, see Miller, "Making the Paris Polyglot Bible."

196 "si l'on ne rencontre premierement quelqu'un qui chante sans jamais avoir ouy chanter." Mersenne, *Harmonie universelle*, "Livre troisiesme des genres,"148. On Rameau's position, see Bloechl, *Native American Song*, 179

197 "si nous sommes estimez fols par les Barbares, il nous importe fort peu, car outre que nous sommes prês de déffendre nos coûtumes de vivre, nous avons la loy naturelle, & la divine, qui nous guide à tout ce qui appartient à nostre salut." Mersenne, *La verité*, 21.

the naturalness of the diatonic scale. Although he never said so explicitly, he seems to have suggested that cultures that depart from these ratios stray away from the natural laws that God gave to creation. When music from outside Europe was described as noise or absent of any harmony, this was not just an aesthetic judgment. Noise was not just something related to taste that could be debated or discussed. Rather, it was an entirely different ontological category.[198]

Mersenne remained convinced that natural and divine laws guided him and his French contemporaries and that these laws of harmony could be codified. Other musical cultures could be discussed as a testimony to the work of his patron or presented as a curiosity, but they could never shake the foundations of his rational and divine science. Ultimately, Mersenne stated, "I therefore exhort all the musicians of the world to employ their compositions only to sing the praises of God, and to excite one another to praise Him by these words of the Royal Prophet: *Come, bless the Lord, all you servants of the Lord.*"[199] A central characteristic of Mersenne's project was an emphasis on unity, expressing the need for a single will and single religion that would shake the entire world in concord. In contrast to many contemporaries, Mersenne considered the unison the most perfect consonance, exactly because of its demonstration of the central role of unity. During this process, Mersenne's interests did not remain static. Whereas in previous books he emphasized the need to study ancient sources to reform modern music, in his *Harmonie universelle* he showed little interest in the humanist endeavor of reviving antique practices. He did not expect Ottoman or ancient musicians to outperform his contemporaries. And if modern composers would work from his descriptions, they would know more than all other musicians. Instead of trying to reintroduce the octave species of the ancients, Mersenne increasingly saw such efforts as futile. Practical implementation of the quarter tones of the enharmonic genus would take a long time, and in his view, the diatonic-chromatic scale used "in all kingdoms" sufficed: "it [the diatonic-chromatic genus] is sufficient for the recreation of ears and the imagination, and for singing the praises of God."[200] At the same time, Mersenne's relationship to the role of custom and taste was complex. The next chapter will further flesh out the many tensions between mathematical rules and individual listeners present in his work.

198 For this point, see Bloechl, "Wendat Song and Carnival Noise."
199 "J'exhorte donc tous les Musiciens du monde à n'employer leurs compositions quà chanter les loüanges de Dieu, & à s'exciter les uns les autres à le loüer par ces paroles du Prophete Royal: *Ecce nunc benedicite Dominum omnes servi Domini.*" Mersenne, *Les preludes*, 186.
200 "tous les Royaumes: lequel suffit pour la recreation des oreilles, & de l'imagination, & pour chanter les loüanges de Dieu." Mersenne, *Harmonie universelle*, "Livre sixiesme de l'art de bien chanter," 439.

Bibliography

Manuscript sources

Carpentras, Bibliothèque Inguimbertine, MS 1831. Papers of Nicolas-Claude Fabri de Peiresc.

Carpentras, Bibliothèque Inguimbertine, MS 1874. Letters of Nicolas-Claude Fabri de Peiresc.

Paris, Bibliothèque nationale de France, MS Dupuy 668. A volume of seventeenth-century letters by various authors.

Paris, Bibliothèque nationale de France, MS fr. 9531. Papers of Nicolas-Claude Fabri de Peiresc.

Paris, Bibliothèque nationale de France, MS fr. 9532. Papers of Nicolas-Claude Fabri de Peiresc.

Printed primary sources

Belon, Pierre. *Les observations de plusieurs singularitez et choses memorables trouvées en Grèce, Asie, Judée, Egypte, Arabie et autres pays étrangèrs*. Antwerp: Steelsius, 1555.

De Caus, Salomon. *Institution harmonique*. Frankfurt: Jan Norton, 1615. Reprint, Geneva: Minkoff, 1980.

De Lery, Jean. *Histoire d'un voyage faict en la terre du Brésil, autrement dite Amerique*. Geneva: s.n., 1585.

Della Valle, Pietro. "Della musica dell'età nostra." In Giovanni Battista Doni, *Lyra Barberini*, vol 2 of 2 vols., 249–64. Florence: Stamperia imperiale, 1763.

Gaffurius, Franchinus. *Theorica musicae*. Milan: Mantegatium, 1492.

Galilei, Vincenzo. "Discorso di Vicentio Galilei intorno all'uso dell enharmonio." In Frieder Rempp, *Die Kontrapunkttraktate Vincenzo Galileis*, 163–280. Cologne: Arno Volk Verlag, 1980.

Gassendi, Pierre. *The Mirrour of True Nobility and Gentility Being the Life of the Renowned Nicolaus Claudius Fabricius, Lord of Pieresk, Englished by W. Rand*. London: Streater for Moseley, 1657.

Kepler, Johannes. *The Harmony of the World*. Translated by E. J. Aiton, A. M. Duncan, and J. V. Field. Philadelphia: American Philosophical Society, 1997.

Kircher, Athanasius. *Musurgia universalis sive ars magna consoni et dissoni in X libros digesta*. Rome: Corbelletti, 1650.

La Mothe le Vayer, François de. "Des doutes raisonnes." In *Oeuvres des François de la Mothe le Vayer, conseiller d'estat ordinaire*, vol 12 of 15 vols., 309–317. Paris: Billaine, 1669.

La Mothe le Vayer, François, de. "Discours sceptique sur la musique." In Marin Mersenne, *Questions harmoniques. Dans lesquelles sont contenuës plusieurs choses remarquables pour la physique, pour la morale, et pour les autres sciences*, 84–165. Paris: Villery, 1634.

La Mothe le Vayer, François de. "La physique du Prince." In *Oeuvres des François de la Mothe le Vayer, Tome Premier*, vol. 1 of 2 vols., 933–1015. Paris: Courbe, 1662.

Le Jeune, Claude. *Dodecacorde contenant douze pseaumes de David, mis en musique selon les douze modes, approuvez des meilleurs autheurs anciens & moderns.* La Rochelle: Haultin, 1598.

Lescarbot, Marc. *Histoire de la Nouvelle France. Contenant les navigations, découvertes, & habitations faites par les François és Indes Occidentales & Nouvelles-France* .... Paris: Milot, 1609.

Lodewijcksz, Willem. *Prima pars descriptionis itineris navalis in Indiam Orientalem* .... Amsterdam: Nicolai, 1598.

Mersenne, Marin. *Correspondance du P. M. Mersenne*. Edited by Paul Tannery, Cornelis de Waard, Bernard Rochot, Armand Beaulieu, and René Pintard. Paris: Centre National de la Recherche Scientifique et al., 1932–88.

Mersenne, Marin. *Harmonicorum libri XII*. Paris: Baudry, 1648. Reprint, Geneva: Minkoff, 1972.

Mersenne, Marin. *Harmonie universelle contenant la théorie et la pratique de la musique (Paris, 1636). Édition fascimilé de l'exemplaire conserve à la Bibliothèque des Arts et Métiers et annoté par l'auteur.* Introduction by François Lesure. Paris: Éditions du Centre National de la Recherche Scientifique, 1963.

Mersenne, Marin. *Les preludes de l'harmonie universelle ou questions curieuses. Utiles aux predicateurs, aux theologiens, aux astrologues, aux medicins & aux philosophes*. Paris: Guenon, 1634.

Mersenne, Marin. *Quaestiones celeberrimae in Genesim, cum accurata textus explicatione: in hoc volumine athei, et deistae impugnantur & expugnantur, & vulgata edition ab hereticorum calumnijs vindicator. Graecorum, & Hebraeorum musica instauratur.* Paris: Cramoisy, 1623.

Mersenne, Marin. *Questions harmoniques. Dans lesquelles sont contenuës plusieurs choses remarquables pour la Physique, pour la Morale, et pour les autres sciences.* Paris: Villery, 1634.

Mersenne, Marin. *Traité de l'harmonie universelle*. Paris: Baudry, 1627.

Mersenne, Marin. *La verité des Sciences, contre les sceptiques ou pyrrhoniens*. Paris: Du Bray, 1625.

Montaigne, Michel de. *The Complete Works of Montaigne*. Translated by Donald Frame. Stanford: Stanford University Press, 1958.

Nicot, Jean. *Thresor de la langue françoyse, tant ancienne que moderne*. Paris: Douceur, 1606.

Peiresc, Nicolas Claude-Fabri de. *Les correspondants de Peiresc, XV: Lettres inédites écrites de Tunis à Peiresc*. Vol. 15 of 21 vols. Edited by Philippe Tamizay de Larroque. Algiers: Adolphe Jourdain, 1889.
Peiresc, Nicolas-Claude Fabri de. *Lettres a Claude Saumaise et a son entourage (1620–1637)*. Edited by Agnès Bresson. Florence: Olschki, 1992.
Peiresc, Nicolas-Claude Fabri de Peiresc. *Lettres de Peiresc*. Vol. 4 of 7 vols.: *Lettres de Peiresc à Borrilly, à Bouchard et à Gassendi. Lettres de Gassendi à Peiresc, 1626–1637*. Edited by Philippe Tamizey de Larroque. Paris: Imprimerie Nationale, 1893.
Peiresc, Nicolas-Claude Fabri de. *Lettres de Peiresc aux frères Dupuy. Tome troisième: janvier 1634–juin 1637*. Vol. 3 of 7 vols. Edited by Philippe Tamizay de Larroque. Paris: Imprimerie Nationale, 1892.
Peiresc, Nicolas-Claude Fabri de. *Lettres de Peiresc. Tome septième: lettres à divers, 1602–1637*. Vol. 7 of 7 vols. Edited by Philippe Tamizey de Larroque. Paris: Imprimerie National, 1898.
Postel, Guillaume. *De la republique des Turcs*. Poitier: Enquibert de Marnef, 1560.
Praetorius, Michael. *Syntagmatis musici Michaelis Praetorii tomus secundus De Organographia*. Vol. 2 of 3 vols. Wolfenbüttel: Holwein, 1618.
Sagard, Gabriel. *Histoire du Canada et voyages que les Freres Mineurs Recollects y ont faicts pour la conversion des infidelles*. Paris: Sonnius, 1636.
Sextus Empiricus. *Outlines of Scepticism*. Edited by Julia Annas and Jonathan Barnes. Cambridge: Cambridge University Press, 2000.
Stelluti, Francesco. *Persio tradotto in verso sciolto e dichiarato*. Rome: Mascardi, 1630.
Taveau, Laurence. "Le manuscrit 2884 du P. Mersenne à la bibliothèque de l'Arsenal: Étude et édition critique." PhD diss., Université de Paris-Sorbonne, Laboratoire IREMUS, 2018.
Thévenot, Jean. *Relation d'un voyage fait au Levant: dans laquelle il est curieusement traité des estats sujets au Grand Seigneur* [...]. Paris: Joly, 1664.
Trigault, Nicholas. *Histoire de l'expédition chrestienne au royaume de la Chine* .... Lyon: Horace Cardon, 1616.
Tyard, Pontus de. *Solitaire second*. Edited by Cathy M. Yandell. Geneva: Droz, 1980.
Zarlino, Gioseffo. *Dimostrazioni harmoniche del R. M. Gioseffo Zarlino da Chioggia*. Venice: Francesco Franceschi, 1571.

Secondary literature

Barker, Andrew. *Greek Musical Writings II: Harmonic and Acoustic Theory*. Vol. 2 of 2 vols. Cambridge: Cambridge University Press, 1989.
Behar, Cem. "The Ottoman Musical Tradition." In *The Cambridge History of Turkey*, edited by Suraiya N. Faroqhi, 319–407. Cambridge, Cambridge University Press, 2006.

Ben-Zaken, Avner. *Cross-Cultural Scientific Exchanges in the Eastern Mediterranean, 1560–1660*. Baltimore: The Johns Hopkins University Press, 2010.

Bertoglio, Chiara. *Reforming Music: Music and the Religious Reformations of the Sixteenth Century*. Berlin: De Gruyter, 2017.

Bevilacqua, Alexander. *The Republic of Arabic Letters: Islam and the European Enlightenment*. Cambridge, MA: The Belknap Press of Harvard, 2018.

Bianchi, Eric. "Bad Latin, Bad Manners: Giovanni Battista Doni, Marin Mersenne, and Literary Style in Seventeenth-Century Music Theory." *Music & Letters* 96, no. 2 (2015): 167–84.

Bloechl, Olivia. *Native American Song at the Frontiers of Early Modern Music*. Cambridge: Cambridge University Press, 2008.

Bloechl, Olivia. "The Pedagogy of Polyphony in Gabriel Sagard's *Histoire du Canada*." *The Journal of Musicology* 22, no. 3 (2005): 365–411.

Bloechl, Olivia. "Wendat Song and Carnival Noise in the Jesuit Relations." In *Native Acts: Indian Performance, 1603–1832*, edited by Laura L. Mielke and Joshua David Bellin, 117–43. Lincoln, NE: University of Nebraska Press, 2012.

Brentjes, Sonja. *Peiresc's Interests in the Middle East and Northern Africa in Respect to Geography and Cartography*. Berlin: Max Planck Institute for the History of Science, 2004.

Burnett. Charles. "European Knowledge of Arabic Texts Referring to Music: Some New Material." *Early Music History* 12 (1993): 1–17.

Burnett, Charles. "Musical Instruments as Conveyors of Meaning from One Culture to Another: The Example of the Lute." In *The Power of Things and the Flow of Cultural Transformations: Art and Culture between Europe and Asia*, edited by Lieselotte E. Saurma-Jeltsch and Anja Eisenbeiß, 156–69. Berlin: Deutscher Kunstverlag, 2010.

Dakhlia, Jocelyne. "Une archéologie du même et de l'autre: Thomas-Osman d'Arcos dans la Méditerranée du XVIIe siècle." In *Les Musulmans dans l'histoire de l'Europe*. Vol. 2 of 2 vols.: *Passages et contacts en Méditerranée*, edited by Jocelyne Dakhlia and Wolfgang Kaiser, 61–163. Paris: Albin Michel, 2013.

Dew, Nicholas. *Orientalism in Louis XIV's France*. Oxford: Oxford University Press, 2009.

Duncan, David Allen. "Mersenne and Modern Learning: The Debate over Music." In *The Rise of Modern Philosophy: The Tension between the New and Traditional*, edited by Tom Sorell, 89–106. Oxford: Clarendon Press, 1993.

Ghrab, Anas. "Livre de la générosite dans la connaisance des modes: Edition et traduction." Diplome d'etudes approfondies, Université Lumiere-Lyon, 2002.

Goodman, Glenda. "Sounds Heard, Meaning Deferred: Music Transcription as Imperial Technology." *Eighteenth-Century Studies* 52, no. 1 (2018): 39–45.

Greenblatt, Stephen. *Marvelous Possessions: The Wonder of the New World*. Chicago: University of Chicago Press, 1991.

Hamilton, Alistair. "'To divest the east of all its manuscripts and all its rarities': The Unfortunate Embassy of Henri Gournay de Marcheville." In *The Republic of Letters and the Levant*, edited by Alastair Hamilton, Maurits H. van den Boogert, and Bart Westerweel, 123–50. Leiden/Boston: Brill, 2005.

Hamlin, William. "On Continuities between Skepticism and Early Ethnography: Or, Montaigne's Providential Diversity." *Sixteenth Century Journal* 31 (2000): 361–80.

Haug, Judith I. *Ottoman and European Music in 'Ali Ufuḳī's Compendium, MS Turc 292: Analysis, Interpretation, Cultural Context.* Münster: Universitäts- und Landesbibliothek Münster, 2019.

Irving, David. "Ancient Greeks, World Music, and Early Modern Constructions of Western European Identity." In *Studies on a Global History of Music: A Balzan Musicology Project*, edited by Reinhard Strohm, 21–41. London/New York: Routledge, 2018.

Irving, David. "Comparative Organography in Early Modern Empires." *Music and Letters* 90, no. 3 (2009): 372–98.

Irving, David. "The Dissemination and Use of European Music Books in Early Modern Asia." *Early Music History* 28 (2009): 39–59.

Irving, David. *The Making of European Music in the Long Eighteenth Century.* Oxford: Oxford University Press, 2024.

Kellman, Jordan. "Mendicants, Minimalism, and Method: Franciscan Scientific Travel in the Early Modern French Atlantic." *Journal of Early Modern History* 26 (2022), 10–37.

Latour, Melinda. *The Voice of Virtue: Moral Song and the Practice of French Stoicism, 1574–1652*. Oxford: Oxford University Press, 2023.

Levenberg, Jeffrey. "A Diagram from the Mubarak Shah Commentary." In *The Museum of Renaissance Music: A History in 100 Exhibits*, edited by Vincenzo Borghetti and Tim Shephard, 392–95. Turnhout: Brepols, 2022.

Locke, Ralph P. *Music and the Exotic from the Renaissance to Mozart.* Cambridge: Cambridge University Press, 2015.

Malcolm, Noel. *Aspects of Hobbes.* Oxford: Clarendon Press, 2002.

Miller, Peter. "Making the Paris Polyglot Bible: Humanism and Orientalism in the Early Seventeenth Century." In *Die europäische Gelehrtenrepublik im Zeitalter des Konfessionalismus/The European Republic of Letters in the Age of Confessionalism*, edited by Herbert Jaumann, 59–85. Wiesbaden: Harrassowitz, 2001.

Miller, Peter. *Peiresc's Mediterranean World.* Cambridge, MA: Harvard University Press, 2015.

Miller, Peter. *Peiresc's Orient: Antiquarianism as Cultural History in the Seventeenth Century.* Farnham: Ashgate, 2012.

Ndiaye, Noémie. *Scripts of Blackness: Early Modern Performance Culture and the Making of Race.* Philadelphia: University of Pennsylvania Press, 2022.

Norton-Wright, Jenny. "Innovations in Music Notation in Late Medieval Syria: British Library Manuscript Or. 13019." *British Library* (music blog), June 29, 2020. https://blogs.bl.uk/music/2020/06/innovations-in-music-notation-in-late-medieval-syria-british-library-manuscript-or-13019.html.

Obelkevich, Mary Romen. "Turkish Affect in the Land of the Sun King." *The Musical Quarterly* 63, no. 3 (1977): 367–38

Palisca, Claude V. *Humanism in Italian Renaissance Musical Thought*. New Haven: Yale University Press, 1985.

Prins, Jacomien. *Echoes of an Invisible World: Marsilio Ficino and Francesco Patrizi on Cosmic Order and Music Theory*. Leiden/Boston: Brill, 2012.

Prins, Jacomien. "Patrizi's and Mersenne's Critiques of Ficino's Interpretation of the Harmony of the Spheres." In *Perfect Harmony and Melting Strains*, edited by Cornelia Wilde and Wolfram R. Keller, 59–80. Berlin: De Gruyter, 2021.

Psychoyou, Théodora. "De la musique universelle à l'oreille des autres: altérités culturelles dans le discours sur la musique en France au XVIIe siècle." In *La musicologie francophone et la circulation des savoirs dans un contexte multiculturel: interpénétration et résilience*, edited by Achille Davy-Rigaux, Catherine Deutsch, Hamdi Makhlouf, and Anas Ghrab, 43–65. Tunis: Centre des Musiques Arabes et Méditerranéennes, 2022.

Quint, David. "A Reconsideration of Montaigne's *Des cannibales*." *Modern Language Quarterly* 51, no. 4 (1990): 459–90.

Randel, Don M. "Al- Fārābī and the Role of Arabic Music Theory in the Latin Middle Ages." *Journal of the American Musicological Society* 29/2 (1976): 173–188.

Reiss, Timothy. "Montaigne, the New World, and Precolonialisms." In *The Oxford Handbook of Montaigne*, edited by Philippe Desan, 196–214. Oxford: Oxford University Press, 2016.

Scherpereel, Joseph. "Peiresc et la musique." In *Peiresc ou la passion de connaître: colloque de Carpentras, novembre 1987*, edited by Pierre Constabel and Anne Reinbold, 153–85. Paris: Vrin, 1990.

Shiloah, Amnon. *Music in the World of Islam: A Socio-Cultural Study*. Farnham: Ashgate, 1995.

Shiloah, Amnon, and Annie Berthier. "A propos d'un 'petit livre arabe de musique.'" *Revue de Musicologie* 71, no. 1–2 (1985): 164–77.

Smith, Justin E. H. *Nature, Human Nature, and Human Difference: Race in Early Modern Philosophy*. Princeton: Princeton University Press, 2015.

Stoessel, Jason, and Denis Collins. "'La grandezza del numero sonoro': Canonic Techniques, Combinatorics, and Early Scientific Thought in Seventeenth-Century Rome." In *Music and Science from Leonardo to* Galileo, edited by Rudolf Rasch, 155–86. Turnhout: Brepols, 2022.

Taylor, Kathryn. "Ancients and Moderns in Sixteenth-Century Ethnography." *History of European Ideas* 46, no. 2 (2020): 113–130

Tolbert, Jane. "Ambiguity and Conversion in the Correspondence of Nicolas-Claude Fabri de Peiresc and Thomas d'Arcos, 1630–1637." *Journal of Early Modern History* 13, no. 1 (2009): 1–24.

Tomlinson, Gary. *The Singing of the New World: Indigenous Voice in the Era of European Contact.* Cambridge: Cambridge University Press, 2007.

Van der Miesen, Leendert. "Songs at the Docks: Nicolas-Fabri de Peiresc and Musical Entanglements in Seventeenth-Century Marseille." *Music and Letters* 105, no. 1 (2024): 27–51.

Van Orden, Kate. "Hearing Franco-Ottoman Relations circa 1600: The Chansons Turcquesques of Charles Tessier (1604)." In *Seachanges: Music in the Mediterranean and Colonial Worlds, 1550–1800*, edited by Kate van Orden, 33–68. Florence: Officina, 2021.

Varwig, Bettina. *Music in the Flesh: An Early Modern Musical Physiology.* Chicago: University of Chicago Press, 2023.

Wegman, Rob C. "Sense and Sensibility in Late-Medieval Music: Thoughts on Aesthetics and 'Authenticity.'" *Early Music* 23, no. 2 (1995): 298–312.

Weiss, Gillian. "Infidels at the Oar: A Mediterranean Exception to France's Free Soil Principle." *Slavery & Abolition*, 32, no. 3 (2011): 397–412.

Welch, Ellen. "The Specter of the Turk in Early Modern French Court Entertainments." *L'Esprit Créateur* 53, no. 4 (2013): 84–97.

Wright, Owen. "The Sight of Sound." *Muqarnas: An Annual on the Visual Culture of the Islamic World* 21 (2004): 359–71.

Wright, Owen. "Turning a Deaf Ear." In *The Renaissance and the Ottoman World*, edited by Anna Contadini and Claire Norton, 143–65. Farnham: Ashgate, 2013.

Wymeersch, Brigitte van. "Peiresc et la musique." In *Peiresc (1604–2004): actes du colloque international 26–30 août 2004*, edited by Jean Dhombres and Agnès Bresson, 111–34. Turnhout: Brepols, 2006.

Yates, Frances. *The French Academies of the Sixteenth Century.* London: Warburg Institute, 1947.

5. Counting and Composing

Abstract: This chapter brings the study full circle by looking at the applicability of rules and rationality in the work of Marin Mersenne. The alignment of what is rational with what is perceived as beautiful was a central concern in the early modern period. Attention was increasingly paid to the capabilities of the subjective listener, and thinkers of the period were conflicted on whether art should follow a prescribed set of rules. This chapter investigates how these problems were articulated in the work of Mersenne and his contemporaries, pointing to an increasing tension between rules, taste, mind, and ear.

Keywords: combinatorics, composition, taste, rules, aesthetics

Can one compose a song that everyone will recognize as the most beautiful? Of all the musical intervals, which gives the most pleasure to hear? What is the role of individual taste in the appreciation of music? For Mersenne, these questions were central to his effort to investigate the powers and confines of the universal harmony mentioned in his many books. They also point to a larger problem in determining the extent to which the domain of reason can speak to aesthetic experience. At times to the great annoyance of some of his correspondents, Mersenne continued to ask whether it was possible to find a rational explanation for beauty. The previous chapter already pointed to the increasing emphasis on taste and custom within musical writings of the early modern period. By the second half of the seventeenth century, having "good taste" had become essential for all those who wanted to participate in high society. Whereas some writers continued to argue that music follows immutable rules, others emphasized the role of custom and individual taste. This tension between mathematical rules and taste continued to be articulated throughout the late seventeenth and eighteenth centuries. By showing how Mersenne accounted for taste and indeterminacy, this chapter argues that some of these tensions were already present in the early seventeenth century. As I will demonstrate, Mersenne attempted to

Van der Miesen, Leendert. *Marin Mersenne and the Study of Harmony: From Sound to Music.* Taylor & Francis Group, 2025.
DOI 10.5117/9789048564132_CH05

give his own answer to these questions, situating music somewhere between immutable rules and indeterminacy. In his view, certain aspects of music are capable of rationalization, but ultimately reason cannot fully explain music and musical experience.

The question of the extent to which musical experience could be rationally explained, either by physical theories or mathematical ratios, was being asked with greater frequency in the early seventeenth century. Mersenne's correspondent René Descartes wrote in his *Compendium musicae* that each sense is capable of pleasure, but this pleasure requires a certain proportionality of the object to the sense that is addressed.[1] He later wrote in his letters that the science of music should be considered morally, not mathematically or physically.[2] In contrast, Baptiste Masoyer Deshommeaux wrote to Mersenne in 1642 that "the science of music is entirely physical [*physicale*]. It consists of the body of air sounding pleasant to human ears."[3] These opposing ideas would continue to be uttered throughout the early modern period, part of a movement that Timothy Reiss has called "aesthetic rationalism." Reiss sees this as an effort to determine quantitative measurements of the emotional and moral effect of the arts, with music playing an important role, as musical proportions could be quantified and experienced through the senses.[4] More generally, musicians and composers were increasingly seeking a relaxation of contrapuntal rules, aligning with the introduction of new styles of composition in the early seventeenth century. In a letter on Italian music, the French viola da gamba player André Maugars pleaded for a looser application of musical rules, as he found that his musical contemporaries kept themselves "religiously enclosed in pedantic categories."[5] Mersenne's correspondent Bredeau wrote that compositions often have more grace when they include licenses from contrapuntal rules rather than when following the rules exactly.[6] And the French court composer Anthoine Boësset wrote that the Dutch composer and theorist Joan Albert Ban "studied too much

1 Descartes, *Musicae compendium/Leitfaden der Musik*, 2–3.

2 "nec mathematicas, nec physicas, sed morales tantum mihi videri." Descartes to Ban, December 1640, in Descartes, *Correspondance*, 3:843.

3 "Il me semble que la science de musique et toutte physicale. Elle consydere le corps de l'air sonnant agreablement aux oreilles humaines." Deshommeaux to Mersenne, 7 July 1642, CM 11, 199. I abbreviate the *Correspondance du P. Marin Mersenne* with CM and the number of the volume.

4 Reiss, *Knowledge, Discovery, and Imagination*, 196. On the application of such ideas in the work of Descartes see Van Orden, "Descartes on Musical Training."

5 "religieusement renfermez dans des cathegories pedantesques." Maugars, *Reponse faite à un curieux*, 5.

6 Bredeau to Mersenne, 12 December 1625, CM 1, 318.

philosophy to have been able to acquire the title of great musician."[7] Rules and mathematical reason would be seen increasingly as obstacles in the way of tasteful music rather than the central path laid out by theorists that practitioners should follow.

Mersenne's own position on these questions was complex and changed over time; his approach was remarkably flexible. As cited in the introduction, in his *La verité* the figure of the skeptic called music "an appearance" due to the existence of differences in taste, concluding that "what I find pleasant, someone else will find discordant according to their mood."[8] At other moments, Mersenne came remarkably close to this position himself. His manuscript presenting combinatory exercises at the Bibliothèque nationale states, "Hearing is like taste, some like sour and spicy flavors, others sweet and insipid."[9] And in his *Harmonie universelle*, he adopted a comparison from one of his correspondents, writing that "everybody has their own taste."[10] Whereas in *La verité* and *Questions harmoniques*, such statements are only uttered by the skeptic in order to be refuted, *Harmonie universelle* adopts the position that Mersenne had previously combatted. This position does not offer evidence that rules were of no use for Mersenne. It demonstrates that taste and licenses should not be excluded but rather accounted for, under certain conditions and by certain composers.

Although himself not a composer, Mersenne was a close observer of the music composed and performed in Paris. The library of the Minim convent had many contemporary editions of French and Italian music that seemed to have been specifically collected by Mersenne, such as works by Claude Le Jeune, Orlando di Lasso, Eustache Du Caurroy, and Antoine Boësset.[11] And throughout his works, there are many references to contemporary musicians and performers. He praised musicians like the organist and composer Charles Racquet and the lute player René Mezangeau, but little

7 "Et ledict Sr Bannius a trop estudié en philosophie pour avoir peu acquerir le tiltre de grand musicien." Boesset to Constantijn Huygens, 15 November 1640, CM 10, 252.

8 "qu'en apparence, puis que ceque je treuve agreable, un autre le treuve discordant." Mersenne, *La verité*, 32.

9 "dont les uns se plaisent aux saveurs aigres, et picquantes, et les autres aux douces, et aux fades." Paris, Bibliothèque nationale de France, MS fr. 24246, fol. 347v. Transcription from Coumet, "Mersenne," 37.

10 "chacun a son goust." Mersenne, *Harmonie universelle*, "Livre second, des chants," 106. Adam Fix has recently argued that "real tastes, passions, and effects of sound were pushed out of Mersenne's science." Fix, "The Acoustical Paradox," 406. I contend that Mersenne's response to taste and the listener's response to sound should not be seen solely as a rejection of taste but rather a process of accounting for it within his broader framework, investigating its limits.

11 Guillo, "Sous la main du père Mersenne."

is known about Mersenne's own listening experiences. He wrote that those who had listened to Louis XIII's ensemble of violin players had "never heard anything more ravishing or more powerful," but not if he himself had actually heard the ensemble.[12] He also observed that one could listen to plainchant for hours without getting bored.[13] In a late letter to Constantijn Huygens, he wrote on listening to "our choirs," but only to remark that one could hear an overtone of a twelfth in some of the lower voices, not whether he found the musical austerity of the Minims' services especially beautiful.[14] It is possible that Mersenne himself occasionally composed, such as when he sent a piece transcribed in numbers to Titelouze or sent a motet to the physician Villiers, but he only rarely noted specific compositions that he heard.[15]

Despite his many connections and contacts, he often disparaged contemporary musicians throughout his writings, describing them as lewd and disparaging their indifference toward religious music. In the introduction to *Harmonie universelle*, he writes,

> It is a strange thing that, of the thousands of players of the lute and other instruments, one does not meet ten who take pleasure in singing and expressing the divine canticles; they prefer to play hundreds of courantes, sarabandes, and allemandes, rather than a spiritual song, so that they seem to have devoted all their work to vanity, which they channel into their heart by the ears, like funnels.[16]

To remedy this, Mersenne suggests establishing a music academy for musicians to discuss "the reasons of harmony." In offering alternatives to concerts of two or three hours, he proposes shortening the time for music and allowing time for "several honnête hommes" to discuss the reasons why certain compositions are more pleasant than others, why certain sounds appeal more, how to apply dissonances, etc. These instructors would also

12 "qu'ils n'ont jamais rien ouy de plus ravissant ou de plus puissant." Mersenne, *Harmonie universelle*, "Livre quatriesme des instrumens a chordes," 177.

13 Mersenne, *Questions harmoniques*, 10–11.

14 "nos choeurs," Mersenne to Constantijn Huygens, 12 January, CM 15, 44. For music within the order of Minims, see Vaast, "L'esthétique musicale," as well as the introduction of this book.

15 Titelouze to Mersenne, 26 March 1628, CM 2, 44–45; Villiers to Mersenne, 7 July 1640, CM 9, 480.

16 "Certes c'est une chose estrange que de mille joüeurs de Luth, & des autres instrumens, l'on n'en rencontre pas dix qui prennent plaisir à chanter, & à exprimer les Cantiques divins; & qui n'ayment mieux joüer une centaine de courantes, de sarabandes, ou d'Allemandes, qu'un air spirituel: de sorte qu'il semble qu'ils ayent voüé tout leur travail à la vanité, qu'ils entonnent dans le coeur par les oreilles, comme par autant d'entonnoirs." Mersenne, *Harmonie universelle*, n.p. ("Preface generale au lecteur").

consider the eschatological, or how "to make the practice and theory of music useful for salvation."[17] Mersenne found that contemporary musicians lacked such knowledge and that they needed a re-education to be encouraged to praise the divine. With its experiential and mathematical dimensions, music was very suitable for the contemplation of what lies beyond immediate perception.

Rather than focusing on Mersenne's own musical experiences, this chapter asks what Mersenne wanted to offer the practitioner and where the aims and methods of music as a science and of music as an art diverge. In doing so, it brings together several trajectories of Mersenne's writings, such as his plea to practitioners to consider the theoretical foundations of their art and his quest to align beauty and reason. In each of my case studies, there is a different configuration regarding the role of numbers, taste, custom, and pleasure. The unifying problem was the extent to which the realm of reason could say something about the realm of (sensual) experience and the extent to which the world of musical listening and practice could provide evidence for the harmonies in the universe at large. Within musicological studies, the seismic shifts of musical objectives and discourse have already received much attention. Musicologists such as Brigitte van Wymeersch, Ivana Rentsch, and Thomas Christensen have explored how early modern authors reordered the classical hierarchy between sense and reason, anticipating a paradigm shift from music as universal harmony to musical practice.[18] Contemporary courtly literature emphasized the sensual engagement with music, positioning music in a courtly setting where discourse and gallant conversation mattered more than mathematics.[19]

The question of the extent to which theory can explain practice looms large over Mersenne's works. A large part of the *Harmonie universelle* is devoted to the correct use of counterpoint, the different tonal species, and the treatment of dissonances and artistic licenses. The compositional rules and music theoretical principles of Mersenne's book have been thoroughly discussed in studies by Herbert Schneider, Wilhelm Seidel, and Elisabeth Honn Hoegberg. Together, these studies point to the central role of Mersenne in seventeenth-century French music theory but also observe a lack of clarity

17 "raisons de l'harmonie ...plusieurs honnestes hommes ... de rendre la pratique, & la theorie de la Musique utile au salut." Mersenne, *Harmonie universelle*, n.p. ("Preface generale au lecteur"). On this passage, see also Rentsch, *Die Höflichkeit musikalischer Form*, 125–32.

18 Wymeersch, *Descartes*; Rentsch, "Sinn statt Verstand"; Christensen, "Sensus, Ratio, and Phtongos."

19 Rentsch, *Die Höflichkeit musikalischer Form*, 157–210. On Mersenne's position in this history, see also Christensen, "Rules, License and Taste."

in the Minim's writing. Herbert Schneider holds that *Harmonie universelle* reads more like a journal than a logically structured piece of writing.[20] This also concerns the question of what the role of music theory should be for the practicing composer and musician. As Honn Hoegberg concludes, "depending on what section of the *Harmonie universelle* is examined, Mersenne either agrees or disagrees with the blending of theory and practice."[21] In contrast to theory, which was considered immutable and a noble philosophical endeavor, Mersenne considered musical practice fallible and limited by human capabilities.[22] While some have argued that Mersenne completely separated the study of sound as something "objective and experimentally verifiable" from the realm of music as something relying on the imagination and inborn sensibilities, this chapter argues that aesthetic and scientific questions were very much entangled for Mersenne.[23] Although Mersenne would indeed write that there were no infallible mathematical rules for musical pleasure, he nonetheless searched for scientific and rational foundations of music, as it would improve the status of music as a science and stimulate his contemporaries to consider its divine origins. Mersenne wanted to educate his contemporaries "both by reason and by examples."[24] Although these realms did not completely overlap, reason and empirical evidence were both utilized to demonstrate and instruct.

Theory and Practice

The relation to music practice for polymaths like Mersenne was complex. Studies on such figures have often sidelined developments in music practice. As Penelope Gouk has pointed out, for polymaths like Mersenne, "it was the possession of higher academic degrees in philosophy and theology, rather than any practical training in performance, that qualified them to write authoritatively on music theory."[25] In his study of musical authorship and social status, Eric Bianchi has pointed to the existence of continuity between the worlds of practitioners and polymaths but has warned against positioning them too closely together: "it would be somewhat misleading to view the

20 Schneider, *Die Französische Kompositionslehre*, 49.

21 Hoegberg, "From Theory to Practice," 123.

22 Seidel, "Französische Musiktheorie," 69.

23 As for example is suggested in Cohen, *Quantifying Music*, 159.

24 "à monstrer tant par raison que par exemples." Mersenne, *Harmonie universelle*, "Livre quatriesme de la composition de musique," 197.

25 Gouk, "The Role of Harmonics," 228.

various social and professional groups as forming a single intellectual (or even music-theoretical) landscape."[26] Although scholars like Mersenne drew sharp lines between their own intellectual or "scientific" work and the art of the practitioner, the relations between these fields were manifold: musical scientists were skilled amateurs themselves, musicians cooperated or performed for scholars, and methods derived from practice were written down and transformed by observers.[27] With so many relations between the world of theory and practice, the question of what Mersenne wanted to offer to the practitioner demands an answer.

The status of the music theorist was increasingly contested around 1600. For example, in his *La sambuca lincei* (Naples, 1618), naturalist Fabio Colonna criticizes the writings of "arithmetical mathematicians" who have not written their books as "practical musicians."[28] In his *L'entretien des musiciens* (Auxerre, 1643), a collection of fictitious letters discussing the ins and outs of the French music scene, Annibal Gantez makes clear that he speaks about music primarily as a musician. He announces that he will not speak about music as "astrologers" and "arithmeticians" do and that he will leave for "Père Mercene," who, Gantez writes, "can speak better on the reasoning of a motet that he would not know how to make."[29] The chapel master and composer Gantez saw himself as practitioner ranked below theorists, something that he derided in his writings when he stated that it is much easier to talk than to do.[30] Others defended musical scholars with little practical experience themselves, such as the Spanish scholar Juan Caramuel y Lobkowitz. Lobkowitz wrote that Mersenne was an ingenuous "musicus," meaning musical philosopher, even though he could not sing well.[31] The rumors about Mersenne's bad voice must have been circulating, as in 1645 the Genevan merchant Jean-Louis Calandrini made a similar remark to Constantijn Huygens when he praised Mersenne's theoretical knowledge but also mentioned his deficiency in practice due to his inadequate voice.[32] Could one be a good theorist without practical experience?

26 Bianchi, "Scholars, Friends, Plagiarists," 65.

27 For such an investigation of these relationships in seventeenth-century England, see Gouk, *Music, Science and Natural Magic*, 23–65.

28 "scritta da Mathematici Arithmetici, & non Musici prattici." Colonna, *La sambuca lincea*, 15.

29 "je laisse cela pour le Pére Mercene qui dira mieux les raisons d'un mottet qu'il ne les sçauroit faire." Gantez, *L'entretien des musiciens*, 106.

30 For this point, see Bianchi, "Asses at the Lyre," 186–87.

31 Quoted and translated in Bianchi, "Scholars, Friends, Plagiarists," 98.

32 Calandrini to Constantijn Huygens, 5 July 1645, CM 13, 456.

This question had ramifications beyond the realm of music theory or the status of the theorist. In his attack on the certain foundations of music, La Mothe le Vayer raised the issue of who had the authority to make judgments: Who is the better judge of the food at the table? The guests who dine yet are ignorant of the preparation and seasoning, or the cook who made and prepared the meal?[33] This question was an old one. Classical authors like Aristotle maintained that makers were themselves not good judges of their art, and disparaging comments about practitioners had long been part of music-theoretical discourse.[34] Most famously, Boethius argued that the "musicus" possesses and employs reason, in contrast to performers and poets; he held that what is practiced by reason is always more honorable than what is practiced by the hand.[35] Such statements were repeated throughout the sixteenth century. Mersenne's correspondent Giovanni Battista Doni, for example, found that practitioners hardly knew anything of music. He complained regularly in his letters to Mersenne about the "the ignorance, laziness, and malignancy of most of our practitioners," who were seemingly incapable of adopting his theoretical insights.[36] Writing about the composer and organist Girolamo Frescobaldi, Doni stated that "for the setting of words, he is very ignorant and lacks judgment, so one could say he has all his knowledge [*science*] at the ends of his fingers."[37] Such dismissive rhetoric was also socially motivated. As Lorenzo Bianconi writes, seventeenth-century musicians were treated at court "on a level with cup-bearers and cooks" and in the church "as an equal to minor clerics."[38]

Mersenne often insisted that there was a lack of rules and reason within musical practice, describing most composers and musicians as unfit to describe their choices in music making. He writes in his *Harmonie universelle*, "The most excellent masters prove every day by experience that they have no established rules for making good songs, then they encounter them most often by outbursts and by chance, as they confess themselves."[39]

33 La Mothe le Vayer, "Discours sceptique sur la musique," 145.

34 Aristotle, *Politics*, 1282a20, 226–27.

35 Boethius, *The Fundamentals of Music*, I.34, 50. For this tradition, see Harne, "Unstable Embodiments."

36 "de l'ignorance, paresse et malignité de la pluspart de noz praticiens." Doni to Mersenne, 26 February 1640, CM 9, 147.

37 "mais pour accomoder les paroles il est fort ignorant et despourveu de jugement, de façon qu'on peut dire qu'il ait toute sa science aux bouts des doits." Doni to Mersenne, 22 July 1640, CM 9, 488. See also the letter Doni to Mersenne, 7 August 1638, CM 8, 17–18.

38 Bianconi, *Music in the Seventeenth Century*, 88.

39 "En effet les plus excellens Maistres preuveut tous les jours par experience qu'ils n'ont point de regles asseurees pour faire de bons chants, puis qu'ils ne les rencontrent le plus souvent que

Musicians could at times try for days without composing anything that satisfied them, while other times they produced many melodies in a short time, all "according to the different dispositions of their imagination and their health."[40] The compositional rules that existed were often broken, and composers themselves could not explain why they made certain decisions. Therefore, Mersenne writes, "one must always follow reason rather than their [the practitioners'] opinion, since its [reason's] light outshines experience."[41] His work aimed to formulate a discourse that, in his view, could elevate musical composition to the level of reason. Mersenne generally remained negative about the practitioners of his day. In his *Questions harmoniques*, he writes that he met some "very honest and very knowledgeable men who touch the lute, or sing," but acknowledges that many musicians are ignorant and scandalous.[42] In his *Les preludes*, he argues that musicians themselves are not to be blamed, as "they use their trade to support their families ... and everyone can enjoy some of the innocent pleasure that comes from their sounds and their harmony."[43] But these innocent pleasures could not be compared to those musicians who sing praises to God, who could not be praised enough, and for that reason, it was permitted to spend day and night on music to find new songs and "raise all mortals to the contemplation of divine things."[44]

Like Boethius and his predecessors, Mersenne valued intellectual labor than any activity using one's hands. In his *Traité de l'harmonie universelle*, he writes that the theoretician or "speculative musician" should be regarded as higher than the practicing musician. Here he compares the practitioners to the material world, the body, and the realm of the senses because "he [the practitioner] has no other light or guidance than custom, usage and the ear, without reason, and judges the goodness of concerts and the beauty of songs

par boutades, & par hazard, comme ils confessent eux-mesmes." Mersenne, *Harmonie universelle*, "Livre second, des chants," 97. On this passage, see also Duncan, "Persuading the Affections."

40 "suivant les differentes dispositions de leur imagination, & de leur santé." Mersenne, *Harmonie universelle*, "Livre second, des chants," 97.

41 "consequemment qu'il faut tousiours plustost suivre la raison que leur opinion, puis que sa lumiere surpasse l'experience." Mersenne, *Harmonie universelle*, "Livre premier des consonances," 20.

42 "tres-honnestes, & de tres-sçavants, qui touchent le luth, ou qui chantent." Mersenne, *Questions harmoniques*, 68.

43 "qu'ils se servent de leur industrie pour entretenir leurs familles ... & que chacun peut recevoir quelque partie du plaisir innocent, qui procede de leurs sons, & de leur harmonie." Mersenne, *Les preludes*, 185.

44 "pour élever tous les mortels à la contemplation des choses divines." Mersenne, *Les preludes*, 186.

according to his phantasy, his passions, and impressions."[45] Consequently, the composer is seen as more excellent than the singer for the same reasons. The theorist must be regarded as the highest of all: "the theorist rejects all kinds of affections, embraces no passion, and follows not the darkness of the imagination but the light of reason alone."[46] Mersenne remained conflicted on whether these realms should be mixed in the same individual. He hoped that musicians could be re-educated and disciplined in the sciences, but at other times he advocated for a separation of theory and practice in his own work:

> I do not want to mix [the parts on practice] with theory, in order to imitate the wisdom and goodness of God, who from all eternity has employed himself in the theory and contemplation of his essence, before coming to the practice that is made manifest in the harmony of the universe.[47]

Since theory is the more ideal and divine form of knowledge, it would be wrong to mix it with practice, as practice is automatically imperfect and considered to be lower than theory. It is thus necessary to contemplate first the existence of eternal truths before understanding how this essence appears in the harmony of the universe. Even in parts of *Harmonie universelle* devoted to musical practice, Mersenne made statements such as "it is nevertheless absolutely true that theory is more excellent and more noble than practice."[48] In doing so, Mersenne placed practical knowledge at the bottom of the hierarchy, reflective of a wider socio-political structure that privileged philosophical over productive knowledge.

At other moments, Mersenne wants to bring practitioners closer to reason. In his *Questions harmoniques*, he cites optical experiments to demonstrate the fallibility of the senses and the necessity to impose reason, as discussed chapter 1 of this book. Mersenne here compares the need for rules in music with those of architecture and perspective. Just as architecture pleases

45 "le Praticien est semblable à l'appetit sensitif, au corps & au monde elementaire; car il n'a autre lumiere ny conduite que la coustume, l'usage & l'oreille, sans la raison, & juge de la bonté des concerts, & de la beauté des chants selon sa phantaisie, ses passions, & impressions qu'il a receuës de ses Maistres, ou qu'il a acquises par l'usage." Mersenne, *Traité*, 92.

46 "le Theoricien rejette toutes sortes d'affections, n'épouse aucune passion, & ne suit point les tenebres de l'imagination, mais la seule lumiere de la raison." Mersenne, *Traité*, 92.

47 "je ne veux pas mesler avec la Theorie, afin d'imiter la sagesse & la bonté divine qui s'est employee de toute eternité à la theorie & à la contemplation de son essence, avant que de venir à la pratique qu'elle a fait paroistre dans l'harmonie de l'Univers." Mersenne, *Harmonie universelle*, "Livre premier des consonances," 10.

48 "il est neantmoins tres-asseuré que la theorie est plus excellente & plus noble que la pratique." Mersenne, *Harmonie universelle*, "Livre premier des consonances," 110.

the onlooker more when the doors and windows are proportioned and symmetrical, musical parts need to be orderly and balanced.[49] Similarly, in his writing on composition, Mersenne emphasizes the need for rules and reason to create and judge musical composition, but as this chapter will later show, these rules could be broken in some cases. Mersenne seems convinced that those who think otherwise can be persuaded:

> As for those who find that concerts are more pleasant when they are lighter and less full, and even more imperfect in the opinion of the masters, it is easy to bring their minds back to reason by showing them that they are mistaken, and that they have neither noticed nor discovered what is most excellent in concerts, being in this similar to those who, having found two boxes, make more of the heavier one, which is only filled with lead, than of the lighter one that is full of diamonds, or other things of great price; but as soon as they have opened them and seen the riches of the lighter one, they change their minds.[50]

Rather than advocating for a complete separation of the field of musical aesthetics and musical science, Mersenne wishes to bring the practitioners "back to reason." Here words like "demonstration" and "proof" are used in a musical context, suggesting that although music does not have the certainty of a mathematical discipline, irrefutable demonstrations do exist. In Mersenne's view, it would be a mistake to leave music to the judgment of taste, opinion, or custom. Rather, reason and experience should guide them. For example when describing the musical judgment of music masters, he writes, "The light of his demonstration, which like the light of the sun dissipates the fog of erroneous opinions and makes the darkness of ignorance vanish."[51] And when he speaks about the enharmonic and chromatic genera, he suggests that practitioners need to be elevated to reason, although they

49 Mersenne, *Questions harmoniques*, 204–5.

50 "Quant à ceux qui trouvent que les concerts sont plus agreables quand ils sont plus legers & moins remplis, & mesme plus imparfaits au jugement des maistres, il est facile de ramener leur esprit à la raison, en leur faisant voir qu'ils se trompent, & qu'ils n'ont pas remarqué ny découvert ce qui est le plus excellent dans les concerts, estant en cela semblables à ceux qui ayant trouvé deux boëttes font plus d'estat de la plus pesante, qui n'est remplie que de plomb, que de la plus legere qui est pleine de diamans, ou d'autres choses de grand prix; mais si tost qu'ils les ont ouvertes, & qu ils ont veu les richesses de la plus legere, ils changent d'avis." Mersenne, *Questions harmoniques*, 196–97.

51 "ils voyent la lumiere de sa demonstration, qui comme la lumiere du Soleil dissipe les brouillards des opinions erronées, & fait évanouïr les tenebres de l'ignorance." Mersenne, *Questions harmoniques*, 200.

tend to flee from the light of reason in the same way in which owls flee from the sun.[52] Here he chastises those composers who do not dare to expand their harmonies to the enharmonic or chromatic genera:

> But as we have experienced, musicians who have no other reason than their fantasy, and some old routine they have learned from their masters are so preoccupied with authority, or custom, that there is no room in their minds for reason, and they condemn certain passages, because they dare not use them, or do not know how to use them as they should, even though they greatly enrich the composition and are judged to be very excellent and received as rarities of music by those over whom reason and demonstration have more force than custom.[53]

Here custom and imitation of the masters are signaled as important impediments to the advancement of music, as musicians follow them instead of reason. The rhetoric on the light of reason present in Mersenne's writings on the certainty of the sciences is now used here in a musical context.

A more certain approach to music would also lead to a higher standing of the musician, Mersenne argued. The low status of the musician was one of the most important reasons for some of his contemporaries to dismiss music. Mersenne aimed for the immediate edification of contemporary musicians, but he also had a higher goal in mind. In his *Les preludes*, he lists all the sciences that musicians should know, as the ideal musician should be versed in grammar, medicine, rhetoric, history, philosophy, physics, ethics, and theology. Although an ambitious task, it would also put practitioners in a better position alongside their contemporaries: "they will elevate music, and render the musician more creditable, as one will see his science better supported by knowledge … instead of having been, as it is still seen at present, reduced to the routine of three or four miserable harmonies."[54] By striving toward this goal, "this science may no longer be so despised as it has been until

52 Mersenne, *Les preludes*, 169–70.

53 "Mais comme l'on experimente que les Musiciens qui n'ont autre raison que leur fantasie, & quelque vieille routine, qu'ils ont apprise de leurs maistres, sont tellement prevenus de l'authorité, ou de la coustume, qu'il n'y a plus de place dans leur esprit pour la raison, & qu'ils blasment certains passages, à raison qu'ils n'en osent pas user, ou qu'ils ne les sçavent pas employer comme il faut, quoy qu'ils enrichissent grandement la composition, & qu'ils soient jugez tres-excellens, & receuz pour des raretez de la Musique, par ceux, sur qui la raison, & la demonstration ont plus de force que la coustume." Mersenne, *Les preludes*, 175–76. Mersenne did write later in his *Harmonie universelle* that in the field of music, no perfect demonstrations exist; see chapter 1.

54 "elles releveront la Musique, & mettront le Musicien d'autant plus en credit, que l'on verra sa science mieux appuyée de toutes les connoissances … au lieu qu'ayant esté, comme on la

now." If so, music would "able to enter the homes of princes and kings, and finally [be] worthy of being presented to the supreme author of all things."[55]

There is only limited information about the musical practitioners who encountered his books. The composer Jean Titelouze wrote Mersenne that he hoped that musicians would see Mersenne's work and "take the trouble to follow its precepts."[56] As quoted before, Christoph de Villiers wrote that Mersenne turned music from "a vilified and misunderstood science" to "a science that will forever be known as a model of excellence with all others."[57] But later Villiers had to acknowledge that Mersenne desired too much from the practitioners. In 1640 he wrote to him,

> If you had taken from your book what is simply useful for music, both theoretical and practical, and reduce all that can be done by rules, as you are very accustomed and clear in your books, this would be a book or compendium that would be well received by apprentice musicians as well as the masters, at least for theory, as your large volume [*Harmonie universelle*] treats infinite matters for the curious only.[58]

Villiers saw a danger in Mersenne's approach, as he included many things that had no immediate use for the practitioners, but rather "only for the curious." The type of book Villiers had in mind would be published more and more in the second half of the century, often with attempts to distance the work from more theoretical and scholarly literature. When the composer Guillaume Gabriel Nivers wanted encapsulate the rules of music used by the greatest masters of the day, he wrote that he skipped "these useless and obscure questions of theory."[59] Some decades later, Charles Masson

void encore à present, reduite à la routine de trois ou quatre miserables accords." Mersenne, *Les preludes*, 141.

55 "ceste science ne soit plus si méprisée comme elle a esté jusques à present … capable d'entrer chez les Princes & les Roys, & finalement qu'elle soit digne d'estre presentée au Souverain Autheur de toutes choses." Mersenne, *Les preludes*, 144.

56 "prendront paine d'en suivre les preceptes." Titelouze to Mersenne, 24 December 1629, CM 2, 366.

57 "d'une science avile et mesprisee vous en avez fait une qui aura pour tousjours le renom d'entree en parangon avec toutes les autres." Villiers to Mersenne, 13 May 1634, CM 4, 119–20.

58 "Si vous aviez tiré de vostre livre ce qui sert simplement à la musique, tant theorique que pratique, et reduit tout ce qui s'en peut par reigles, comme vous estes fort familier et intelligible en vos livres, certainement ce seroit un livre ou abregé autant bien receu des musiciens aprentifs que des maistres, du moins pour la theorique, que vostre gros volume, où il se traite des matieres infinies pour les curieux seulement." Villiers to Mersenne, 7 July 1640, CM 9, 463.

59 "toutes ces questions inutiles & obscures de Theorie." Nivers, *Traité de la composition*, n.p. ("Preface").

wrote that his *Nouveau traité* (Paris, 1705) contains "neither curiosities, nor difficult and confusing terms of the ancients, but only what is useful in practice."[60] Although these authors catered to different audiences than Mersenne's, such statements can be seen as critique of the genre of music theory that he put forward.

Interestingly, these tensions were already present in the immediate response to Mersenne's books. The lawyer Claude Bredeau commented perhaps most harshly on Mersenne's *Traité de l'harmonie universelle*, writing in 1627,

> For those who do not know how to sing or compose music by routine, having only learned that one must guard against a false fifth and a false relation or similar thing, will understand little of your book, being versed in neither arithmetic, geometry, nor physics, from which you derive your theorems and demonstrations, unless they are taught by intelligent people. But mathematicians and philosophers will take pleasure in it, as they will see what these sciences have in common with music, and which cannot be properly discovered or perfectly understood without having read and heard them. I therefore believe that the two books [i.e., *Traité de l'harmonie universelle*] with which you wanted to make a first attempt will be well received and that there will not be a learned musician or anyone who wishes to understand it perfectly who does not read them and have an insatiable thirst to see the fourteen that remain to be printed.[61]

Bredeau wrote this about Mersenne's publication from 1627, which had much less practical interest than his later 1636 *Harmonie universelle*. Later Bredeau reiterated this point, writing that it is easier to learn from

60 "ni curiositez, ni termes des Anciens, difficiles & embarrassans; mais seulemens ce qui est utile dans la Pratique." Masson, *Nouveau traité*, n.p. ("Avertissement"). On these changing outlooks, see Seidel, "Französische Musiktheorie," 88–90.

61 "Car ceux qui ne sçavent que chanter ou composer la musique par routine, ayant seulement appris qu'il se faut garder d'une faulse quinte et d'une faulse relation ou chose semblable, entendront peu à vostre livre, n'estant versez ny en l'arithmetique, ny en la geometrie, ny en la physique, d'où vous tirez vos theorems et demonstrations, sinon qu'ils s'en facent enseigner par personnes intelligentes. Mais les mathematiciens et philosophes y prendront plaisir, comme voyans à decouvert ce que ces sciences ont de commun avec la musique, et qu'elle ne se peut pas bien decouvrir, ny parfaictement cognoistre sans les avoir leuës et entenduës. Je croy donc que les deux Livres, par lesquels vous avez voulu faire l'essay, seront bien receus, et qu'il n'y aura docte musician ou personne qui desire l'entendre parfaictement qui ne les lise, et n'ayt comme une soif insatiable de voir les quatorze, qui restent à imprimer." Bredeau to Mersenne, 2 November 1627, CM 1, 588–89.

examples in music notation, something that Mersenne would expand in his later publications.[62] But his comments do point at a problem: What would somebody—who in Bredeau's words only knows that one should avoid a diminished fifth—do with the hundreds of pages of theological analogies metaphors, acoustic measurements, and long digressions into the smallest movement of a string? In Bredeau's words, although "mathematicians and philosophers" will appreciate seeing in Mersenne's work the shared principles of the sciences of arithmetic, geometry and physics, what would musicians do with that understanding? As mentioned in the introduction, by 1627 Mersenne already seemed to have received such criticism; he writes at the end of his *Traité*,

> I am astonished that some ignoramuses who know how to play the lute or the organ, have the audacity to criticize this music treatise on the pretext that it is philosophical, or that they do not understand it.[63]

He writes that with his efforts in translating Euclid and Bacchius from Greek, it is now possible to understand ancient music in one or two hours. Moreover, Mersenne believed that he had the socially and morally superior class of people on his side: "it is enough for me to have all the *honnêtes hommes* on my side, whom I very willingly admit as competent judges of this *Traité*."[64] In writing his *Traité*, it was not so much the musicians' judgment he was after but rather that of a social group that was esteemed higher, a heterogeneous group of courtiers, aristocrats, and upper gentry. As mentioned in the introduction, this group utilized music to conceptualize models for harmony, interaction with the divine, and disciplining the body. Music was frequently referred to by moralists and pedagogues and was perceived as a more virtuous alternative for spending idle time than drinking or playing card games, an attitude to which Mersenne would have been happy to contribute.[65]

62 Bredeau to Mersenne, 12 March, 1628, CM 2, 39.

63 "Je m'étonne de ce quelques ignorans qui sçavent un peu joüer du Luth ou de l'Orgue, ont la hardiesse de blasmer ce Traité de Musique sous pretexte qu'il est philosophique, ou qu'ils ne l'entendent pas." Mersenne, *Traité*, n. p. ("Advertissement de l'autheur"). The editors of the *Correspondance du P. Marin Mersenne* maintain that Mersenne added this part in a later stage in response to some of the criticism he received, see CM 2, 41–42.

64 "Mais il me suffit d'avoir tous les honnestes hommes de mon costé, que j'admets tres-volontiers pour Juges competens de ce Traité." Mersenne, *Traité*, n. p. ("Advertissement de l'autheur"). Mersenne also wrote that the book had been seen and approved by Jacques Mauduit, who passed away in the summer of 1627.

65 Van Orden, *Music, Discipline and Arms*; Brooks, *Courtly Song*.

The Most Beautiful Song

At the end of his *Les preludes*, Mersenne asks what the perfect song could be. He immediately dismisses secular songs, as they only consist of flattery, vanity, and lies and thus could never be perfect. Mersenne thus insists here that the most perfect song must be religious but is unable to say much more about it.[66] Instead, he saves this for a later work, meaning his *Harmonie universelle*, where the topic of musical perfection is discussed at multiple moments, for example in inquiries on the most perfect tuning systems and the rules for making melodies. One of the most remarkable methods through which Mersenne tried to answer the question of music's perfection was the art of combinations, most visible in a manuscript that has received little attention. Manuscript fr. 24256 at the Bibliothèque nationale in Paris consists of 672 pages filled with music notation (fig. 5.1).[67] Although three smaller anonymous compositions are included, most of the volume lacks any songs or tunes. Rather, the manuscript is a combinatorial exercise: page after page is filled with all the possible combinations of an eight-note series. Without any repetition or rhythmical alteration, the total comes to 40,320 possibilities. With sixty melodies per page, the total amounts to more than 600 pages of music. Mersenne boasted in his *Harmonie universelle* and *Harmonicorum libri XII* that he himself filled these pages. Addressing the reader in his preface, he writes, "If anyone wants to see the 4,320 [*sic*] songs composed of the 8 notes of the octave, I have made an entire volume."[68] Manuscript fr. 24256 affirms Mersenne's empirical approach to combinatorics, as he writes down every combination with a visually overwhelming result.

The manuscript was not solely an intellectual exercise. At the end of the manuscript, he explains how knowing the 40,320 combinations of the eight notes could help the practicing musician. Here it becomes clear that the manuscript must be read not only as an exercise in the art of combinations but also as a tool for the musical practitioner. The

66 Mersenne, *Les preludes*, 223–24.

67 Paris, Bibliothèque nationale, MS fr. 24256. The manuscript is also discussed in Coumet, "Mersenne"; Coumet, *Oeuvres, Vol 2*; Redwood, "Combinatorics."

68 "si l'on veut voir les 4320 Chants composez des 8 notes de l'Octave, j'en ay fait un Volume entier." Mersenne, *Harmonie universelle*, n.p. ("Preface generale au lecteur"); Mersenne, *Harmonicorum*, "De cantibus," 123. Doni wrote to Mersenne that he did a disservice to his mind and the time God had given him by writing down all the combinations himself: "it would be sufficient to make a first trial and leave the rest to someone else, who would not be capable of better things" ("veu qu'il suffisoit d'en mettre quelque petit essay et laisser parachaver le rest à quelqu'autre, qui ne seroit capable de choses meilleures"). Doni to Mersenne, May 1636, CM 6, 72.

Figure 5.1: Demonstration of the first sixty possible combinations of eight notes from a total of 40,320 combinations. Paris, Bibliothèque nationale, MS fr. 24256, fol. 4r.

invention of new melodies using the art of combinations was closely related to Mersenne's efforts to discuss the poetics and inspiration of musical composition.[69] As André Redwood has shown for Mersenne's discussion of combinatorics in *Harmonie universelle*, the endless permutations served to demonstrate musical abundance for musicians to pursue themselves.[70] The goal of Mersenne's laborious task of writing down 40,320 melodies was to show the number of possibilities in music left to explore, arriving at the (nearly) endless. Mersenne's theological understanding of music as preparation for the afterlife becomes clear here; human ears receive a mere shadow of the perfection that awaits the blessed, but the plenitude of possibilities in this life already shows the richness of creation.[71] The musical practitioner should absorb this realization and utilize musical science to push his art further.

Besides providing information for the practitioner, Mersenne also utilized combinatorics to investigate the limits of the human taste and the uncertainties in music. One of Mersenne's earliest discussions of combinatorics can be found in his *La verité des sciences* of 1625. In this work, he aims to fight the arguments of the "Skeptics and Pyrrhonists," as the full title makes clear. The work is structured as a conversation between a skeptic, an alchemist, and a Christian philosopher, who voices Mersenne's opinions. The skeptic begins with a discussion of the uncertainty of the sciences, discussing each discipline, including music. In this view, there is no reasonable explanation as to why the octave, fifth, and fourth are considered consonances while the seventh and semitone are dissonances. The diversity of taste among humans and animals is formulated as a real anxiety concerning the foundations of knowledge.[72] Throughout his *La verité*, Mersenne argues against the skeptic's claims, coming to a position that is often called "mitigated skepticism."[73] Although he agrees with skeptical thinkers that the senses are feeble and difficult to trust, he did not accept the conclusion that most knowledge is therefore uncertain: the senses could be verified and controlled. And this is precisely what the majority of *La verité* aims to articulate: under the right circumstances, certain knowledge is attainable, even in the field of music.

69 Klotz, *Kombinatorik*, 70.

70 Redwood, "Mersenne and the Art of Delivery"; Redwood, "Combinatorics."

71 For example: Mersenne, *Harmonie universelle*, "Livre des consonances," 24.

72 "d'autres ayment le cri des pourceaux, que l'accord des orgues." Mersenne, *La verité*, 32. Like François de La Mothe le Vayer's skeptical approach to music in the *Questions harmoniques*, this was based on the Ten Modes of Sextus Empiricus to challenge dogmatic beliefs, the first of which concerns variation among animals.

73 Popkin, *The History of Scepticism*, 112–27.

Mersenne turned to combinatorics to demonstrate the existence of certain knowledge in *La verité*, relying mostly on the Jesuit mathematician Christoph Clavius. Clavius had investigated how many times certain sets can be combined, asking how many elements can be formed from the four qualities: hot, cold, dry, and wet. He also discussed sets such as the alphabet and musical notes.[74] Mersenne adopted these examples and seemed especially interested in the calculation of musical notes. As he establishes in *La verité*, two notes can be combined in two different ways without doubling them, but six notes can already be combined 720 times. To bring his point home, he gives a table with the possibilities of fifty different musical notes, coming to a number with sixty-six digits.[75] Although Mersenne acknowledges that any set of signs could be used, such as letters or soldiers, music serves as an important motivator for his interest in combinatorics. Writing specifically on music, Mersenne states that the great number of possibilities demonstrate "that one is marvelously far removed from the perfection of music theory and music practice, which one will never reach until one can demonstrate which song is the most beautiful."[76] He writes in *Harmonie universelle* that he considers avoiding repetition and finding new melodies one of the greatest difficulties facing composers.[77] The goal of these calculations was to establish that there are many songs still remaining to be composed and sung. Calculating how many ways one could vary a song of three notes with a 8-course lute in *La verité*, Mersenne comes to 1,307,674,368,000 possibilities and observes that some of these possibilities are "perhaps more pleasant than the songs that have been used until the present day."[78] Calculating how many songs are possible with twelve notes, Mersenne comes to a number greater than the amount of minutes that have passed since the creation of the world.[79] The basic components of music provide more combinations of new melodies than there is time in history.

After introducing the method of combining, Mersenne asks whether the most beautiful song possible could be found using the art of combinations. Here combinatorics serves as a first step, since knowing all the possibilities

74 Mersenne, *La verité*, 527–80; Dear, *Mersenne*, 194–95.

75 Mersenne, *La verité*, 551.

76 "qu'on est merveilleusement éloigné de la perfection de la Musique Theorique, & de la Pratique, laquelle on n'atteindra jamais jusques à ce qu'on puisse demontrer quel chant est le plus beau." Mersenne, *La verité*, 551.

77 Mersenne, *Harmonie universelle*, "Livre sixiesme de l'art de bien chanter," 362.

78 "sont peut estre meilleurs, & plus agreables que tous ceus desquels on s'est servi jusques à present." Mersenne, *La verité*, 552.

79 Mersenne, *La verité*, 553.

is necessary to identify which song is the most beautiful.[80] The musician must consider each one after the other before ultimately being able to decide which is the most beautiful or excellent. It is uncertain whether Mersenne truly believed that one could indeed find the most beautiful melody through mathematical rules; in his *La Verité*, he acknowledges that perfection in music is impossible and that such a thing would be attainable only with God's help. But by investigating all the different possibilities, some of this perfection could be approached, as musicians could weigh each of the combinations and thus get a sense of which melody could be considered the most beautiful.[81]

Although Mersenne recognized early on the difficulties of finding the most beautiful melody with the help of combinatorics, many of his correspondents criticized him for even suggesting the possibility. Giovanni Battista Doni, for example, wrote Mersenne that he could have spent his time more wisely than by writing out the many possibilities of combinations.[82] The lawyer Robert Cornier responded that beauty in music depends on "taste and respectively on the persons who listen."[83] In response to Mersenne's questions, Descartes wrote that it was not possible to connect a particular interval with a particular sensation. There was no universal law concerning musical effect. Growing slowly exasperated with Mersenne's questions, Descartes wrote, "I do not know any qualities of the consonances that respond to the passions."[84] Music was in this sense similar to food, the philosopher found: everyone knows that honey is sweeter than olives, but many people still prefer to eat olives.[85] In March 1630, Descartes wrote that it is impossible to determine what is beautiful, since it depends on the point of view of the subject and "the judgments of man are so different."[86] He summarized in a different letter, "To determine what is more pleasant, we must suppose the capacity of the listener, which changes, like taste, from person to person."[87]

80 Mersenne, *La verité*, 557.

81 Mersenne, *La verité*, 558–59. On Mersenne's discussion of combinatorics and the compositional process, see also Klotz, *Kombinatorik*, 70–74.

82 Doni to Mersenne, May 1636, CM 6, 72.

83 "le goust, est respective aux persones qui l'escoutent." Cornier to Mersenne, 18 August 1625, CM 1, 264.

84 "Je ne connois point de qualitez aux consonances qui répondent aux passions." Descartes to Mersenne, 4 March 1630, CM 2, 408.

85 Descartes to Mersenne, 4 March 1630, CM 2, 407–8. On these letters, see also Mace, "Marin Mersenne"; Buzon, "The *Compendium Musicae*."

86 "les jugemens des hommes sont si differens." Descartes to Mersenne, 18 March 1630, CM 2, 417.

87 "pour déterminer ce qui est plus agreable, il faut supposer la capacité de l'auditeur, laquelle change comme le goust, selon les personnes." Descartes to Mersenne, mid-January 1630, CM 2, 371.

Mersenne's correspondents pointed out to him that each person has a different disposition and taste (*goût*); what is perceived as beautiful depends on the listener. One more critic must be mentioned here: somewhere around 1627, Mersenne received a letter from the physician and professor of medicine René Moreau, which he quoted extensively in *Harmonie universelle*.[88] Moreau's letter is well known for its discussion of rhetoric, comparing music to the art of oratory. Eventually Moreau came to a conclusion similar to that of Cornier and Descartes: there is no possible way to reconcile mathematical and individual taste. More than the other correspondents, Moreau argued that there was a fundamental difference between notated melody as it exists on paper—as in Mersenne's immense manuscript—and a melody as heard. He emphasized a performative element that Mersenne hardly touched upon. One needed to be able to distinguish between how a melody was judged on paper and how it was judged by the ear. A melody that could be considered perfect using the rules of composition might still not be pleasing to its listeners. Moreau made the comparison to rhetoric by stating that a speech always needs to have the right *decorum*—that is, it needs to address the right audience.[89] Moreau argued that even when an oration was composed using all the textbook rules, "it will possibly not be the most pleasing to the ear, nor the most persuasive."[90] Later Moreau offered the following comparison: "finally, it is like food: everybody has their own taste; or to put it better, it is like eloquence: one loves it plain and diffuse, the other concise and nervous."[91]

Moreau's reference to rhetoric demonstrates the importance of the discipline for all the sciences in seventeenth century, and music in particular. Although Mersenne had a distaste for the elevated style of certain rhetoricians, he emphasized the utility of rhetoric for musicians and the understanding of music at large. The musician needed to be able to accentuate, articulate, and elevate certain words in a way that persuaded the listener. In his *Traité*, Mersenne deepened the connection, comparing

88 Mersenne, *Harmonie universelle*, "Livre second, des chants," 106. On Moreau's letter, see also Mace, "Marin Mersenne," 10–11; Dear, "Marin Mersenne," 183–84; Redwood, "Combinatorics," 7.

89 In his *De oratore*, Cicero continually emphasizes that eloquence is based on both rational and irrational principles. Some parts of making a convincing speech cannot be strictly formulated in rules but must instead rely on the emotional and persuasive capabilities of the orator to convince and move the audience.

90 "il ne sera pas ny le plus agreable à l'oreille peut-estre, ny le plus persuasif." Moreau to Mersenne, CM 1, 634.

91 "Enfin c'est comme des viandes: chascun a son goust, ou pour mieux dire, c'est comme de l'eloquence: l'un l'aime plaine et diffuse, l'autre concise et nerveuse." Moreau to Mersenne, CM 1, 637.

the use of intervals, melodies, and accents in music to public speaking, exhorting musicians to use the art of rhetoric to move their listeners.[92] The increasing concern with music's emotional power gave medical language, imagination, and temperament a central role within musical discourse, slowly supplanting the role of mathematical proportions.[93] The number of references to rhetoric is indicative of the increasing effort to situate music closer to the language arts. This also influenced the way musical taste and judgment were conceptualized. With music being treated increasingly as an art or language, it was considered an issue of individual taste and shared opinion. For example, La Mothe Le Vayer wrote that "just as the goal of the orator is to persuade his listeners, that of the musician is to please the majority."[94]

Ultimately, the use of combinatorics to find the most beautiful song was a conundrum that Mersenne could not solve. Throughout *Harmonie universelle*, he adopts some of the criticisms he received, for example citing Moreau's letter without naming the author. Here Mersenne compares music to oratory: even when the orator follows all the rules of rhetoric and precepts of the art, there is no certainty that it will be "the most pleasant to the ear nor be the most persuasive."[95] Like the orator, the composer can never compose a work perfectly, since one cannot account for different tastes, temperaments, and expectations. In his description of how his manuscript of all the 40,320 combinations should be used, he had come to a similar position when explaining why none of the combinations would please all possible listeners:

> This is because hearing is like taste: some people like sour and spicy flavors, and others sweet and insipid, since we experience those songs that use large intervals please some [listeners] more than those that use small steps. This can be applied to the other external senses as well, since different eyes prefer different colors, and the smell that agrees with one offends another.[96]

92 Mersenne, *Traité*, 199–202; Duncan, "Persuading the Affections"; Redwood, "Mersenne and the Art of Delivery."

93 Palisca, *Humanism*, 333–38; Moyer, "Musical Scholarship in Italy," 185–202.

94 "comme la fin de l'Oreteur est de persuader ses auditeurs, celle du Musicien est de plaire à la multitude." La Mothe le Vayer, "Discours sceptique sur la musique," 146.

95 "encore qu'il ne semble pas le plus agreable à l'oreille, ny le plus persuasif." Mersenne, *Harmonie universelle*, "Livre second, des chants," 105.

96 "Car il est des oreilles comme des gousts, dont les uns se plaisent aux saveurs aigres, et picquantes, et les autres aux douces, et aux fades, puisque l'on experimente que les chants qui

Reflecting on the difficulty of finding the most beautiful melody with the help of combinatorics, Mersenne came to an unusual position for himself: hearing can be compared to the sense of taste, and what pleases one listener might offend the other. Here he took the position that he had previously contested, allowing and incorporating individual taste within a universalistic framework.

In his writings after his combinatory manuscript, Mersenne continued to return to this problem but never again made the comparison between musical listening and taste so strongly. Although the precise reasons why certain consonances sound better were unknown, Mersenne argued that music was more capable than other sciences of determining what is pleasant and what is not. As quoted in chapter 1, Mersenne found that scholars in optics still disagreed with one another on how many colors there are but that music theorists had established at least some certainties regarding what is pleasing and what not.[97] But in looking at his contemporaries, Mersenne had to acknowledge that passions and fantasies seemed to guide musical listeners rather than reason.[98] He was thus ultimately conflicted, as the pleasure in musical beauty was more rational and intellectual than the other sensory delights, although it remained something out of reach.

It is unknown whether Mersenne showed the manuscript to musicians in Paris, but he must have shown it to the many visitors to his cell. In August 1635, his patron Nicolas-Claude Fabri de Peiresc suggested that Mersenne take the manuscript with him on his travels so that others might benefit from seeing it. If he went to Italy or Germany, Peiresc wrote, Mersenne might find a printer who was willing to print the 600 pages of music, since printing was cheaper and less cumbersome there than in Paris.[99] Taking the volume with him did not appeal to Mersenne, who called it a "work for the drawer that will never see the light of day."[100] It is difficult to say whether the musicians who visited his cell at the Minim convent were actually impressed and inspired by the plethora of possibilities and combinations. But the visual and physical potencies of the book are clear: much remained to explore in the field of music.

usent de grands intervalles plaisent plus à quelques uns que ceux qui vont par degrez conjoincts; ce que l'on peut aplicquer aux autres autres sens exterieurs, puisque les yeux differents font choix de differentes couleurs, et que l'odeur qui agree à l'un blesse l'autre." Paris, Bibliothèque nationale, MS fr. 24246, fol. 347v. Transcription from Coumet, "Mersenne," 37.

97 Mersenne, *Harmonie universelle*, "Livre premier des consonances," 88, and chapter 1.

98 "mais parce que l'on mesure plutost le plaisir de la Musique par la passion & par la fantaisie des auditeurs que par la raison." Mersenne, *Harmonie universelle*, "Livre premier des consonances," 54.

99 Peiresc to Mersenne, 20 August 1635, CM 5, 357.

100 "car c'est une piece de cabinet qui ne verra jamais le jour." Mersenne to Peiresc, around 1 September 1635, CM 5, 372.

Tables for Composing

In several of his works, Mersenne played with the idea of standardizing or mathematizing the compositional process, with the goal of edifying his contemporaries. In the *Questions inouyes* of 1634, for example, he discussed how one could learn to compose music within an hour or less, concluding that this should be possible, since music consists only of arranging and ordering consonances.[101] In a letter to Peiresc from 1635, he described the lack of rules for judging good concerts as "solid and invariable as the *Elements* of Euclid."[102] If composers had immutable rules, then they could write songs at any time of the day, much as mathematicians could demonstrate and draw lines at any moment according to infallible rules. At the same time, he seemed skeptical as to whether this was truly possible. In his *Harmonie universelle*, he considered the question of whether "it is possible to discover and prescribe infallible rules and maxims for making good songs" one of the greatest difficulties in all of music.[103]

The compositional rules that Mersenne collected in his books were drawn from a variety of sources. As Herbert Schneider has shown, Mersenne included many basic rules for making counterpoint that were widely shared among sixteenth-century theorists.[104] Although he expressed his distaste for the work of Gioseffo Zarlino at several points, many rules can be matched to his *Le istitutioni harmoniche* (Venice, 1558) and other sixteenth-century music literature. He also based his work on the many composers and musicians in his circle, such as Mauduit, Racquet, and Titelouze. In his emphasis on the close relationship between music and rhetoric, Mersenne advised composers to think about phrasing, sections, and ornamentation, much as orators should attend to these matters in their speeches.[105] Most of his compositional advice in *Harmonie universelle* is quite general. For example, in the "Livre second des chants," he aims to include several rules for making good melodies. He writes that the music masters already established several "certain rules" in the field of music. Pointing to the possibility of improvement, he added that if a great number of "wise and judicious men" would employ their time to music, "we may hope for certain rules for making good

101 Mersenne, *Questions inouyes*, 160–72.

102 "fermes et invariables que celle des *Elemens* d'Euclide." Mersenne to Peiresc, 1 September 1635, CM 5, 373.

103 "A sçavoir si l'on peut treuver & prescrire des regles & des maximes infallibles selon lesquelles on fasse de bons Chants." Mersenne, *Harmonie universelle*, "Livre second, des chants," 97.

104 Schneider, *Französische Kompositionslehre*, 89–99.

105 Duncan, "Persuading the Affections."

songs."[106] The ultimate rules discussed here, however, were more difficult to formulate. First, he writes that "it is necessary to follow and imitate the movement of the passions that one wishes to excite with the listeners."[107] The second rule pertains to the use of intervals and degrees, which must follow the passions as well. Perhaps realizing that it is more difficult to formulate invariable rules in music, Mersenne then proceeds to compare the colors, intervals, and four elements, emphasizing that compositions should be measured and organized and eliminate elements that may hurt the ears.[108] Whereas in other moments Mersenne considered music superior to optics in that its harmonies could be quantified, he now emphasized the comparison as a tool for inspiration.

Although some of his correspondents and contemporaries found that Mersenne's books contained too many digressions and curiosities to be useful to practitioners, the theorist did indeed address his books to musicians and composers. The extent to which he took steps to make his work appealing to practitioners differed in each part, but the many repetitions and reformulations that one finds across *Harmonie universelle* are certainly partially explained by his effort to ensure that readers who might skip the more theoretical parts on sound and mechanics would still encounter these ideas in other books. Statements from the first books on sound and the mechanics of the string, for example, are often repeated in the books on composition and instruments.

One mode of addressing the practitioner utilizes tables and figures that specifically depict the ways in which knowledge could be systematized or quantified. As Domenico Bertoloni Meli has shown, Mersenne seemed to relish columns and rows of numbers.[109] It seems that presenting work in a tabular form had an aesthetic goal. This also becomes clear from Mersenne's work on music, which is filled with tables. In his *Traité*, for example, he adds a "universal table for composing any piece of music desired."[110] With this table, one could compose "with ease and without error."[111] It consists of numerous musical rules

106 "grande multitude d'hommes sçavans et judicieux ... je croy que l'on pourra esperer des regles certaines pour faire de bons chants." Mersenne, *Harmonie universelle*, "Livre second, des chants," 98.

107 "qu'il faut suivre & imiter le mouvement de la passion à laquelle on veut exciter les auditeurs." Mersenne, *Harmonie universelle*, "Livre second, des chants," 99.

108 Mersenne, *Harmonie universelle*, "Livre second, des chants," 103.

109 Bertoloni Meli, "The Role of Numerical Tables," 177. Several of the tables considered here are also discussed in Honn Hoegberg, "From Theory to Practice."

110 "Table universelle pour composer telles pieces de Musique qu'on voudra." Mersenne, *Traité*, 184.

111 "avec facilité sans erreur." Mersenne, *Traité*, 183.

or guidelines written out in words, for example: "if the soprano is in unison with the tenor, and the bass is a third under the tenor, the *haute-contre* will make a fifth or sixth over the bass."[112] With these guidelines in mind, composers should have no trouble in composing pleasant harmonies, the theorist argues.

This line of thought continues in his *Harmonie universelle*, where Mersenne writes that an arithmetician can easily learn "without a master" and states that "there is no science so easy, since the best ratios consist solely in counting 1, 2, 3, 4, 5, 6, 7, etc. and relating these numbers with each other."[113] He immediately mentions that he is speaking here of theory, not practice. This tension between an ideal theoretical realm that is fully intelligible in numbers and a messier practice is a common thread running through his work, as Mersenne ultimately did not think music could be fully rationalized or encapsulated in rules. But he did believe that practitioners could be stimulated to understand the rational foundations behind craft and that, with the help of more education, musical amateurs and listeners could reflect on the beauty of musical numbers. For these purposes, Mersenne often presented his findings in tabular form. One example of such a table is the "table of one hundred consonances" (fig. 5.2). This table is grouped together with the "table of fifty dissonances," and together they list 150 musical intervals with their ratios: unison 1:1, minor third 6:5, major third 5:4, all the way to an interval of fourteen octaves with the ratio 98304:5.[114] The reader can thus look up every possible interval and its corresponding mathematical proportion, here expressed in lengths of two strings. Mersenne acknowledged that the table is not necessary for making music, as it employs enormous numbers and intervals that are never used in practice. Rather, it was important to demonstrate that mathematical relations exist, governing proportions from the smallest interval to the largest. Composers should still memorize these proportions, Mersenne argued, "since nature is more powerful than art, and since we can understand by reason that these octaves are possible, it is reasonable for the musician to know all the proportions, however large they may be."[115] Even with no practical use in sight, he found

112 "Si le dessus fait / l'unisson avec la taille. Et la basse / La tierce sous la taille. La haute-contre fera / La quinte ou la sexte sur la basse." Mersenne, *Traité*, 184.

113 "Il s'ensuit semblablement qu'un Arithmeticien peut apprendre la Musique sans maistre, & qu'il n'y a nulle science si aisee, puis que ses meilleures raisons consistent seulement à conter 1, 2, 3, 4, 5, 6, 7, &c. & à comparer ces nombres les uns aux autres." Mersenne, *Harmonie universelle*, "Livre premier des consonances," 105.

114 Mersenne, *Harmonie universelle*, "Livre premier des consonances," 108–9.

115 "Mais puis que la nature est plus puissante que l'art, & que nous pouvons comprendre par la raison que ces Octaves sont possibles, il est raisonnable que le Musicien connoisse toutes

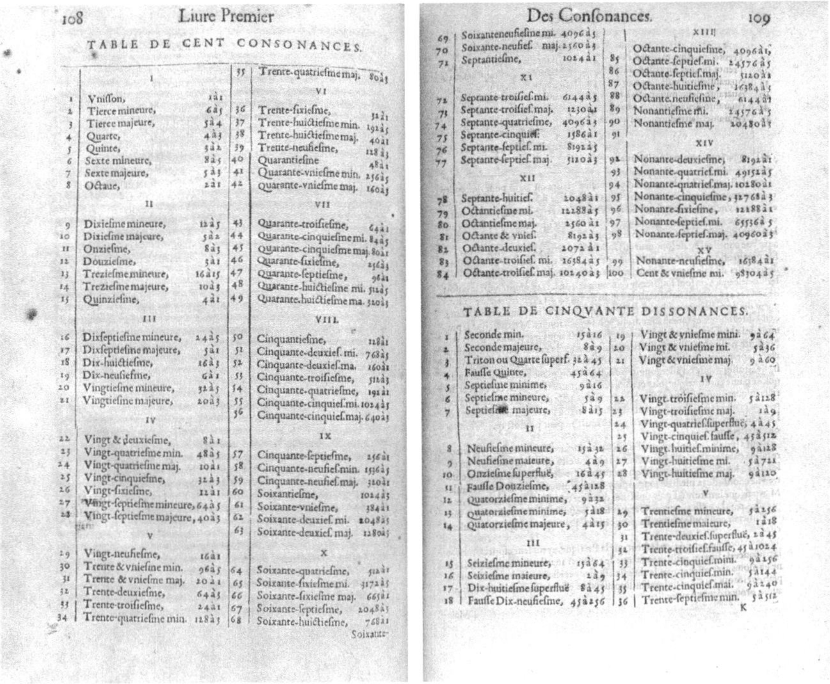

108 Liure Premier

TABLE DE CENT CONSONANCES.

	I	
1	Vniſſon,	1 à 1
2	Tierce mineure,	6 à 5
3	Tierce majeure,	5 à 4
4	Quarte,	4 à 3
5	Quinte,	3 à 2
6	Sexte mineure,	8 à 5
7	Sexte majeure,	5 à 3
8	Octaue,	2 à 1
	II	
9	Dixieſme mineure,	12 à 5
10	Dixieſme majeure,	5 à 2
11	Onzieſme,	8 à 3
12	Douzieſme,	3 à 1
13	Trezieſme mineure,	16 à 15
14	Trezieſme majeure,	10 à 3
15	Quinzieſme,	4 à 1
	III	
16	Dixſeptieſme mineure,	24 à 5
17	Dixſeptieſme majeure,	5 à 1
18	Dix-huictieſme,	16 à 3
19	Dix-neufieſme,	6 à 1
20	Vingtieſme mineure,	32 à 5
21	Vingtieſme majeure,	20 à 3
	IV	
22	Vingt & deuxieſme,	8 à 1
23	Vingt-quatrieſme min.	48 à 5
24	Vingt-quatrieſme maj.	10 à 1
25	Vingt-cinquieſme,	32 à 3
26	Vingt-ſixieſme,	12 à 1
27	Vingt-ſeptieſme mineure,	64 à 5
28	Vingt-ſeptieſme majeure,	40 à 3
	V	
29	Vingt-neufieſme,	16 à 1
30	Trente & vnieſme min.	96 à 5
31	Trente & vnieſme maj.	20 à 1
32	Trente-deuxieſme,	64 à 3
33	Trente-troiſieſme,	24 à 1
34	Trente-quatrieſme min.	128 à 5
35	Trente-quatrieſme maj.	80 à 3
	VI	
36	Trente-ſixieſme,	32 à 1
37	Trente-huictieſme min.	192 à 5
38	Trente-huictieſme maj.	40 à 1
39	Trente-neufieſme,	128 à 3
40	Quarantieſme	48 à 1
41	Quarante-vnieſme min.	256 à 5
42	Quarante-vnieſme maj.	160 à 3
	VII	
43	Quarante-troiſieſme,	64 à 1
44	Quarante-cinquieſme mi.	84 à 5
45	Quarante-cinquieſme maj.	80 à 1
46	Quarante-ſixieſme,	256 à 3
47	Quarante-ſeptieſme,	96 à 1
48	Quarante-huictieſme mi.	512 à 5
49	Quarante.huictieſme ma.	320 à 3
	VIII.	
50	Cinquantieſme,	128 à 1
51	Cinquante-deuxieſ. mi.	768 à 5
52	Cinquante-deuxieſ.ma.	160 à 1
53	Cinquante-troiſieſme,	512 à 3
54	Cinquante-quatrieſme,	192 à 1
55	Cinquante-cinquieſ.mi.	1024 à 5
56	Cinquante-cinquieſ.maj.	640 à 3
	IX	
57	Cinquante-ſeptieſme,	256 à 1
58	Cinquante-neufieſ.min.	1536 à 5
59	Cinquante-neufieſ.maj.	320 à 1
60	Soixantieſme,	1024 à 3
61	Soixante-vnieſme,	384 à 1
62	Soixante-deuxieſ. mi.	2048 à 5
63	Soixante-deuxieſ. maj.	1280 à 3
	X	
64	Soixante-quatrieſme,	512 à 1
65	Soixante-ſixieſme mi.	3172 à 5
66	Soixante-ſixieſme maj.	665 à 1
67	Soixante-ſeptieſme,	2048 à 3
68	Soixante-huictieſme,	768 à 1

Soixante-

Des Conſonances. 109

69	Soixanteneufieſme mi.	4096 à 5
70	Soixante-neufieſ. maj.	2560 à 3
71	Septantieſme,	1024 à 1
	XI	
72	Septante-troiſieſ.mi.	6144 à 5
73	Septante-troiſieſ.maj.	1230 à 1
74	Septante-quatrieſme,	4096 à 3
75	Septante-cinquieſ.	1586 à 1
76	Septante-ſeptieſ. mi.	8192 à 5
77	Septante-ſeptieſ. maj.	5120 à 3
	XII	
78	Septante-huitieſ.	2048 à 1
79	Octantieſme mi.	12288 à 5
80	Octantieſme maj.	2560 à 1
81	Octante & vnieſ.	8192 à 3
82	Octante-deuxieſ.	2072 à 1
83	Octante-troiſieſ. mi.	16384 à 5
84	Octante-troiſieſ. maj.	10240 à 3
	XIII	
	Octante-cinquieſme,	4096 à 1
85	Octante-ſeptieſ.mi.	24576 à 5
86	Octante-ſeptieſ.maj.	5120 à 1
87	Octante-huitieſme,	16384 à 3
88	Octante.neufieſme,	6144 à 1
89	Nonantieſme mi.	24576 à 5
90	Nonantieſme maj.	20480 à 3
91		
	XIV	
92	Nonante-deuxieſme,	8192 à 1
93	Nonante-quatrieſ.mi.	49152 à 5
94	Nonante-qnatrieſ.maj.	10280 à 1
95	Nonante-cinquieſme,	32768 à 3
96	Nonante-ſixieſme,	12288 à 1
97	Nonante-ſeptieſ.mi.	65536 à 5
98	Nonante.ſeptieſ.maj.	40960 à 3
	XV	
99	Nonante-neufieſme,	16384 à 1
100	Cent & vnieſme mi.	98304 à 5

TABLE DE CINQVANTE DISSONANCES.

1	Seconde min.	15 à 16
2	Seconde majeure,	8 à 9
3	Triton ou Quarte ſuperf.	32 à 45
4	Fauſſe Quinte,	45 à 64
5	Septieſme minime,	9 à 16
6	Septieſme mineure,	5 à 9
7	Septieſme majeure,	8 à 15
	II	
8	Neufieſme mineure,	15 à 32
9	Neufieſme maieure,	4 à 9
10	Onzieſme ſuperfluë,	16 à 45
11	Fauſſe Douzieſme,	45 à 128
12	Quatorzieſme minime,	9 à 32
13	Quatorzieſme minime,	5 à 18
14	Quatorzieſme majeure,	4 à 15
	III	
15	Seizieſme mineure,	15 à 64
16	Seizieſme maieure,	2 à 9
17	Dix-huitieſme ſuperfluë	8 à 45
18	Fauſſe Dix-neufieſme,	45 à 256
19	Vingt & vnieſme mini.	9 à 64
20	Vingt & vnieſme mi.	5 à 36
21	Vingt & vnieſme maj.	9 à 60
	IV	
22	Vingt-troiſieſme min.	5 à 128
23	Vingt-troiſieſme maj.	1 à 9
24	Vingt-quatrieſ.ſuperfluë,	4 à 45
25	Vingt-cinquieſ. fauſſe,	45 à 512
26	Vingt.huitieſ.minime,	9 à 128
27	Vingt-huitieſme mi.	5 à 721
28	Vingt-huitieſme maj.	9 à 120
	V	
29	Trentieſme mineure,	5 à 256
30	Trentieſme maieure,	1 à 18
31	Trente-deuxieſ.ſuperfluë,	2 à 45
32	Trente-troiſieſ.fauſſe,	45 à 1024
33	Trente-cinquieſ.mini.	9 à 256
34	Trente-cinquieſ.min.	5 à 144
35	Trente-cinquieſ.mai.	9 à 240
36	Trente-ſeptieſme min.	5 à 512

K

Figure 5.2: Table of a hundred consonances and fifty dissonances. Mersenne, *Harmonie universelle*, "Livre premier des consonances," 108–9. Paris, Bibliothèque nationale de France, département Arsenal, FOL-S-1449.

it important for practitioners to reflect on the scientific foundations of their art, as the theory of music is a worthwhile pursuit on its own.

Mersenne gave a slightly more practical table in the first book composition, entitled the "universal table" of consonances. This table collects all the movements that are possible in two-voice counterpoint toward and from a specific consonance; the example given here shows the unison (fig. 5.3).[116] The idea is similar to that of the table presented in his *Traité* but is made significantly more intricate by his complex system of notation. The table utilizes numbers for the different consonances (1 for unison, 3 · for a minor third, 3 for a major third, etc.), letters for the degree of the genera (D for diesis, S · for a minor semitone, T for a major tone, etc.), and numbers for the air vibrations emitted to create each tone (for example the number 15 for a lower note and 18 for a

les proportions pour grandes qu'elles puissent estre." Mersenne, *Harmonie universelle*, "Livre premier des consonances," 110.

116 Mersenne, *Harmonie universelle*, "Livre quatriesme de la composition," 234–36. The designation of this table as a "table universelle" appears on p. 233.

234 Liure Quatriesme

PREMIERE. TABLE.

L'on passe à l'Vnisson.

		BCCA
1	De la 3̇, l'vn passant par le T, & l'autre par le S	15.16.16.18
2	ou de la 3, par le Ṫ & par le T	8. 9. 9.10
3	ou par le Ṡ, & par la 3	20.24.24.25
1	De la Quarte par la 3̇ & par le Ṫ	9.10.10.12
2	ou par le S, & par la 3	15.16.16.20
3	ou par la 3', & par le S Pythagorique	3888.4096.4096.4121
1	De la V par le T, & la 4	8.9.9.12
2	ou par la 3̇ & 3	10.12.12.15
1	De la 6̇ par la V, & par le S	10.15.15.16
2	ou par la 4, & par la 3̇	5.6.6.8
3	ou par le T', & par la 4'	
1	De la 6 par la V, & par le Ṫ	6.9.9.10
2	ou par la 4 & la 3	3.4.4.5
3	ou par la V', & par le Ṡ	

II

L'on passe à la 3̇

1	De l'Vnisson comme deuant par le T & le S	15.16.16.18
1	De la 3 par le Ṡ, ny ayant qu'vne partie qui se meut	24.25.30
1	De la 4 par le Ṡ & le S	24.25.30.32
2	ou par la D, & deux Ṡ	
1	De la V par le Ṫ & T	40.45.54.60
2	ou par la 3̇ & le Ṡ	24.25.30.36
1	De la 6̇ par le Ṡ, & par la 3	75.80.96.120
2	ou par la 3̇, & par le Ṫ	25.30.36.40
1	De la 6 par le Ṡ, & par la 4	24.25.30.40
2	ou par la 3, & par le Ṫ	9.10.12.15
1	De l'8 par la V, & par le Ṫ	9.10.12.18
2	ou par la 4, & la 3	4.5.6.8
3	ou par la 6̇, & par le Ṡ	24.25.30.48

III.

On va à la Tierce maieure.

1	De l'Vnisson par le Ṫ, & le T	8.9.9.10
2	De la 3̇ comme deuant par le Ṡ	24.25.30
1	De la 4 par le S	12.15.16
2	ou par la D, & par le Ṡ	96.100.125.128
1	De la V par le S, & le T	32.36.45.48
2	ou par le Ṡ & le T'	
3	ou par la Tierce mineure diminuee, & par la D	

1 De

Figure 5.3: Universal table of transitions from the unison. Mersenne, *Harmonie universelle*, "Livre quatriesme de la composition de musique," 234. Paris, Bibliothèque nationale de France, département Arsenal, FOL-S-1449.

higher note). The difficulty is increased because Mersenne also switches the lower and higher numbers when they designate string lengths. The numbers 18 and 16 could thus also designate the proportions between two strings in which the higher number represents the lower note. The table is divided into eight parts, accounting for the first eight consonances. The table is a testament to Mersenne's effort to record different options using empirical evidence rather than attempt to develop a single rule.

All the motions in this table are either contrary or conjunct. In the next proposition of *Harmonie universelle*, Mersenne details how one could judge these passages: for example, a passage is good when it moves from an imperfect consonance to a perfect one (minor third to the unison, major sixth to the octave). The first example from the table is ultimately judged as a good passage. Mersenne thus shows how to judge passages, which movements are deemed agreeable, and which movements are considered worse, either because they have no variety, move in steps that are too great, or offend the listener. This table is translated into musical notation slightly later in the book when he gives all the different passages that can be made in two-part compositions between the unison and a different consonance.[117] Although he again focuses on two-part compositions, this could also be applied to works with more voices. Compositions with only two voices were governed by stricter rules regarding dissonance treatment than compositions in six or seven parts.

As the next system, he gives the composition according to his so-called "universal tablature of harmonic ratios" (fig. 5.4).[118] It is a method of composing that does not require any music notation and uses Mersenne's nineteen-tone octave division, ranging four octaves, which Mersenne considered the range of the human voice. The table connects each individual note to a number that forms a ratio with the lowest individual pitch (C ut), here with the number 1800 and the highest note C ut (28800). Mersenne emphasizes that this method is "nothing other than the practice of those who compose *airs* and motets."[119] The tablature also contains small intervals, such as the comma (4050/4000 in the third comma), distinguishes between a minor and major semitone, and includes the diesis (for example between *c and Xd in the third column with the ratio of 6750/6912).

117 Mersenne, *Harmonie universelle*, "Livre cinquiesme de la composition," 307–8.

118 "la Tablature universelle des raisons Harmoniques." Mersenne, *Harmonie universelle*, "Livre quatriesme de la composition," 245.

119 "n'est autre chose que la pratique de ceux qui composent des Airs, & des Motets." Mersenne, *Harmonie universelle*, "Livre quatriesme de la composition," 245.

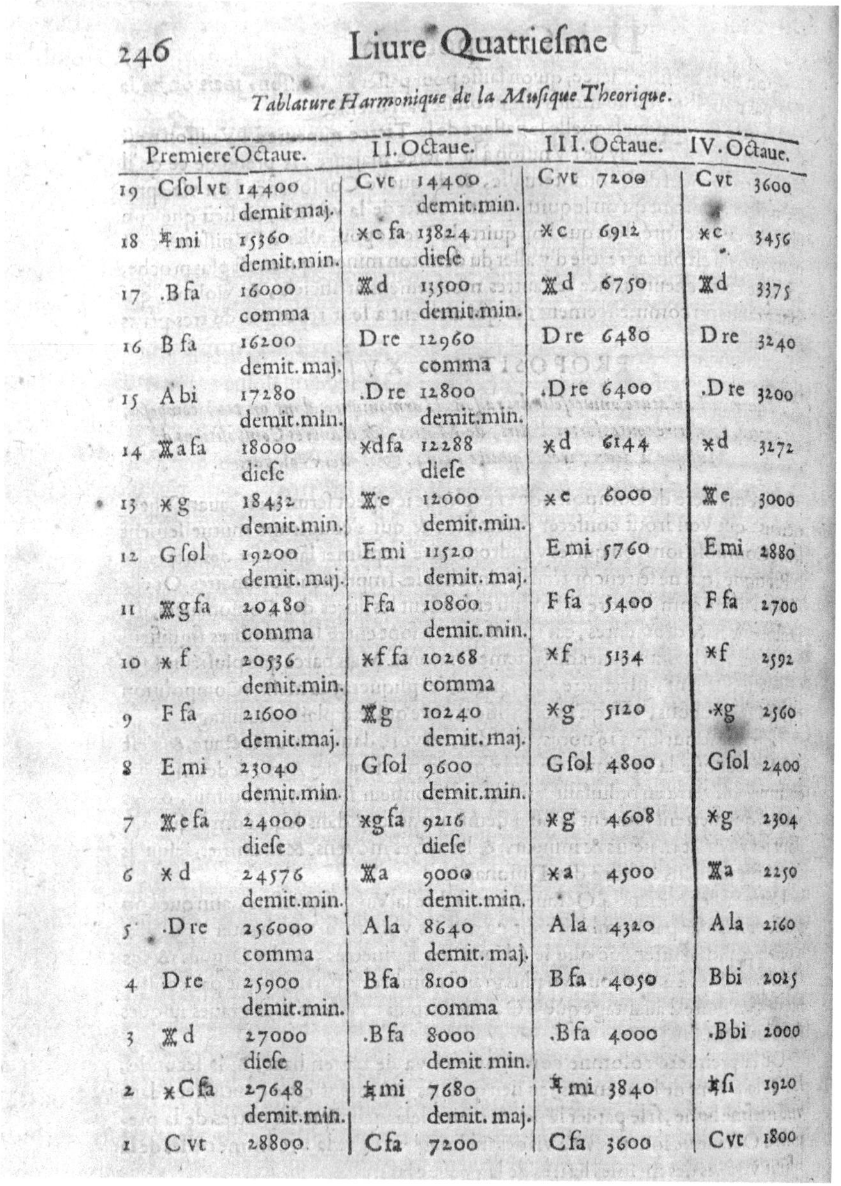

246 Liure Quatrieſme

Tablature Harmonique de la Muſique Theorique.

	Premiere Octaue.			II. Octaue.			III. Octaue.		IV. Octaue.	
19	Cſol vt	14400	demit maj.	C vt	14400	demit.min.	C vt	7200	C vt	3600
18	♯mi	15360	demit.min.	×c fa	13824	dieſe	×c	6912	×c	3456
17	.B fa	16000	comma	♯d	13500	demit.min.	♯d	6750	♯d	3375
16	B fa	16200	demit. maj.	D re	12960	comma	D re	6480	D re	3240
15	A bi	17280	demit.min.	.D re	12800	demit.min.	.D re	6400	.D re	3200
14	♯a fa	18000	dieſe	×d fa	12288	dieſe	×d	6144	×d	3272
13	×g	18432	demit.min.	♯e	12000	demit.min.	×e	6000	♯e	3000
12	G ſol	19200	demit. maj.	E mi	11520	demit. maj.	E mi	5760	E mi	2880
11	♯g fa	20480	comma	F fa	10800	demit. min.	F fa	5400	F fa	2700
10	×f	20536	demit.min.	×f fa	10268	comma	×f	5134	×f	2592
9	F fa	21600	demit.maj.	♯g	10240	demit. maj.	×g	5120	.×g	2560
8	E mi	23040	demit.min.	G ſol	9600	demit.min.	G ſol	4800	G ſol	2400
7	♯e fa	24000	dieſe	×g fa	9216	dieſe	×g	4608	×g	2304
6	×d	24576	demit.min.	♯a	9000	demit.min.	×a	4500	♯a	2250
5	.D re	256000	comma	A la	8640	demit. maj.	A la	4320	A la	2160
4	D re	25900	demit.min.	B fa	8100	comma	B fa	4050	B bi	2025
3	♯d	27000	dieſe	.B fa	8000	demit min.	.B fa	4000	.B bi	2000
2	×C fa	27648	demit.min.	♯mi	7680	demit. maj.	♯mi	3840	♯ſi	1920
1	C vt	28800		C fa	7200		C fa	3600	C vt	1800

Figure 5.4: "Harmonic tablature of theoretical music." Mersenne, *Harmonie universelle*, "Livre quatriesme de la composition," 246. Paris, Bibliothèque nationale de France, département Arsenal, FOL-S-1449.

These tables collect scale divisions or movements in counterpoint that musicians could use in their practice to develop better melodies and harmonies but do not fundamentally alter the practice of composing. Mersenne does give other suggestions that would be more tied to compositional practice, offering composers alternative writing methods that turn compositions themselves into rows of numbers or letters. Throughout the book, he presents over a dozen such methods, most of them using a single composition by Eustache du Caurroy, a small "Misericordias Domini" based on psalm 88 ("The mercies of the Lord I will sing forever") in fauxbourdon style and thus in homophonic texture with parallel harmonies. In his first discussion of the piece (fig. 5.5), Mersenne prints the notes together with numbers above each note, indicating the position of the note according to the bass. Thus, a 12 represents a twelfth above the bass. According to Mersenne, such numbers could be used in practice, but they had the clear drawback that the progression of the individual notes was not clear, as the bass itself is moving. He explains that they are helpful in a different sense:

> They are useful, however, to indicate the consonance of each note, in order that those who do not know how to analyze [*partir*] might have the pleasure of reflecting upon the consonances that they hear by considering these numbers with which I mark the consonances that the three parts of this example form with the bass.[120]

The numbers thus could be used to reflect on the music rather than to change compositional practice. The words suggest that Mersenne thought that there was also a particular type of pleasure in reflecting on the numbers of the harmonies, combined or separate from the listening experience. As they depict the relationships with reference to the bass, it functions almost as a sort of proto-figured bass, but as a method of analyzing counterpoint, it is ultimately useless.[121]

On the next page, Mersenne also gives the same composition in the numbers of his universal table (fig. 5.6). Although no rhythm is given, small marks could be added above the numbers to indicate the rhythm, but here he was content having each syllable with the same duration to

120 "quoy qu'ils soient utiles pour marquer les Consonances de chaque note, afin que ceux qui ne sçavent pas partir, ayent le plaisir de faire reflexion sur le Consonances qu'ils, en considerant lesdits nombres, avec lesquels je marque les Consonances, que font les trois parties de cet exemple avec le Basse." Mersenne, *Harmonie universellle*, "Livre quatriesme de la composition," 247. I am following Elizabeth Honn Hoegberg's insightful analysis of this passage; see Honn Hoegberg, "From Theory to Practice," 162–64.

121 Figured bass was first codified in Italy around 1600 but was only used in France in the second half of the seventeenth century. Mersenne would include the technique first in his *Cogitata* of 1644. See Schneider, *Kompositionslehre*, 94.

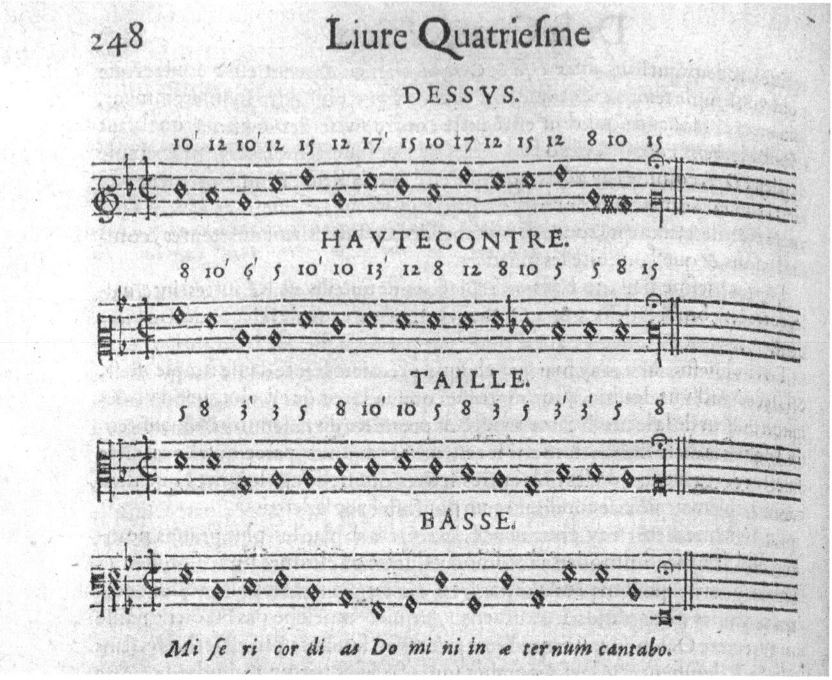

Figure 5.5: Compositional table. Mersenne, *Harmonie universelle*, "Livre quatriesme de la composition," 248. Paris, Bibliothèque nationale de France, département Arsenal, FOL-S-1449.

"reduce harmony to its ancient simplicity."[122] As a final example, he gives his method of writing music in the number of air vibrations already briefly discussed in chapter 1. He calls this "the most excellent" way of writing music, since "it shows the nature of sound."[123] By extending a string and counting the number of vibrations, Mersenne quantifies the movement of air and the pitches produced. Through this method, he argues, a piece of music could be sung at the same pitch around the world. Here it becomes a method to write down music for circulation and performance, again utilizing the same composition by Du Caurroy, but this time he only prints the bass (fig. 5.7). According to his calculations, the first G in the bass would need to be around 144 vibrations, but Mersenne thought that the air was in fact one hundred times faster than his observations. He concludes that "each will sing all kinds of musical pieces at the same tone

122 "reduire l'Harmonie à son ancienne simplicité." Mersenne, *Harmonie universelle*, "Livre quatriesme de la composition de musique," 249.

123 "qu'elle monstre la nature du son." Mersenne, *Harmonie universelle*, "Livre quatriesme de la composition de musique," 249.

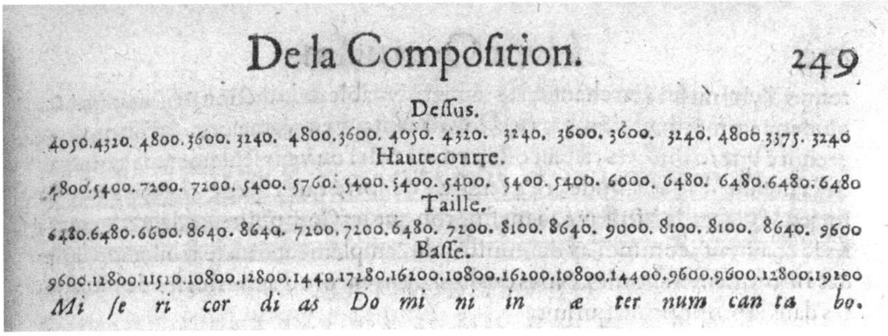

De la Compoſition. 249

Deſſus.

4050.4320. 4800.3600. 3240. 4800.3600. 4050. 4320. 3240. 3600. 3600. 3240. 4800 3375. 3240

Hautecontre.

4800.5400. 7200. 7200. 5400. 5760. 5400. 5400. 5400. 5400. 5400. 6000. 6480. 6480.6480.6480

Taille.

6480.6480.6600. 8640. 8640. 7200. 7200. 6480. 7200. 8100. 8640. 9000. 8100. 8100. 8640. 9600

Baſſe.

9600.12800.11520.10800.12800.14400.17280.16200.10800.16200.10800.14400.9600.9600.12800.19200

Mi ſe ri cor di as Do mi ni in æ ter num can ta bo.

Figure 5.6: Compositional table. Mersenne, *Harmonie universelle*, "Livre quatriesme de la composition," 249. Paris, Bibliothèque nationale de France, département Arsenal, FOL-S-1449.

14400.	19200.	17280.	16200.	19200.	21600.	25600.	24000.
MI	SE	RI	COR	DI	AS	DO	MI-
16200.	24000.	16200.	21600.	14400.	14400.	19200.	28800.
NI	IN	Æ	TER	NVM	CAN	TA	BO.

Figure 5.7: Compositional table. Mersenne, *Harmonie universelle*, "Livre quatriesme de la composition," 249. Paris, Bibliothèque nationale de France, département Arsenal, FOL-S-1449.

that composers desire that they be sung," but it is hard to see how this could this work in practice.[124]

Mersenne gives two more methods here that represent the notes as ratios of consonances, and thus the opening harmony of G, D, G, Bb is represented by the numbers 10, 15, 20, 24 (figs. 5.8 and 5.9). Through this method, it is easy to see the mastery of Du Caurroy in fauxbourdon, Mersenne argues, as the different harmonies could be more easily observed than in standard notation.[125] Mersenne holds that the second example was in fact practiced by an unknown music master and includes the ratio that each note forms with the final note of the bass. Interestingly, in the next theorem, Mersenne presents the table of different Arabic and Hebrew characters, discussed in chapter 4 (fig. 4.3). According to the theorist, this table could be used in musical communication across the globe and teach other cultures the diatonic system. Methods of musical composition, analysis, and circulation were thus closely aligned.

Mersenne returns to the topic of notational systems in book sixteen of *Harmonie universelle*, which is devoted to singing. Here he introduces

124 "de sorte que chacun chantera toutes sortes de pieces de Musique au mesme ton que les Compositeurs desirent qu'elles se chantent." Mersenne, *Harmonie universelle*, "Livre quatriesme de la composition de musique," 250.

125 Mersenne, *Harmonie universelle*, "Livre quatriesme de la composition de musique," 250.

Deſſus	24.	15.	12.	12.	40.	6.	24.	8.	5.	5.	12.	30.	30.	20.	5.	4.
Hautecontre	20.	12.	8.	6.	24.	5.	16.	6.	4.	3.	8.	24.	15.	15.	4.	4.
Taille	15.	10.	6.	5.	15.	4.	12.	5.	3.	2.	5.	15.	12.	12.	3.	2.
Baſſe	10.	5.	5.	4.	10.	2.	5.	2.	2.	1.	4.	10.	10.	10.	2.	1.
	Mi	*ſe*	*ri*	*cor*	*di*	*as*	*Do*	*mi*	*ni*	*in*	*æ*	*ter*	*num*	*can*	*ta*	*bo.*

Figure 5.8: Compositional table. Mersenne, *Harmonie universelle*, "Livre quatriesme de la composition," 250. Paris, Bibliothèque nationale de France, département Arsenal, FOL-S-1449.

Deſſus	24.	9.	4.	16.	6.	4.	16.	24.	9.	6.	16.	16.	6.	4.	15.	4.
	5.	2.	1.	3.	1.	1.	3.	5.	2.	1.	3.	3.	1.	1.	4.	4.
Hautecontre	4.	15.	8.	8.	15.	10.	15.	15.	15.	15.	15.	16.	4.	4.	4.	4.
	1.	4.	3.	3.	4.	3.	4.	4.	4.	4.	4.	5.	1.	1.	1.	1.
Taille	3.	3.	2.	9.	9.	8.	8.	3.	8.	12.	9.	2.	12.	12.	9.	2.
	1.	1.	1.	4.	4.	3.	3.	1.	3.	5.	4.	1.	5.	5.	4.	1.
Baſſe	2.	3.	5.	9.	3.	4.	9.	6.	9.	6.	9.	4.	2.	2.	3.	1.
	1.	2.	3.	5.	2.	3.	8.	5.	5.	5.	5.	3.	1.	1.	2.	1.
	Mi	*ſe*	*ri*	*cor*	*di*	*as*	*Do*	*mi*	*ni*	*in*	*æ*	*ter*	*nũ*	*can*	*ta*	*bo.*

Figure 5.9: Compositional table. Mersenne, *Harmonie universelle*, "Livre quatriesme de la composition," 251. Paris, Bibliothèque nationale de France, département Arsenal, FOL-S-1449.

different ways of depicting music by means of letters that would benefit the understanding and circulation of music, again departing from compositions by Du Caurroy (fig. 5.10). He describes the different methods as "literary music" (*musique literaire*), as these systems rely on letters of the alphabet instead of numbers. It is notable that Mersenne expands the number of voices, arriving at compositions with six, seven, and eight parts. Here he argues once more that such methods would have a range of benefits; for example, it would be easy to observe different intervals between the voices. Knowing the order of the alphabet would be enough, as one could see that an "f" and "g" together form a small interval. In his eyes, a musician could easily spot a seventh, as it would be seven letters apart. From these rules, Mersenne argues, it is possible "to learn to compose correctly in a short time."[126] With regard to rhythm, Mersenne also introduces a variety of figures above the letters, shown in the two-voice "O seigneur" (fig. 5.11). If no character is added, the duration will be a semibreve. For flats and

126 "Apprendre à composer correctement en Musique dans peu de temps." Mersenne, *Harmonie universelle*, "Livre sixiesme de l'art de bien chanter," 351.

Compoſition du meſme auteur à 8. parties.

I Deſſus	l *m m m m* m z *u* ſ *r*	*u u u u u u m m m.*
II Deſſus	mm m m m u s s s s	ſ m l l l l m z m.
I Haute-contre	LU L Z M U Z m r r	m m m m L m l ſ l.
II Haute-contre	UL L S' S L Z S Z Z	ſ u u u u u L z u.
Haute Taille	LM M M Z M S V R R	u S L L L L u m m.
Baſſe Taille	MM V Z Z V R M S S	M M M F M M M M M.
I Baſſe	l*m m m m* m z *u* f r	S S V V V V V Z L.
II Baſſe	LL L M M L S S S S	v v L L F L L M L.
	Ecce quàm bonum & quàm iucundum	ha bita re fratres in vnum.

Figure 5.10: Compositional table. Mersenne, *Harmonie universelle*, "Livre sixiesme de l'art de bien chanter," 350. Paris, Bibliothèque nationale de France, département Arsenal, FOL-S-1449.

Duo du 5. Mode ou du Mode de MI.

Deſſus	e, f, e, *a*, *g*, *a*, *b*, *c*, *b'*, *a*, g, a, f, e,
Baſſe ou Concordant.	a, a, c, cb'a, c, d, d, a, b', d, e, c, d, e,
	O Seigneur mon Dieu que tu es grand en tou te grandeur.

Figure 5.11: Compositional table. Mersenne, *Harmonie universelle*, "Livre sixiesme de l'art de bien chanter," 351. Paris, Bibliothèque nationale de France, département Arsenal, FOL-S-1449.

crosses, he sought out symbols found in the printer's shop and concludes, "Now this way of writing music, as well as the preceding one, teaches so easily to know all the consonances, and the dissonances, and even the art of analysis [*partir*], and to compose with as many parts as one would like so that it does not take one hour to understand the system [*l'artifice*]."[127]

These methods offered a way of circulating printed music without specialized music type from the printer, but they also had other benefits, such as providing a way to quickly scan over harmonies. Throughout his description of the benefits, Mersenne uses the phrase "a blink of the eye" [*clin d'oeil*] several times, such as when discussing the method of assigning numbers to each of the voices: "since in the blink of an eye you can see all the consonances made by all the parts."[128] Regarding the method of using harmonic numbers, he describes the effect as follows:

127 "Or cette maniere d'escrire la Musique, aussi bien que la precedente, apprend si aysement à connoistre toutes les Consonances, & les dissonances, & mesmes l'art de partir, & de composer à tel nombre de parties que l'on voudra, qu'il ne faut pas une heure pour en comprendre l'artifice." Mersenne, *Harmonie universelle*, "Livre sixiesme de l'art de bien chanter," 346.

128 "dautant que l'on void en un clin d'oeil toutes les Consonances que font toutes les parties." Mersenne, *Harmonie universelle*, "Livre quatriesme de la composition de musique," 275.

> one sees in the twinkling of an eye all the cadences that accommodate it and bind together agreeably, and those which are inimical: so that if the mind followed the eye in this matter, one could learn to compose in music, at the same time as one would have a look at the said table.[129]

Here he is clearly attempting to find a way to depict music that could quickly give an overview of the harmonies used. Even though the mind does not follow the eye immediately, Mersenne saw a benefit in musical information being presented in rows of numbers. The methods thus provide a way to look at the different parts, harmonies, and cadences at once, forming a score. Here Mersenne argues that music has been fully captured by the numbers on the page so that the mind could immediately grasp it, and a glimpse at the table would be enough to learn the entire art of composition. On a more practical level, by assigning numbers to the position according to the bass or tonal system, readers could scan the harmonies and progressions quickly, possibly following musical practice, much as early modern composers used erasable slate tablets to assemble voices before they wrote out the different parts.[130]

That many of the examples were taken from Eustache du Caurroy, who already passed away in 1609, was not by accident. Du Caurroy's career was interwoven with the French court and its royalist agenda, where he held the position of composer of the royal chamber and chapel.[131] Mersenne thought that the composer was especially adept in counterpoint and wrote, "Those who wish to learn to compose in counterpoint, both simple and figurative, will find nothing more advantageous than to read, analyze, and imitate the works of Eustache de Caurroy."[132] In 1643 Gantez similarly argued that musicians should imitate the works of Du Caurroy, Sauvaire Intermet, and Claude le Jeune instead of seeking "new inventions and fashionable movements." If not, they were considered "fools."[133] By using the works of De Caurroy as examples for harmony and voice leading, Mersenne positioned himself in

129 "dans laquelle on void en un clin d'oeil toutes les cadences qui s'accommodent & se lient ensemble agreablement, & celles qui sont ennemies: de sorte que si l'esprit suivoit l'oeil en cette matiere, l'on pourroit apprendre à composer en Musique, en mesme temps que l'on auroit envisagé ladite Table." Mersenne, *Harmonie universelle*, "Livre cinquiesme de la composition de musique," 291.

130 On the possible use of scores by composers, see Owens, *Composers at Work*, 4–5.

131 Alexis-Collins, "Eustache du Caurroy," 205.

132 "ceux qui veulent apprendre à composer en Contrepoint tant simple que figuré, ne trouvent rien plus avantageux que de lire, partir, & imiter les oeuvres d'Eustache de Caurroy." Mersenne, *Harmonie universelle*, "Livre sixiesme de l'art de bien chanter," 363.

133 "fous, pour rechercher de nouvelles inventions et des mouvemens." Gantez, *L'entretien des musiciens*, 148.

a longer tradition that argued that contemporary composers should depart from imitating the composer's works. There is also another meaning to Du Caurroy's position. As Sebastian Klotz has noted, Du Caurroy formulated his compositional approach as a process of problem solving, a rational art that could be understood through study and reason.[134] In the preface to his *Preces ecclesiasticae* (Paris, 1609), a collection of fifty-three sacred motets, Du Caurroy writes, "My eyes were opened to solve more and more the difficulties of this science." He goes on to describe music as a "science, being part of mathematics," that is "demonstrative and very certain."[135] But like Mersenne, Du Caurroy found music being misused, with many authors treating the subject with confusion, "relying on a few common and vulgar rules that do little for true harmony." He hoped that "professors [would] take care to bring it into the reasons and considerations required for such a science."[136] Mersenne's methods attempted to do precisely this in demonstrating the scientific foundations of music and thereby raising its status. He achieved this by doing something that Du Caurroy could not. Vocal works like Du Caurroy's *Preces ecclesiasticae* were printed as partbooks, meaning that each individual voice was printed separately and each musician had only their own part in front of them (fig. 5.12). There was often no single score where all parts were gathered. The methods that Mersenne introduced in his book were all different ways of gathering and comparing different vocal parts, bringing together the different voices to demonstrate the harmonies present in the composition. The consonances and dissonances of the entire composition could be quickly scanned and evaluated. Although it would become quite complex with figured counterpoint, Mersenne was primarily thinking of simple counterpoint, which he considered the basis for each composer.

By turning music into numbers that can be rearranged and decoded, these methods come close to the seventeenth-century interest in universal languages and cryptography. But the exercise was not just theoretical or speculative for Mersenne; he saw his efforts as serving the practical need to preserve musical compositions that would otherwise be lost due to the lack of a sufficiently flexible technique of music printing in France. He

134 Klotz, "Vibration und Vernunft," 288.

135 "ç'a esté les yeux ouverts pour me resoudre de plus en plus sur les difficultez de ceste science ...ceste science faisant partie des Mathematiques, est demonstrative & tres-certain." Du Caurroy, *Preces ecclesiasticae*, n.p. ("Au Roy").

136 "plusieurs se meslent de la traicter avec confusion si grande (s'appuyans de quelques reigles communes & vulgaires qui servent de bien peu à la vraye harmonie) ... les professeurs prennent garde d'y apporter les raisons & considerations requises à une telle science." Du Caurroy, *Preces ecclesiasticae*, n.p. ("Au Roy").

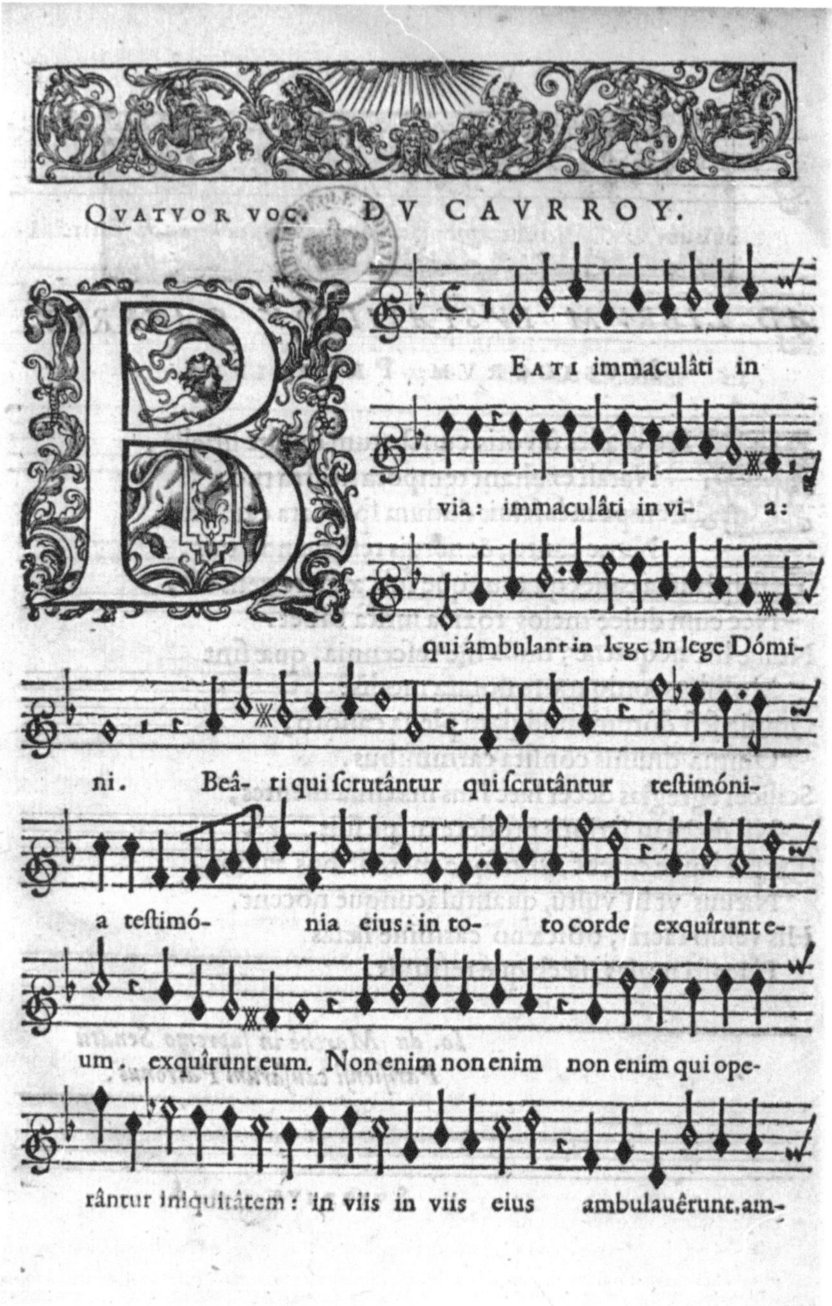

Figure 5.12: Superius (soprano) part of a partbook. Du Caurroy, *Preces ecclesticae*, 1. Paris, Bibliothèque nationale de France, RESVM1-48.

also thought that it might enable composers to print and share their music more widely, since they would no longer rely on the music printer Pierre I Ballard, who held the privilege on music printing in France. That meant that any composer or theorist who wanted to print musical notes had to rely on Ballard. Already in his *Quaestiones*, Mersenne demonstrated an interest in numerical and literary modes of composing, partially motivated by Ballard's monopoly.[137] With such notational symbols made up of letters or numbers, "one can easily write, read, and sing every kind of music with ordinary printing characters," without any need for musical characters "that are very rare in France."[138] He thought that this would greatly benefit music, as compositions tended to disappear and not receive public attention because of Ballard's monopoly. The method thus not only brought music into the realm of quantifiable numbers and "scientific" representation but would also allow music to be considered more universally transmittable.

The extent to which such notational systems served practical purposes remains unclear. Mersenne believed, however, that music could be more easily shared, learned, and transmitted by this method. In his correspondence, he continued to share examples, for example to Theodor Haak in England. Haak in turn shared the method put forth by the Englishman William Braithwaite in 1638, who used the numbers 1 through 7 to represent the different pitches of the scale and provided different symbols for register, duration, and rhythm.[139] Comparing Braithwaite's method to Frenchman Jean le Maire's method to transform the solmization syllables, Mersenne wrote in 1640, "Regarding the music by numbers of your Braithwaite: it is entirely ridiculous, as you say very well, and neither these characters, nor those of Monsieur le Maire give us anything new in the field of theory, which is what I esteem worthy of man, they can only serve practice."[140] Braithwaite's and Le Maire's systems could thus potentially benefit practice

137 Mersenne, *Quaestiones*, col. 1694. For Ballard and music printing in France, see the masterful work by Laurent Guillo in Giullo, *Pierre I Ballard*. For more on Mersenne's relation to Ballard, see Van der Miesen, "'Unbelievably Hard Work.'"

138 "l'on peut aisement escrire, lire & chanter toute sorte de Musique avec les caracteres ordinaires des impressions, sans qu'il soit besoin des notes ... qui sont fort rares en France." Mersenne, *Harmonie universelle*, "Livre sixiesme de l'art de bien chanter," 349.

139 Braithwaite, *Methodus nova docendi*.

140 "Pour la Musique par nombres de vostre Br(aithwaite): elle est entièrement ridicule, comme vous dites fort bien; et ni ces Caractères, ni ceux du Sr. Le Maire ne nous donnent rien de nouveau dans le fond de la Theorie, qui est ce(lle) que j'estime digne de l'homme, elles ne peuvent servir qu'à la prattique." Mersenne to Haak, 23 November, 1640, CM 11, 421. See also the previous letter: Mersenne to Haak, 4 September 1640, CM 11, 413. On Le Maire, see Cohen, "Jean le Maire"; Knowlson, "Jean Le Maire."

by making printed music notation more widely available than had been the case in France and by being easier to memorize, but Mersenne's interest was deeply rooted in the field of theory. This is why he preferred his own devised systems, as they simultaneously transmit musical information in a written form and also elucidate musical relations, progressions, and harmonies in a visual format.

The Tyranny of Rules

Ultimately, Mersenne considered rules unable to capture what makes a composition excellent. While in some letters he played with the idea of immutable rules in music similar to those from the field of geometry, in his *Harmonie universelle* he emphasized the inner genius of the composer. Here he writes, "the quality of beauty ... depends more on the excellent genius of the composer than the strict observance of the rules." He continues, "one cannot form a certain science from it."[141] Science and reason cannot capture what makes a good composition, as this depends on the skill and imagination of the composer. Although the composer uses rules and certain principles in his works, the true composer should be above the rules:

> But the learned musician should be so far above all rules, so that he does not impose any laws on himself harmful to beautiful songs and the movements of each part, which form the principal attractions of music. ... for the rules of harmony are not like those of geometry, which compel the mind of all those who have common sense to embrace them, for they depend upon the ear and on custom. And many [composers] use two or three successive fifths, which implies that they have a good effect, which does not matter at all, as long as the composer and the listener are satisfied.[142]

141 "la qualité du beau ...dépend plustost de l'excellent genie du Compositieur, que de l'estroite observance des regles; c'est pourquoy je n'en parle pas icy, dautant que l'on n'en peut former une certaine science." Mersenne, *Harmonie universelle*, "Livre quatriesme de la composition de musique," 256.

142 "Mais le docte Musicien doit tellement estre par dessus toutes les regles, qu'il ne s'impose nulle loy qui prejudicie aux beaux chants, & aux mouvemens de chaque partie, qui font les principaux charmes de la Musique ... car les regles de l'Harmonie ne sont pas comme celles de la Geometrie, qui contraignent l'esprit de tous ceux qui ont le sens commun à les embrasser: elles dépendent de l'oreille, & de la coustume, & plusieurs font deux ou trois Quintes de suite, qui soustiennent qu'elles ont un bon effet: ce qui n'importe nullement, pourveu que le Compositeur, & l'Auditeur soient satisfaits." Mersenne, *Harmonie universelle*, "Livre quatriesme de la composition," 261.

The true musician stands above rules, Mersenne argues, retracting his previous comparison of music to mathematics. Rather, he emphasizes that the composer must have "favorable genius" and "natural inclinations."[143] Composers acquire these inclinations by nature, and art could never reach the same level of genius as nature. In fact, rules could constrain the composer, as he found with the composer Claude Le Jeune. According to Mersenne, Le Jeune possessed a "natural good" that surpassed all reasonable explanations, as there were no infallible musical principles upon which everybody could agree.[144] Upon publication of *Harmonie universelle*, Mersenne no longer believed that there were absolute rules within the realm of music making. The rules of harmony, he suggested, "depend upon the ear and custom." Whereas someone like Salomon de Caus wrote that the rules of music are more certain and natural than geometry, Mersenne arrived at the opposite conclusion: music is not like geometry, and it instead depends on the disposition and natural inborn talent of the composer.

Mersenne's mitigated standpoint becomes especially clear in his debate with the Dutch composer and theorist Joan Albert Ban. Throughout the 1630s and 1640s, the Haarlem composer developed his own idea of musical composition, which he called "zin-roerende zangh" or "musica flexamina" (soul-moving songs). He worked on an eighteen-keyed tuning for harpsichord and on a Dutch musical vocabulary that could be used instead of Latin music terminology (introducing the tongue twister "zanghswelschikkelykheidt" for "harmonia concinnitas").[145] Ban wrote several treatises on music, such as the *Dissertatio epistolica de musicae natuae* (Leiden: 1637) and an introduction to his *Zangh-bloemzel* (Amsterdam, 1642), in which he explains his musical principles.[146] Mersenne and Ban came in contact in 1637 through the mediation of Descartes, and the two exchanged numerous letters. Ultimately both theorists judged each other negatively; Ban saw Mersenne's *Harmonie universelle* as leading in the wrong direction, and Mersenne concluded that Ban had little talent or affinity for music.

143 "le genie heureux, & l'inclination naturelle." Mersenne, *Harmonie universelle*, "Livre sixiesme de l'art de bien chanter," 364.
144 "dont le bon naturel surpassoit toute leur science, laquelle n'a pas encore esté establie par des principes infallibles, dont tous les hommes puissent tomber d'accord." Mersenne, *Harmonie universelle*, "Livre quatriesme de la composition," 205. On Mersenne's opposition of the composers Eustache du Caurroy and Claude Le Jeune, see Schneider, *Französische Kompositionslehre*, 112.
145 Ban, *Zangh-bloemzel*, 48.
146 The competition has been discussed in numerous studies, see especially Walker, "Mersenne's Musical Competition of 1640"; Duncan, "Persuading the Affections"; Rasch, "*Me veux-tu voir mourir.*"

The two men began their exchange when Mersenne sent some questions to Ban via the theologian André Rivet.[147] In response, Ban explained his theories regarding the consonances. According to Ban, the minor third should be considered soothing, the major third rousing [*incitativus*], and the fourth vehement or excited [*vehementior*].[148] In his particular idiosyncratic style, Ban formulated his findings as "rules and axioms" and emphasized that all musicians, composers, and listeners will experience the power of his music.[149] In his words, a "soul-moving music" will be heard "if the singers and players follow the rules without any failures."[150] As he repeatedly emphasized, music must not only come from the mind of the musician but be calculated with reason and follow clear rules. According to Ban, no music theorist had been able to formulate the rules of musical practice [*musica practica*].[151] He saw his own efforts as a significant step in the right direction. Anyone who denied the existence of his rules either had no musical ear or did not know mathematics. As he recounts in *Zang-bloemzel*, his own interest in music began when he found "immutable reasons and rules" in other disciplines, such as language, surveying, and geometry, and thought that these must exist in music as well.[152] Ban has long been ridiculed by later historians for his self-assured style of writing and attempt to codify musical text expression by means of ornamentation and word-painting or madrigalisms. According to Ban, the text had to be expressed in both the melody and harmony, with particular feeling being assigned to each interval. But his efforts to unite the mathematical proportions of the consonances with their emotional effects were part of a larger tendency of the early modern period to align the emotional power of music with its mathematical foundations. Whereas some doubted that such a relationship even existed, Ban was vehemently convinced that he had found it.

The tensions with Mersenne's position became especially clear during the musical competition between Ban and Antoine Boësset that the friar organized. In 1640 Mersenne challenged Ban to a competition with the

147 Mersenne's original letter is lost. Ban's answer has survived as a letter to the English diplomat William Boswell. The response is found in Ban to Boswell, 15 December 1637, CM 7, 1–5.

148 Ban to Mersenne, 31 October 1638, CM 8, 166. For more on Ban's view on the choice of intervals and mode, see Rasch, "*Me veux-tu voir mourir.*"

149 "regulis axiomatisque." Ban to Mersenne, 31 October 1638, CM 8, 168; Ban to Doni, 1 January 1639, CM 8, 261.

150 "Sequeste regole delli cantori et sonatori siano bene observato senza fallo si sentirà la Musica flexanima." Appendix to Ban to Doni, 1 March 1639, CM 8, 346.

151 "Musica practica a nemine hactenus, quod sciam, demonstrata est, nec rite explicata." Ban to Mersenne, 31 October 1638, CM 8, 164.

152 "vaste redenen, ende Regulen." Ban, *Zangh-bloemzel*, n.p. ("Tot den leezer").

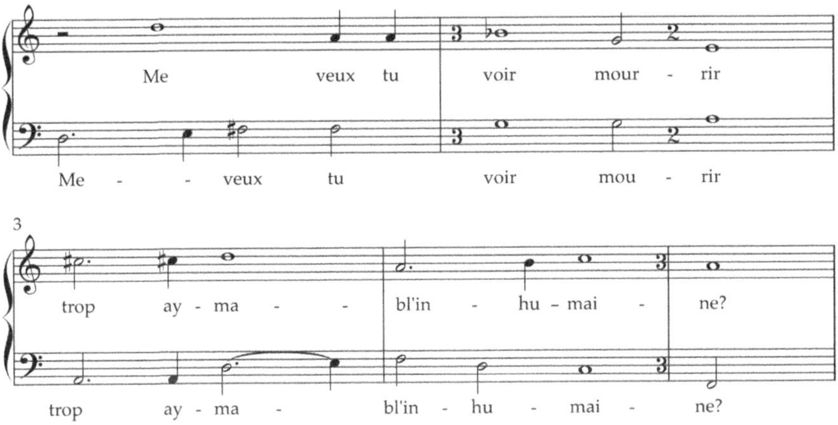

Example 5.1: Antoine Boësset, "Me veux mourir," opening. From the letter of Joan Albert Ban to Anna Maria van Schurman, 20 August 1640, *Correspondance du P. Marin Mersenne*, vol. 10, 31.

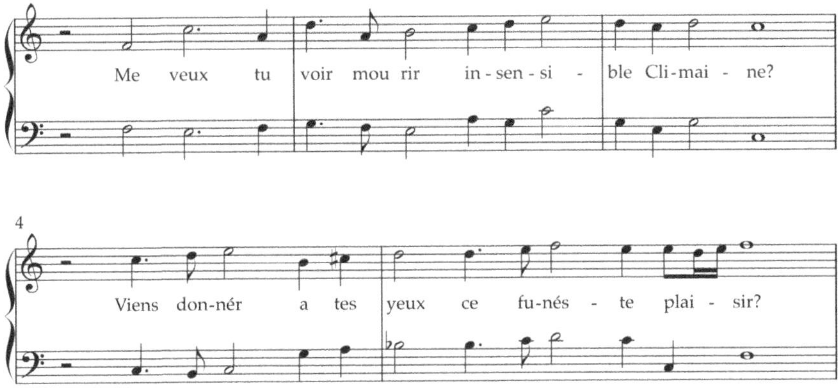

Example 5.2: Joan Albert Ban, "Me veux te mourir," opening. From the letter of Joan Albert Ban to Anna Maria van Schurman, 20 August 1640, *Correspondance du P. Marin Mersenne*, vol. 10, 36.

Parisian composer. He selected the poem "Me veux-tu voir mourir" (Do you want to see me die) but did not tell Ban that Boësset had already composed a piece on the same text. In May of that year, Mersenne circulated both Ban's and Boësset's composition around Europe. Most of his correspondents (such as Christoph de Villiers and Constantijn Huygens) preferred Boesset's composition, which led Ban to defend his compositional principles in a letter to Anna Maria van Schurman.[153] Ban saw numerous flaws in Boësset's composition. He himself had interpreted the opening sentence "Me veux tu voir mourir" as an exclamatory question that should be depicted with a

153 Ban to Van Schurman, 20 August 1640, CM 10, 18–46. The letter was written after the encouragement of Huygens.

rising movement. He described the half tone step on "tu voir" in Boësset's composition as sweet, whereas it should sound desperate (example 5.1).

The upward movement on the word "aymable" was similarly unsuitable according to Ban. As he emphasized in his letter, this was not just textual interpretation but could be demonstrated by the sciences of mathematics and physics. As he wrote to Van Schurman, each interval has its own effect, that could be demonstrated mathematically and experienced by the senses.[154] Later he again argued that all these matters are "mathematically and physically demonstrable."[155] Ban goes exactly against the viewpoint of Descartes who argued around the same time that music could not be explained mathematically and physically.[156] Whereas more and more of his contemporaries found that it was impossible to unite the rational and sensual powers of music, Ban argued that they simply did not know the methods.

Mersenne found Ban's composition harsh and violent, not suitable for French' sensibilities (example 5.2). Music, and the genre of the *air de cour* in particular, was meant "to charm the soul and the ear," and Ban's music did nothing to that end.[157] He valued the composition of Boësset much more, pointing to the beauty of the descending minor thirds in the opening phrase. Whereas Ban argued that the phrase "aymable inhumaine" was not appropriately set to music, Mersenne found that the chosen intervals were very suitable to represent the protagonist's love (measures 3–4 in example 5.1).[158] In several places, Boësset's musical style departed from the compositional advice of *Harmonie universelle* (such as avoiding musical intervals that are difficult to sing), but these licenses were allowed; empirical perfection in the realm of music is impossible to achieve. Contrasting with his earlier statements, Mersenne disagreed with Ban that the obedience to rules would create a perfection composition. He found that Boësset's setting had a particular quality that was difficult to put into words, but its character was instantly recognizable. As he wrote to Constantijn Huygens,

154 "quaelibet enim aliam atque aliam delectandi, movendique habet potestatem; imo quaelibet cujusque species propriam habet, pro diversa graduum constitutione, energiam: mathematice quoad rationem, et phisice quoad sensum demonstrabilem." Ban to Van Schurman, 20 August 1640, CM 10, 30. See also the remark cited in Walker, "Mersenne's Musical Competition of 1640," 82.

155 "Haec omnia mathematicè et physicè demonstrabilia sunt." Ban to Van Schurman, 20 August 1640, CM 10, 37.

156 "nec mathematicas, nec physicas, sed morales tantum mihi videri." Descartes, *Oeuvres complètes*, 3:843.

157 "particulierement et principalement pour charmer l'esprit et l'oreille." Mersenne to Constantijn Huygens, 14 November 1640, CM 10, 237; Mersenne to Constantijn Huygens, 14 November 1640, CM 10, 247–48. This phrase is also discussed in Duncan, "Persuading the Affections."

158 Mersenne to Constantijn Huygens, 14 November 1640, CM 10, 244.

this was the music of the most important court composer, not "a shepherd or villager."[159] To take away Boësset's liberty in choosing which phrasing or setting he preferred "would be to tyrannize harmony and render the muses slaves."[160]

Mersenne's evaluation of Boësset's and Ban's compositions was clearly biased in favor of his fellow Frenchman: the whole competition can be described as a ruse to force Ban to admit the superiority of French composers. The composition does offer insight into Mersenne's response to individual compositions, something he rarely expressed in writing. And his disagreement with Ban makes his mitigated position clear: certain musico-mathematical rules existed for Mersenne, but on the level of the individual response, it was impossible to establish axiomatic principles. Compositions are based on the ear and custom, and both undergo changes. Mersenne praised the works of Boësset and Le Jeune, whose music broke several compositional rules of the sixteenth century, arguing that these composers had certain affinities for music that could not be articulated in words or rules. Like his discussion of the consonances, Mersenne's writings on composition demonstrate that he no longer believed that musical experience could be fully captured in rules, pointing to the increasing acknowledgment of musical style and taste in the later seventeenth century.

It is clear that Mersenne's opinions were not static: they changed over the years. In his 1634 *Les preludes*, for example, he singled out the popular genre of *air de cour*. According to Mersenne, these songs contain only profane and vain subjects, and he argued that they could never achieve the perfection of religious music.[161] Two years later, he was less critical and considered melancholic airs with secular texts true expressions of human sadness after the fall of man.[162] Although it goes too far to suggest that he would have reconsidered his entire musical outlook, he clearly shifted his priorities. Eventually Mersenne was able to incorporate these different elements into his eclectic *Harmonie universelle*, the last parts of which were published only three years before the contest of Ban. Although he defined the goal of music as pleasing the ear, there is little evidence that he reconsidered his hope for his contemporaries to join in "the archetypal harmony," as humans would

159 "d'un berger ou d'un vilageois." Mersenne to Constantijn Huygens, 14 November 1640, CM 10, 241, 246.

160 "ce seroit tyrannizer l'harmonie, et rendre les Musses trop esclaves." Mersenne to Constantijn Huygens, 14 November 1640, CM 10, 246. On Huygens's role in this affair, see Rasch, "*Me veux-tu voir mourir*," 240–41.

161 Mersenne, *Les preludes*, 223.

162 Mersenne, *Harmonie universelle*, "Livre second, des chants," 175.

ultimately join the angels in heaven and encounter musical perfection.[163] If he succeeded in inspiring some divinity among contemporary musicians, he considered his time wisely spent: "I beg God's goodness to inspire them [musicians] with this thought … and then I believe I have used my time as effectively for these treatises, as I have done the same for theology."[164] If successful, good music theory could be just as effective as theology.

Taste and Its Limits

As Jacques de Gouy declared at the start of his *Airs a quatre parties* from 1650, "we are no longer in the age of Pythagoras."[165] Much changed in the first half of the seventeenth century in France. In the sixteenth century, the members of the Académie de musique et de poésie stated that "where music is disordered, the morals are often depraved, and where it is well ordered, there men are well disciplined."[166] Although generally in line with the viewpoints of the Académie de musique et de poésie, Mersenne himself had to admit that there were excellent men with no particular disposition for music. The Pythagorean or Neoplatonic tradition continued to exert influence, but a strict form of Pythagoreanism within music theory slowly declined over the course of the early modern period as other discursive traditions were introduced in its place. Whereas classical philosophers like Plato and Aristotle excluded taste in the making of judgments, from the Renaissance onward the concept of taste was re-evaluated and increasingly applied. Cultural historians have illustrated the increasing role of taste as a category of thought in seventeenth-century France, closely related to the development of aesthetic understandings of art. Taste was transformed from its literal meaning to a normative judgment of what was considered beautiful (having good taste, or *le bon goût*).[167] Around 1650 the discourse surrounding good taste had become even more established, taking taste as a universal principle rather than something that depends on specific

163 "de l'Harmonie Archetype." Mersenne, *Harmonie universelle*, "Livre premier des consonances," 21.

164 "Je supplie la bonté divine qu'il leur inspire cette pensée … & pour lors je croiray avoir employé le temps aussi utilement à ces Traitez, comme si j'en eusse fait autant de la Theologie." Mersenne, *Harmonie universelle*, "Traitez des instrumens a chordes," n.p. ("Preface au lecteur").

165 "Nous ne sommes plus du temps de Pythagoras." De Gouy, *Airs à quatre parties*, n.p. ("preface").

166 "ou la musique est desordonnée, là volontiers les moeurs sont dépravez, et où elle est bien ordonnée, la sont les hommes bien moriginez." Cited in Yates, *The French Academies*, 319.

167 Koch, *The Aesthetic Body*, 179–234.

customs. This sense of taste did not come from knowledge or reason but was seen as something that it is innate: one could be very knowledgeable but still have bad taste.[168]

In the realm of music, taste was similarly reevaluated and discussed, for example in relation to the tradition of rhetoric or humoral theory. As discussed in chapter 4, geography and humoral complexion were considered as factors of individual differences. The search for rules to account for taste discussed here were thus not in opposition with a randomness, but rather with different traditions that grappled with individual difference. Food was another important cultural marker. In his writings on the tonal systems, Galilei described music as something culturally dependent; what appears beautiful to one group might offend others. The Italian composer compared music to food, medicine, or odors.[169] As mentioned before, Mersenne's correspondent Moreau even compared music to food, for which each individual has a different predilection, and he was no exception in this use of language. The French author Jean Bodin compared the composer's use of dissonances in music to cooks who serve their best meats in "sharpe and unsaverie sauces."[170] The collector Pierre Trichet wrote that "just as vision is recreated by a pleasant variety of colors and taste is stimulated by the object of delicate foods, hearing is tickled by the diversity of harmonious sounds produced by musical instruments."[171] For these writers, there was nothing inherently negative in saying that the appreciation of music depended on the senses and on taste, just as one might have food preferences. In a world with increasing emphasis on taste and custom, the axiom that the simplest proportions are the most beautiful was more and more difficult to maintain. Mersenne's skeptical counterpart La Mothe le Vayer would write that "everybody has their passions and particular taste."[172] Somewhat later, the polymath Christiaan Huygens would write that "not all tastes agree completely," noting that the ancient Greeks did not consider the third of sixth to be consonances.[173] As Georgia Cowart has shown, music criticism in the seventeenth century focused increasingly on music as dependent upon

168 Flandrin, "Distinction through Taste"; Tsien, *The Bad Taste of Others.*

169 Galilei, "Discorso di Vicentio Galilei," 164–65.

170 Bodin, *The Six Bookes*, 791.

171 "comme la veue se recrée par l'agréable variété des couleurs et comme l'appétit s'esguise par l'object des viandes délicates, de l'ouïe est chatouillé par la diversité des sons harmonieux que produisent les instruments de musique." Trichet, *Traité de instruments*, 16.

172 "Chacun suit sa passion, & a son goust particulier icy comme ailleurs." La Mothe le Vayer, "Discours sceptique sur la musique," 146.

173 "les gouts ne s'accordent pas tout." Huygens, "Origine de chant," 36.

individual taste and the individual response of the listener until "absolute reason" and "relative taste" would arrive at a temporary compromise in the eighteenth century.[174]

As this chapter has demonstrated, the tensions between the realms of taste and immutable rules can already be found in the work of Mersenne and his contemporaries. Although the friar continuously emphasized the role of mathematics and reason, taste was not necessarily antithetical to his project. Individual taste in music exists, but within bounds—never so much that a listener might enjoy a diminished fifth. The different examples have demonstrated how in certain aspects, Mersenne's work remains committed to rules and reason, albeit in a mitigated way, while at the same time allowing room for individual preference. This allowed Mersenne to come to his own distinctive position within a tradition of investigating universal harmony. Rationality and empiricism could not be fully united within the realm of music. A thousand listeners might have a thousand opinions, but as Descartes once wrote to Mersenne, this does not mean that music lacks solid foundations.[175]

Bibliography

Manuscript sources

Paris, Bibliothèque nationale de France, MS fr. 24256. Manuscript with melodies of eight notes.

Printed primary sources

Aristotle. *Politics: With an English Translation by H. Rackman*. London: Heinemann, 1959.

Ban, Joan Albert. *Zangh-bloemzel & kort sangh-bericht*. With an introduction by Frits Noske. Amsterdam: Knuf, 1973.

Bodin, Jean. *The Six Bookes of a Common-Weale. Out of the French and Latine copies, Done into English, by Richard Knolles*. London: Islip, 1606.

Boethius, Anicius Manlius Severinus. *The Fundamentals of Music*. Translated by Calvin M. Bower, edited by Claude V. Palisca. New Haven: Yale University Press, 1989.

Braithwaite, William. *Methodus nova docendi discendique veterem (ut ita dicam) musicam* [...]. London: Norton, 1639.

174 Cowart, *The Origins of Modern Music Criticism*.

175 "ce qui n'empesche pas que la musique n'ait ses demonstration tres assures." Descartes to Mersenne, October or November 1631, CM 3, 212.

Colonna, Fabio. *La sambuca lincea overo dell'istromento musico perfetto. Con annotazioni critiche manoscritte di Scipione Stella (1618–1624) (rist. anast. Napoli, 1618)*. Edited by Patrizio Barbieri. Lucca: Libreria musicale italiana, 1991.

De Gouy, Jacques. *Airs à quatre parties, sur la paraphrase des pseaumes de Messire Antoine Godeau, évesque de Grasse*. Paris: Ballard, 1650.

Descartes, René. *Musicae compendium/Leitfaden der Musik*. Edited by Johannes Brockt. Darmstadt: Wissenschaftliche Buchgesellschaft, 1992.

Descartes, René. *Oeuvres completes III: Correspondance, janvier 1640–juin 1643*. Vol. 3 of 11 vols. Edited by Charles Adam and Paul Tannery. Paris: Vrin, 1975.

Du Caurroy, Eustache. *Preces ecclesticae ad numeros musices redactae*. Paris: Ballard, 1609.

Galilei, Vincenzo. "Discorso di Vicentio Galilei intorno all'uso dell enharmonio." In *Die Kontrapunkttraktate Vincenzo Galileis*, edited by Frieder Rempp, 163–80. Cologne: Gerig, 1980.

Gantez, Annibal. *L'entretien des musiciens*. Auxerre, 1643. Reprint, Geneva: Minkoff, 1971.

Huygens, Christiaan. "Origine de chant." In *Oeuvres complètes: Musique et Mathématique*, vol. 20 of 22 vols., edited by Johan Adriaan Volgraff, 30–37. The Hague: Nijhoff, 1940.

La Mothe le Vayer, François, de. "Discours sceptique sur la musique." In Marin Mersenne, *Questions harmoniques. Dans lesquelles sont contenuës plusieurs choses remarquables pour la physique, pour la morale, et pour les autres sciences*, 84–165. Paris: Villery, 1634.

Masson, Charles. *Nouveau traité des règles pour la composition de la musique*. Paris: Ballard, 1705.

Maugars, André. *Response faite à un curieux, sur le sentiment de la musique d'Italie. Escrite à Rome le premier octobre 1639*. N.p.: s.n., ca. 1639.

Mersenne, Marin. *Correspondance du P. M. Mersenne*. Edited by Paul Tannery, Cornelis de Waard, Bernard Rochot, Armand Beaulieu, and René Pintard. Paris: Centre National de la Recherche Scientifique, et al., 1932–88.

Mersenne, Marin. *Harmonicorum libri XII*. Paris: Baudry, 1648. Reprint, Geneva: Minkoff, 1972.

Mersenne, Marin. *Harmonie universelle contenant la théorie et la pratique de la musique (Paris, 1636). Édition fascimilé de l'exemplaire conserve à la Bibliothèque des Arts et Métiers et annoté par l'auteur*. Introduction by François Lesure. Paris: Éditions du Centre National de la Recherche Scientifique, 1963.

Mersenne, Marin. *Les preludes de l'harmonie universelle ou questions curieuses. Utiles aux predicateurs, aux theologiens, aux astrologues, aux medicins & aux philosophes*. Paris: Guenon, 1634.

Mersenne, Marin. *Quaestiones celeberrimae in Genesim, cum accurata textus explicatione: in hoc volumine athei, et deistae impugnantur & expugnantur, &*

vulgata edition ab hereticorum calumnijs vindicator. Graecorum, & Hebraeorum musica instauratur. Paris: Cramoisy, 1623.

Mersenne, Marin. *Questions harmoniques. Dans lesquelles sont contenuës plusieurs choses remarquables pour la physique, pour la morale, et pour les autres sciences.* Paris: Villery, 1634.

Mersenne, Marin. *Questions inouyes, ou recreation des scavans.* Paris: Villery, 1634.

Mersenne, Marin. *Traité de l'harmonie universelle.* Paris: Baudry, 1627.

Mersenne, Marin. *La verité des sciences, contre les sceptiques ou pyrrhoniens.* Paris: Du Bray, 1625.

Nivers, Guillaume Gabriel. *Traité de la composition de musique.* Paris: Ballard, 1667.

Trichet, Pierre. *Traite des instruments de musique (vers 1640).* Edited by François Lesure. Geneva: Minkoff, 1978.

Secondary literature

Alexis-Collins, Marie. "Eustache du Caurroy: Un compositeur français aux confins du XVIe et du XVIIe siècle." *Acta Musicologica* 73, no. 2 (2001): 189–258.

Bertoloni Meli, Domenico. "The Role of Numerical Tables in Galileo and Mersenne." *Perspectives on Science* 12, no. 2 (2004): 164–90.

Bianchi, Eric. "Asses at the Lyre: Latin as Musical Language, and the Benefits of Exclusion." In *Baroque Latinity: Studies on "Baroque" European Neo-Latin Literature*, edited by Jacqueline Glomski, Gesine Manuwald, and Andrew Taylor, 171–93. London: Bloomsbury, 2023.

Bianchi, Eric. "Scholars, Friends, Plagiarists: The Musician as Author in the Seventeenth Century." *Journal of the American Musicological Society* 70, no. 1 (2017): 61–128.

Bianconi, Lorenzo. *Music in the Seventeenth Century.* Translated by David Bryant. Cambridge: Cambridge University Press, 1987.

Brooks, Jeanice. *Courtly Song in Late Sixteenth-Century France.* Chicago: University of Chicago Press, 2000.

Christensen, Thomas. "Rules, License and Taste in 17th-Century French Music Theory: From Mersenne to Rameau." In *Musikalische Norm um 1700*, edited by Rainer Bayreuther, 81–95. Berlin/New York: Walter de Gruyter, 2010.

Christensen, Thomas. "Sensus, Ratio, and Phtongos: Mattheson's Theory of Tone Perception." In *The Work of Music Theory: Selected Essays*, 289–310. London/New York: Routledge, 2016.

Cohen, Albert. "Jean le Maire and La Musique Almérique." *Acta Musicologica* 35, no. 4 (1963): 175–81.

Cohen, H. Floris. *Quantifying Music: The Science of Music at the First Stage of the Scientific Revolution, 1580–1650.* Dordrecht: Reidel, 1984.

Coumet, Ernest. "Mersenne: dénombrements, repertoires, numérotations de permutations." *Mathématiques et sciences humaines* 38 (1972): 5–37.

Coumet, Ernest. *Oeuvres d'Ernest Coumet*, vol. 2 of 2 vols., edited by Catherine Goldstein. Besançon: Presses universitaires de Franche-Comté, 2019.

Cowart, Georgia. *The Origins of Modern Music Criticism: French and Italian Music, 1600–1750*. Ann Arbor: UMI Research Press, 1981.

De Buzon, Frédéric. "The *Compendium Musicae* and Descartes's Aesthetics." In *The Oxford Handbook of Descartes and Cartesianism*, edited by Steven Nadler, Tad M. Schmaltz, and Delphine Antoine-Mahut, 254–68. Oxford: Oxford University Press, 2019.

Dear, Peter. "Marin Mersenne and the Probabilistic Roots of 'Mitigated Scepticism.'" *Journal of the History of Philosophy* 22, no. 2 (1984): 173–205.

Dear, Peter. *Mersenne and the Learning of the Schools*. Ithaca: Cornell University Press, 1988.

Dell'Antonio, Andrew. *Listening as Spiritual Practice in Early Modern Italy*. Berkeley: University of California Press, 2011.

Duncan, David Allen. "Persuading the Affections: Rhetorical Theory and Mersenne's Advice to Harmonics Orators." In *French Musical Thought, 1600–1800*, edited by Georgia Cowart, 149–75. Ann Arbor: UMI Research Press, 1989.

Fix, Adam. "The Acoustical Paradox: How Music Was Unmoored and Set Adrift from the Sciences in the Seventeenth Century." In *Music and Science from Leonardo to Galileo*, edited by Rudolf Rasch, 377–420. Turnhout: Brepols, 2022.

Flandrin, Jean-Louis. "Distinction through Taste." In *A History of Private Life, Volume 3: Passions of the Renaissance*, edited by Roger Chartier, translated by Arthur Goldhammer, 265–307. Cambridge, MA: Belknap Press, 1989.

Gouk, Penelope. *Music, Science, and Natural Magic in Seventeenth-Century England*. New Haven: Yale University Press, 1999.

Gouk, Penelope. "The Role of Harmonics in the Scientific Revolution." In *The Cambridge History of Western Music Theory*, edited by Thomas Christensen, 223–45. Cambridge: Cambridge University Press, 2002.

Guillo, Laurent. *Pierre I Ballard et Robert III Ballard. Imprimeurs du roy pour la musique (1599–1673)*, 2 vols. Sprimont: Mardaga, 2003.

Guillo, Laurent. "Sous la main du père Mersenne: la bibliothèque de musique des Minimes de la place Royal (Paris, XVIe–XVII siècles)." In *Collectionner la musique*. Vol. 3 of 3 vols.: *Erudits collectionneurs*, edited by Denis Herlin, Catherine Massip, and Valérie De Wispelaere, 29–80. Turnhout: Brepols, 2015.

Harne, George A. "Unstable Embodiments of Musical Theory and Practice in the *Speculum musicae*." *Plainsong and Medieval Music* 21, no. 2 (2012): 113–36.

Honn Hoegberg, Elisabeth. "From Theory to Practice: Composition and Analysis in Marin Mersenne's Harmonie universelle." PhD diss., Indiana University, 2005.

Knowlson, James R. "Jean le Maire, the Almérie, and the "musique almérique": A Set of Unpublished Documents." *Acta Musicologica* 40, no. 1 (1968): 86–89.

Klotz, Sebastian. *Kombinatorik und die Verbindungskünste der Zeichen zwischen 1630 und 1780*. Berlin: Akademie Verlag, 2006.

Klotz, Sebastian. "Vibration und Vernunft: Zur experimentellen Agenda in Marin Mersennes *Harmonie universelle* (Paris, 1636)." In *Spektakuläre Experimente: Praktiken der Evidenzproduktion im 17. Jahrhundert*, edited by Helmar Schramm, Ludger Schwarte, and Jan Lazardzig, 279–94. Berlin: De Gruyter, 2006.

Koch, Erec. *The Aesthetic Body: Passion, Sensibility and Corporeality in Seventeenth-Century France*. Newark: University of Delaware, 2008.

Lenoble, Robert. *Mersenne ou la naissance du mécanisme*. Paris: Vrin, 1943.

Mace, Dean T. "Marin Mersenne on Language and Music." *Journal of Music Theory* 14, no. 1 (1970): 2–34.

Moyer, Ann. E. "Musical Scholarship in Italy and the End of the Renaissance, 1500–1650: From Veritas to Verisimilitude." In *History and the Disciplines: The Reclassification of Knowledge in Early Modern Europe*, edited by Donald R. Kelley, 185–202. Rochester: University of Rochester Press, 1997.

Owens, Jessie Ann. *Composers at Work: The Craft of Musical Composition 1450–1600*. Oxford: Oxford University Press, 1997.

Palisca, Claude V. *Humanism in Italian Renaissance Musical Thought*. New Haven: Yale University Press, 1985.

Popkin, Richard. *The History of Scepticism: From Savonarola to Bayle*. Oxford: Oxford University Press, 2003.

Rasch, Rudolf. "*Me veux-tu voir mourir*: Joan Albert Ban versus Antoine Boësset (1640–1641)." In *Music and Science from Leonardo to Galileo*, edited by Rudolf Rasch, 211–49. Turnhout: Brepols, 2022.

Redwood, André. "Combinatorics, Composition, Copia: Mersenne's Permutations as Rhetoric of Abundance." *Music Theory Spectrum* 45, no. 1 (2023): 71–88.

Redwood, André. "Mersenne and the Art of Delivery." *Journal of Music Theory* 59, no. 1 (2015): 99–119.

Reiss, Timothy. *Knowledge, Discovery and Imagination in Early Modern Europe: The Rise of Aesthetic Rationalism*. Cambridge: Cambridge University Press, 1997.

Rentsch, Ivana. *Die Höflichkeit musikalischer Form: Tänzerische und anthropologische Grundlagen der frühen Instrumentalmusik*. Kassel: Bärenreiter, 2012.

Rentsch, Ivana. "Sinn statt Verstand: Johann Mattheson und das Geschmacksurteil im 18. Jahrhundert." In *Gut oder schlecht? Urteil und Werturteil in der Musik*, edited by Claudia Maurer Zenck and Ivana Rentsch, 31–43. Frankfurt am Main: Peter Lang, 2015.

Schneider, Herbert, *Die Französische Kompositionslehre in der ersten Hälfte des 17. Jahrhundert*. Tutzing: Schneider, 1972

Seidel, Wilhelm. "Französische Musiktheorie im 16. und 17 Jahrhundert." In *Enstehung nationaler Traditionen: Frankreich, England* (=*Geschichte der Musiktheorie*, vol. 9), edited by Frieder Zaminer, 56–82. Darmstadt: Wissenschaftliche Buchgesellschaft, 1986.

Tsien, Jennifer. *The Bad Taste of Others: Judging Literary Value in Eighteenth-Century France*. Philadelphia: University of Pennsylvania Press, 2012.

Vaast, Corrine. "L'esthétique musicale du père Mersenne et l'austérité musicale des Minimes." In *Saint François de Paule et les Minimes en France de la fin du XVe au XVIIIe siècle*, edited by Pierre Benoist and André Vauchez, 249–71. Tours: Presses universitaires François-Rabelais, 2011.

Van der Miesen, Leendert. "'Unbelievably Hard Work.' Marin Mersenne's *Harmonie universelle* at the Printers." In *Early Printed Music and Material Culture in Central and Western Europe*, edited by Andrea Lindmayr-Brandl and Grantley McDonald, 231–244. London: Routledge, 2021.

Van Orden, Kate. "Descartes on Musical Training and the Body." In *Music, Sensation, and Sensuality*, edited by Linda Phyllis Austern, 17–38. New York/London: Routledge, 2002.

Van Orden, Kate. *Music, Discipline, and Arms in Early Modern France*. Chicago: University of Chicago Press, 2005.

Walker, Daniel Pickering. "Mersenne's Musical Competition of 1640 and Joan Albert Ban." In *Studies in Musical Science in the Late Renaissance*, 81–110. London: Brill, 1978.

Wymeersch, Brigitte van. *Descartes et l'évolution de l'esthétique musicale*. Liège: Mardaga, 1999.

Yates, Frances. *The French Academies of the Sixteenth Century*. London: Warburg Institute, 1947.

Conclusion

Throughout this book, I have highlighted the evidential and epistemic potential of music and have looked at different ways in which sound was turned into evidence for something beyond its immediate sounding context. Marin Mersenne believed that harmony was found in the material constitution of a trumpet and the number of air vibrations in a composition by court composer Boësset. From experiments with echoes and overtones to the recording of all possible melodies using eight notes, Mersenne was especially concerned with making universal harmony audible, visible, and palpable. A wide variety of experiences were taken as evidence for universal harmony, such as the bodily experience of standing close to an organ or travelogues on music making in the Americas. As a central figure in the correspondence networks across Europe, Mersenne experimented with the circulation of questions and images, probing his correspondents on a wide variety of topics. Paying attention to the geographies of scientific investigation, I have focused on spaces such as the instrument maker's workshop and the cabinet of curiosity and have looked at the different social and epistemic processes that took place within such spaces, and where boundaries between theorists and practitioners were drawn.

Rather than arriving at a singular characterization of Mersenne's status, this investigation has demonstrated that Mersenne's work offers a key case study in charting the configurations of music and science in the early seventeenth century. Each of the preceding chapters has looked at a different constellation of musical scholarship in the early seventeenth century, shedding new light on the development of music as a scientific discipline, cultures of scholarship in early modern Europe, and the uses of sound in scientific contexts. This book positions Mersenne as a central figure in the landscape of early modern science, reconsidering several central developments in musical scholarship. The discipline of music consisted of a vast body of knowledge, involving a wide variety of actors, materials, and practices. Mersenne's work demonstrates how the study of music was reconfigured in the early modern period, responding to a decline in Pythagoreanism in musical scholarship and the availability of new sources, methods, and ways of knowing. At the

Van der Miesen, Leendert. *Marin Mersenne and the Study of Harmony: From Sound to Music.* Taylor & Francis Group, 2025.
DOI 10.5117/9789048564132_CON

same time, his work was an agent of change itself, highlighting the central role of sound as evidence, pushing the boundaries of musical investigations, and bringing different actors and their sources together.

I have attempted to show that the early modern attention to evidence was not solely visual but also aural. Although the study of sound lacked technologies such as the microscope and telescope, objects like a musical string, trumpet, and organ pipe functioned as epistemic tools that were capable of demonstrating something about the natural world. Mersenne argued that the human senses are too weak to perceive the universal harmony always present in nature; it required special instruments, techniques, and investigative modes. Following the increasing role of mechanical objects in early modern culture, Mersenne and his collaborators centered instruments, which mediated between the realms of art and nature. The trumpet, for example, was the key instrument for Mersenne to demonstrate the natural scale, while he ignored the more dissonant trumpet tones. Listening carefully to a gut string, he discovered overtones or "small sounds" that he believed to be consonants, demonstrating that God was the sole author of sound. Although convinced of the existence of an a priori harmony beyond immediate perception, Mersenne struggled with how the world of appearances and materials can give insight into the world that exists beyond sensory perception. For example, the skills of an expert trumpeter were seen as "destroying nature," and in his work on instrument building, he did not believe that theory could be completely united with practice. His efforts to find the rules of what makes a beautiful melody ultimately remained incomplete. These different case studies demonstrate that experiences and testimonies functioned within a larger theological and epistemological system and that, when confronted with opposing evidence, the evidence was rejected, not the larger system.[1]

In addition to my investigation of the material dimension of evidence and its techniques, I have looked at who was permitted to give evidence. Building on previous research that has pointed out that Mersenne hesitated to consider the testimony of instrument builders explicitly, I have considered more broadly the role of builders, musicians, and correspondents. Chapter 3 has shown that in his correspondence, Mersenne especially addressed (amateur) scholars, collectors, and so-called *honnêtte hommes* rather than practitioners. Musicians were thought to be incapable of articulating their

1 Historian of science Hans-Jörg Rheinberger has sketched the reversal of this process as a central characteristic of science from the early modern period onward. See Rheinberger, *Spalt und Fuge*, 8–9.

own practices, and builders were encouraged to use the theories discussed in *Harmonie universelle* to acquire a better social position for themselves. The reasons for this stance on the practitioner were both social and epistemological. Mersenne looked for claims that could be universalized, whereas many forms of practical knowledge functioned in specific workflows and were bound to certain contexts and operations. And like most early modern scholars, Mersenne argued that artisans were for only concerned with their livelihood, accusing them of secrecy and self-interest. In chapter 4, I emphasized how the testimonies of certain travelers and translators were rejected or adopted. Combining these different investigations, it is possible to conclude that Mersenne aimed to establish reason over the sensory world, but he was also aware that the world of reason could not explain all experiences and materials, leading to the ambivalent role of empirical methods, experiences, and testimonies in his work.

This plurality of investigations becomes especially clear in his correspondence, where string experiments and combinatorics are discussed alongside counterpoint rules, non-European musical instruments, and ancient musical theories. Just to mention one more example, in a letter to Giovanni Battista Doni from February 1635, he listed many of his current investigations, including artisans he would seek out, images of rare instruments he received, transcriptions of Indigenous music in North America that he studied, and ancient manuscripts that he still hoped to find in libraries. In between these references, Mersenne relayed that he had thought about a book about military music but that he became repulsed by the subject:

> I do not think I will do the book on military music because some of the experiences I have made disgust me, and I do not have the ability to do the other things necessary for this subject because I would need three or four artilleries and suitable places to do ten or twelve different tests.[2]

What repulsed him specifically is difficult to tell. In his own annotated copy of *Harmonie universelle*, he included several drum patterns, such as those

2 "je ne feray pas, je croy, le livre de la musique militaire, parce que quelques experiences que j'ay faites, m'en desgoustent et que je n'ay pas la commodité de faire les aultres necessaires pour ce subject, car j'aurois besoing de 3 ou 4 artilleries, et des lieux propres pour faire dix ou douze essays differentz." Mersenne to Doni, 2 February 1635, CM 5, 34. I abbreviate the *Correspondance du P. Marin Mersenne* with CM and the number of the volume. Mersenne conducted several experiments in "in the field" ("aux champs"), possibly at the country houses of the French government officials André Lefèvre d'Ormesson and Charles de l'Aubespine, see Mersenne to Gassendi, 17 November 1635, CM 5, 483–491.

used in case of alarm or the chamade to signal to the enemy an upcoming proposition.[3] Giving up on a subject seems to have been foreign to the Minim, but Doni would have probably been relieved that Mersenne dropped the issue. In November 1634, he already advised Mersenne not to deal extensively with the topic, writing, "I would not want to talk about it so much that it would seem that the main aim [of the book] was to deal with mechanics [*mecanique*] and not music."[4] Military music and ballistics had more to do with mechanics than with music, Doni thought. Mersenne was less afraid of engaging with the subject of mechanics, even though he defended music's status as a science rather than a "mechanical art" [*art mechanique*] in his *Questions harmoniques*.[5] The experiments with cannonballs performed at the Bastille were just as much part of his investigation of harmony as they were part of the emerging science of ballistics. Media historian Friedrich Kittler has singled out Mersenne's experiments with cannons as a fundamental break with the Pythagorean tradition, writing, "For what stood as a break between Pythagoras and Mersenne went by the simple name of gunpowder."[6] Although the cannon received a privileged role in his acoustic investigations, Mersenne's attention to "unmusical" sounds was part of his broader Augustinian convictions. The difference between Mersenne and his predecessors can be explained as a matter of scale, moving from the proportions of entire strings to the proportions in individual movements of air. His quest to reduce sound to single vibrations of air questioned the entire notion of commensurability, but this was something that he never fully pursued as it could potentially question his entire argument regarding the order and harmony in the world.

Emphasizing the composite character of musical scholarship in the early modern period, I have explored different constellations of musical learning, such as mathematical, rhetorical, antiquarian, humanistic, experimental, and instrumental forms of knowledge. It was a central characteristic of Mersenne's work to mobilize different knowledge traditions within the framework of universal harmony. He was part of a culture of curious scholars who cared little for disciplinary boundaries but instead compiled many forms of knowledge, a strategy that led to the characteristic extravagance of

3 Mersenne, *Harmonie universelle*, "Livre septiesme des instrumens de percussions," between pp. 56 and 57.

4 "je ne voudrois pas aussi en parler si au long qu'il semblast que le but principal fust traicter de la mecanique et non pas de la musique." Doni to Mersenne, 8 November 1634, CM 4, 388.

5 Mersenne, *Questions harmoniques*, 80.

6 "Denn was als Zäsur zwischen Pythagoras und Mersenne lag, hörte auf den schlichten Namen Schießpulver." Kittler, "Wie der Sound von heute vor 300 Jahren enstand," 148.

his work. Although not an exhaustive investigation of Mersenne's writings nor all aspects of early modern musical science, the book demonstrates the plurality of early modern musical and acoustic investigations, exploring new avenues of research and unidentified connections. His interest in non-European musical examples discussed in chapter 4, for example, was closely related to his interest in proving the naturalness of the diatonic scale discussed in chapter 2. The acoustic investigation of echoes and overtones were in turn part of the aesthetic experience of harmony, emphasizing the divine nature of sound. Mersenne's efforts can be understood as an effort to unite all aspects of musical learning before musical learning and the science of acoustics would depart more and more from each other into the second half of the seventeenth century.

The central goal of Mersenne's work on music was to ground harmony in the natural world. Harmony was not a concept created by humans nor a physical property created by chance; it was divine and found in all objects and things. In this he belonged to a longer Pythagorean and Augustinian tradition in which music was central proof of the harmony of the universe. This idea was revitalized around 1600, with the development of mathematical and experimental methods to establish the harmony of the world more clearly. Mersenne never seemed to have abandoned these beliefs. He remained convinced that the eight consonances proscribed by common practice were part of nature, measuring strings and organ pipes and investigating overtones and overblowing. In these efforts, his work responded to an increasing philosophical skepticism, cultural relativism, and emphasis on taste. Against these trends, Mersenne aimed to defend the status of music as a privileged body of knowledge that possesses special access to the divine.

At the same time, there is a rhetorical quality to Mersenne's work. He wanted to offer his readers, courtiers, preachers, theologians, and philosophers a plethora of metaphors and examples to use in their work and social activities. Convinced that it was impossible to find true answers in earthly life, he emphasized that science should strive to understand a bit more of the divine. In his *Harmonie universelle*, he specifically aimed to show preachers, theologians, and politicians how to use harmony in their daily activities. Here Mersenne offered a wide variety of metaphors, applications, and tools to put his findings into practice. In his discussion of the interval of unison, for example, Mersenne delved into a long discussion of all types of unity. He did this not with the goal of finding any apodictic truths as to why unison was more beautiful than the other intervals but rather to meditate on the meaning of unity in a variety of fields. This aspect of Mersenne has already been touched on by others. In the words of Ferdinand Hallyn, Mersenne was preoccupied with "the associative 'fruitfulness' of the mind, the source of metaphors without referential basis."[7] Michael Fend sees a similar tendency when he describes his work as a "discussion machine."[8] As I have argued in chapter 3, many of these tendencies can be explained by Mersenne's attention to curiosity. His *Questions theologiques* contains central clues concerning Mersenne's goals and epistemological position. Aware that he would not find an ultimate answer, Mersenne stressed that scholars should touch on many different subjects. Here he was inspired by the cabinet of curiosity: Just as a king collects many different curiosities to delight, an individual should strive to collect sensory experiences to perceive a bit of the perfection that will be revealed in the afterlife. All human efforts are thus seen as a step toward the divine, with the ultimate goal of living among the chosen ones.

These different aspects of Mersenne's work led to conflicting arguments and incongruous positions. Whereas in his *Questions theologiques* he wrote that God takes pleasure in variety, in his writings on the consonances in *Harmonie universelle* Mersenne called diversity "the mother of corruption."[9] It was not the aim of this study to unite these different passages or hierarchize them. Rather, I have sought to highlight these very incongruities and demonstrate that Mersenne adopted different positions depending on the argument he wished to make. As I have shown in chapter 3, Mersenne often included explanations or theories that he did not believe himself but that he recognized as a possibility. This makes it difficult to ascribe a singular position to him; exceptions or contradictions abound. Mersenne's books seem to be the work of multiple minds, or at least one conflicted mind. Rather than attempting to streamline Mersenne's writings or account for contradictions, I have argued that it is a central feature of Mersenne's efforts to mobilize different arguments and traditions, often leading to tensions.

The period around 1600 has been long identified as a breaking point, or moment of rupture. Michel Foucault's *The Order of Things* still provides a central framework for interpreting early modern culture. By performing an archaeology of knowledge rather than a traditional history of science, Foucault locates an epistemic shift in the mid-seventeenth century. The era before is characterized by notions of resemblance, sympathy, and analogy, an age where the sign was still attached to the signified. According to Foucault,

7 Hallyn, *The Poetic Structure of the World*, 252.

8 "Diskussionsmaschine." Fend, "Probleme," 327.

9 "la mere de corruption." Mersenne, *Harmonie universelle*, "Traitez des consonances," 19.

the mid-seventeenth century saw the shift to the so-called "classical age," during which representation, the map, and the work of Descartes played a key role. Although Foucault pays little attention to music, several musicologists have applied his archaeological methods and epistemic model to music.[10] Gary Tomlinson, for example, has shown how the later musical compositions of Claudio Monteverdi seem to announce the emergence of a new episteme.[11] Brigitte van Wymeersch, Jairo Moreno, and Veit Erlmann have in different ways explored the work of René Descartes, describing a shift from a universalist cosmic view to a more listener-centered conception of music. Foucault's characterization of early modern science, in turn, has been criticized by many historians, who have shown that the emphasis on analogy and resemblance was less pervasive in the sixteenth century than Foucault had argued. With reference to Mersenne, Georg Huppert has shown that many sixteenth-century scholars combatted analogical reasoning and the idea of a natural language. The "Cartesian Mersenne," Huppert argued, continued to propagate ideas his predecessors already declared invalid.[12]

The emphasis within these works has been on the figures like Descartes and Kircher, whose work can perhaps be more easily situated along epistemic boundaries than Mersenne's. Acknowledging the different aspects of a scholar like Mersenne, this study has investigated his position as a transitional figure who complicates claims regarding epistemic shifts within a single lifetime. Although sensitive to the changing interests of Mersenne's books, I have not observed an absolute shift in his work during the early 1630s. As I have shown, several central themes recur throughout his scholarly work, such as his desire to ground the intervals in nature, a topic that returns throughout his earlier works and *Harmonie universelle*. Although the emphasis on synthesis and similarity was less present in his later works, this study has argued for existence of many continuities in his books. Much of Mersenne's work on music can still be understood in light of the response to skepticism in his earliest works. Even after 1634, Mersenne

10 Foucault, *The Order of Things*. As Gary Tomlinson writes: "Foucault's is a deaf Renaissance, it would seem, endowed with eyes but no ears." See Tomlinson, *Music in Renaissance Magic*, 59.

11 Tomlinson, *Music in Renaissance Magic*, 229–46.

12 Huppert, "Divinatio et Eruditio," 198. Although Mersenne is not extensively discussed by Foucault, George Huppert suggests that Robert Lenoble's work on Mersenne was especially important for Foucault's conceptualization of the Renaissance episteme. Huppert thinks it is likely that Foucault gathered his sources on sixteenth-century science from Lenoble's book, as both Lenoble and Foucault argue for the absence of scientific method before the seventeenth century. Among Foucault's papers at the Bilbiothèque Nationale, there are several pages with excerpts from Mersenne. Paris, Bibliothèque nationale de France, MS NAF 28730.

still remained interested in analogies and resemblances, making him the subject of critique. His rejection of the existence of a natural language, on the other hand, would be placed in the classical age.[13] These tensions demonstrates that clear epistemic lines are difficult to draw, as the science of music consisted of different practices and entangled bodies of knowledge, moving in multiple directions.

Because Mersenne struggled with the evidential role of individual experiences, some commentators go as far as suggesting that Mersenne ultimately was not interested in listening, because he had a preconceived notion of universal harmony. Others again describe him as "a pioneer in the organization of an experimental inquiry."[14] As this study has demonstrated, listening was central to many of Mersenne's musical investigations, but it had its own ambivalences. It needed to be embedded, controlled, and disciplined. When successfully done, listening was ultimately able to provide evidence of the harmonies present in the world. From delicate listening to the overtones described in chapter 1, to his conceptualization of musical listening in chapter 5, this study has argued that Mersenne was engaged in many forms of listening. Music was not solely a science of reason. Although its foundations corresponded in some manner to divinity, based on similar principles and operations, it was just as much an empirical science, involving a wide range of actors and informants. Mersenne was ultimately unable to explain taste, compositional talent, and the construction of instruments as rational endeavors. Taste and custom were issues for which musical science needed to account. In a similar way, he included categories such as "natural good" and "inspiration" in his writings on musical composition. These examples demonstrate that musical experience increasingly became something that could not be fully rationalized, pointing to the development of aesthetic sensibilities in the second half of the seventeenth century.

Paying attention to the aesthetic and scientific roles of images, Claus Zittel has argued that René Descartes developed an education of seeing. According to Zittel, Descartes disciplined the imagination of his readers and at the same time relied on visual arguments to demonstrate his natural philosophy.[15] Similarly, one might speak of an education of hearing in Mersenne's work. While it is true that universalistic scholars like Mersenne had little to say on the qualities of specific compositions or specific listening situations, he

13 Foucault, *The Order of Things*, 47.

14 For Mersenne's supposed disinterest in listening, see: Fix, "The Acoustical Paradox," 402. The citation comes from Crombie, "Marin Mersenne," 407.

15 Zittel, *Theatrum philosophicum*, 21.

admonished his readers to listen for harmony, for example, in his choice of musical examples. As Mersenne himself wrote in his introduction, not all compositions included in his *Harmonie universelle* were "as charming as one could desire." The goal of these musical examples was not to charm the reader; the examples were for "inscribing art and science in the mind and ear of its listeners."[16] The musical examples cited in his book served to remind the listeners and readers of the harmonies present in this life and the next one. His notion of universal harmony was especially a harmony of edification, demonstrating to his readers that there is a divine art and science to the musical tones that they hear in their daily lives. "It is enough to have ears" to know that there is harmony, Mersenne wrote, but listeners had to be guided to where and on how this harmony became evidential.[17]

Although the relationship between music and science in the early modern period has received increasing interest, I would like to suggest that this line of investigation can benefit from closer attention to the different techniques and modes of engaging aurally with the natural and social world. Here the history of musical science and sound could benefit from the history of science and history of knowledge, and vice versa.[18] For example, in his study on the visual culture of early modern science, Dániel Margócsy stresses that there were many different ideas surrounding the power of images.[19] Surekha Davies pushes this idea further, writing, "There was not one early modern scientific epistemology, but many."[20] As Pamela Smith summarizes, much recent research has emphasized "the plurality and coexistence of various modes of interaction with nature."[21] Map makers, merchants, antiquarians, natural philosophers, artisans, and theologians all had different ways of engaging with nature, with distinct techniques, technologies, and protocols. These works have stressed the multiplicity of early modern scholarly practices, building in turn on recent work within the history of knowledge. Knowledge is increasingly understood as something plural, with historians exploring different forms, formats, and techniques of knowing beyond canonical figures such as Galileo or Mersenne.

16 "Et si ses Compositions ne sont pas si charmantes qu'on les pourroit desirer, à raison de leur grande simplicité, dont elle a voulu user pour en faire entrer l'art & la science dans l'esprit, & dans l'oreille des auditeurs." Mersenne, *Harmonie universelle*, "Traitez des consonances," n.p. ("A Monsieur Nicolas Claudy Fabry").

17 "il suffit d'avoir des oreilles." Mersenne, *Traité*, 1.

18 See the valuable contribution by Champion and Stanyon, "Musicalising History."

19 Margócsy, *Commercial Visions*, 6.

20 Davies, *Renaissance Ethnography*, 10.

21 Smith, "Science on the Move," 358.

Although historians of the early modern period have explored these different ways of engaging with nature in a variety of contexts, much of their work remains limited to the visual realm. The emphasis on the plural forms of knowledge, however, can also be fruitfully applied to the realm of listening, namely to investigate the different ways in which sound mediated contact with the natural world. Such a study of listening in early modern science would take account of the different ways of making sense of audible phenomena. Just as there was no single way of capturing nature by images or visual technologies, there were multiple acoustemologies—to use Steven Feld's term for "what is knowable, and how it becomes known, through sounding and listening," and acoustic epistemes in the early modern period.[22] Research within Sound Studies has not remained limited to the present, but has shown how an attention to sound can uncover different histories, as sound was extremely meaningful and culturally charged.[23] Several themes outside the realm of the present study demand further exploration, such as the role of religious orders like the Minims, commerce, colonial exploitation, and pedagogy. Whereas Mersenne's correspondence contains hardly any references to purchasing musical instruments, the letters of his friend Constantijn Huygens mention the growing international commerce in viols and other instruments. The allusions to Maltese songs played in Paris and the presence of Ottoman musicians in southern France suggest a world of Mediterranean music making that has only recently received attention. An investigation that centers materials such as gut, wood, and metal would similarly uncover new narratives and avenues of research in the relationship between artisans, players, and scholars. From the goldsmiths who judged the quality of metals using their ears, noted by the English natural philosopher Walter Charleton, or the many travelers who enjoyed the "divers tunes" in "the fields abroad," seventeenth-century actors continued to made sense of the world through sound.[24] As the interrelation of musical aesthetics, industry, and cultural encounters continues to be explored in studies on the nineteenth and twentieth centuries, the cultural archive of the early modern period offers a wealth of examples to see how these different domains were already in conversation in the early modern period.

22 Feld, "Acoustemology."

23 Atkinson, *The Noisy Renaissance*; Biddle and Gibson, *Cultural Histories*; Knighton and Mazuela-Anguita. *Hearing the City*; Linhart Wood, *Sounding Otherness*; Hammond, *The Powers of Sound and Song*; Wilbourne and Cusick, *Acoustemologies*.

24 Charleton, *Physiologia Epicuro-Gassendo-Charltoniana*, 211; Moryson, *An Itenerary*, 3:10.

The sources surrounding Mersenne's efforts already demonstrate the kaleidoscopic nature of early modern culture, but within these multiple trajectories and forms of knowledge, Mersenne directed his work toward his idiosyncratic notion of harmony. He circled around harmony, opting for various methods, different sounds, and listening techniques to approach this ungraspable phenomenon. Over the course of his works, he seemed increasingly aware of the noise and disorder of the early modern world, but also of his inability to explain certain sounds and acoustic phenomena as harmonious. But even in the roaring of cannons or the noise through the windows of his convent, Mersenne was able to find harmony. He remained convinced of his fundamental tenets: God made the material of sound, and it therefore must be studied to understand the divine. Music, as the study of sound, mediates between the divine and the material world, with harmony as its fundamental instrument.

Bibliography

Manuscript sources

Paris, Bibliothèque nationale de France, MS NAF 28730. Fonds Michel Foucault.

Printed primary sources

Charleton, Walter. *Physiologia Epicuro-Gassendo-Charltoniana* [...]. London: Heath, 1654.

Mersenne, Marin. *Correspondance du P. M. Mersenne*. Edited by Paul Tannery, Cornelis de Waard, Bernard Rochot, Armand Beaulieu, and René Pintard. Paris: Centre National de la Recherche Scientifique et al., 1932–88.

Mersenne, Marin. *Harmonie universelle contenant la théorie et la pratique de la musique (Paris, 1636). Édition fascimilé de l'exemplaire conserve à la Bibliothèque des Arts et Métiers et annoté par l'Auteur.* Introduction by François Lesure. Paris: Éditions du Centre National de la Recherche Scientifique, 1963.

Mersenne, Marin. *Questions harmoniques: Dans lesquelles sont contenuës plusieurs choses remarquables pour la physique, pour la morale, et pour les autres sciences.* Paris: Villery, 1634.

Moryson, Fynes. *An Itenerary written by Fynes Moryson Gent. First in the Latine tongue, and then translated by him into English* [...]. London: Beale, 1617.

Secondary literature

Atkinson, Niall. *The Noisy Renaissance: Sound, Architecture, and Florentine Urban Life*. University Park: Pennsylvania State University Press, 2016.

Biddle, Ian, and Kristen Gibson. *Cultural Histories of Noise, Sound and Listening in Europe*. London/New York: Routledge, 2017.

Champion, Matthew S., and Miranda Stanyon. "Musicalising History." *Transactions of the Royal Historical Society* 29 (2019): 79–103.

Crombie, Alistair C. "Marin Mersenne and Scientific Acceptability." In *Science, Optics and Music in Medieval and Early Modern Thought*, 399–417. London: Hambledon Press, 1990.

Davies, Surekha. *Renaissance Ethnography and the Invention of the Human: New Worlds, Maps, and Monsters*. Cambridge: Cambridge University Press, 2016.

Duncan, David Allen. "Persuading the Affections: Rhetorical Theory and Mersenne's Advice to Harmonics Orators." In *French Musical Thought, 1600–1800*, edited by Georgia Cowart, 149–175. Ann Arbor: UMI Research Press, 1989.

Feld, Steven. "Acoustemology." In *Keywords on Sound*, edited by David Novak and Matt Sakaneeny, 12–21. Durham, NC: Duke University Press, 2015.

Fend, Michael. "Probleme mit der Idee der 'harmonia universalis' in der frühen Neuzeit." *Archiv für Musikwissenschaft* 71, no. 4 (2014): 307–34.

Fix, Adam. "The Acoustical Paradox: How Music Was Unmoored and Set Adrift from Sciences in the Seventeenth Century." In *Music and Science from Leonardo to Galileo*, 377–420. edited by Rudolf Rasch. Turnhout: Brepols, 2022.

Foucault, Michel. *The Order of Things: An Archaeology of the Human Sciences*. London/New York: Routledge, 2002.

Hallyn, Fernard. *The Poetic Structure of the World: Copernicus and Kepler*. New York: Zone Books, 1990.

Hammond, Nicholas. *The Powers of Sound and Song in Early Modern Paris*. University Park: Pennsylvania State University Press, 2020.

Huppert, George. "Divinatio et Eruditio: Thoughts on Foucault." *History and Theory* 13, no. 3 (1974): 191–207.

Kittler, Friedrich. "Wie der Sound von heute vor 300 Jahren enstand." In *Weltbürgertum und Globalisierung*, edited by Norbert Bolz, Friedrich Kittler, and Raimar Zons, 145–55. München: Fink, 2000.

Knighton, Tess, and Ascensión Mazuela-Anguita. *Hearing the City in Early Modern Europe*. Turnhout: Brepols, 2018.

Linhart Wood, Jennifer. *Sounding Otherness in Early Modern Drama and Travel: Uncanny Vibrations in the English Archive*. Cham: Palgrave Macmillan, 2019.

Margócsy, Daniel. *Commercial Visions: Science, Trade, and Visual Culture in the Dutch Golden Age*. Chicago: University of Chicago Press, 2014.

Rheinberger, Hans-Jörg. *Spalt und Fuge: Eine Phänomenologie des Experiments*. Berlin: Suhrkamp, 2021.

Smith, Pamela H. "Science on the Move: Recent Trends in the History of Early Modern Science." *Renaissance Quarterly* 62, no. 2 (2009): 345–75.

Tomlinson, Gary. *Music in Renaissance Magic: Toward a Historiography of Others.* Chicago/London: University of Chicago Press, 1993.
Wilbourne, Emily, and Suzanne G. Cusick. *Acoustemologies in Contact: Sounding Subjects and Modes of Listening in Early Modernity.* Cambridge, UK: Open Book Publishers, 2021.
Zittel, Claus. *Theatrum philosophicum: Descartes und die Rolle ästhetischer Formen in der Wissenschaft.* Berlin: Akademie Verlag, 2009.

Appendices

Appendix A: Composition of *Harmonie universelle*

This table collects the composition of *Harmonie universelle*, including all paratextual material and dedications, constructed from Mersenne's own copy, printed in facsimile by François Lesure (see bibliography). Pages without page numbers are designated by roman numerals, prop. refers to the number of propositions. I would like to thank Rudolf Rasch for his help composing this and the following table.

	Prop.	Page numbers
Harmonie universelle, contenant la théorie et la pratique de la musique		
Harmonie universelle		iii
Première préface générale au lecteur		i–xvi
Table des propositions des dix-neuf livres de l'Harmonie universelle		xvii–xxxii
Table des xix. livres de musique		xliii–li
Première observation		lii
Seconde observation		liii–liv
Traitez de la nature des sons, et des mouvemens de toutes sortes de corps		
A [...] Monseigneur Louis de Valois, Conte d'Alais (dedication)		lii–vi
Préface au lecteur		vii–viii
Livre premier de la nature & des proprietez du son	34	1–84
Livre second du mouvement des corps	22	85–156
Livre troisiesme des mouvemens & du son des chordes	23	157–228
Advertissement au lecteur		[229]
[Gilles Personne de Roberval], Traité de méchanique	3	1–36
Traitez de la voix, et des chants		
A Monsieur [Jacques] Hallé, Seigneur de Boucqueval (dedication)		iii–v
Preface au lecteur		vi–vii
Fautes survenües en l'impression		vii–viii
Livre premier de la voix	53	1–88
Livre second, des chants	27	89–180
Traitez des consonances, des dissonances, des genres, des modes, & de la composition		
A Monsieur Nicolas Claude Fabry, Sieur de Peiresc et de Callas (dedication)		iii–vi
Préface, & advertissement au lecteur		vii–xi
Fautes survenües en l'impression		xii
Traitez des consonances [Livre premier des consonances]	40	1–112

	Prop.	Page numbers
Livre second des dissonances	14	113–40
Livre troisiesme des genres de la musique	19	141–96
Livre quatriesme de la composition	27	197–282
Livre cinquiesme de la composition	12	283–330
Livre sixiesme de l'art de bien chanter	34	331–440
Traité des instrumens à chordes		
A Monsieur [Henri] de Refuge (dedication)		iii–v
Préface au lecteur		vi–viii
Livre premier des instruments	29	1–44
Livre second des instruments à chordes	16	45–100
Livre troisiesme des instrumens à chordes	27	101–76
Livre quatriesme des instrumens à chordes	18	177–228
Livre cinquiesme des instrumens à vent	35	225–308
A Monsieur [Étienne] Pascal (dedication)		i–ii
Préface au lecteur		iii–iv
Livre sixiesme des orgues	45	309–412
Livre septiesme des instrumens de percussion	31	1–62
Éloge de Jacques Mauduit, excellent musicien		63–72
Fautes de l'impression, & quelques advis		73–79
Livre de l'utilité de l'harmonie, & des autres parties des mathématiques	18	1–68
Nouvelles observations physiques et mathématiques	14	1–27
Advertissement		28

Appendix B: Composition of *Harmonicorum libri XII*

This table collects the composition of *Harmonicorum libri XII* from the facsimile published by Minkoff in 1972 (see bibliography) and supplemented with materials from the 1636 edition that do not appear in the 1648 edition. Prop. refers to the number of propositions.

	Prop.	Page numbers
Harmonicorum libri XII		
Illustri Viro Henrico Ludovico Haberto Mommoro (dedication)		i–iv
Henrico Ludovico Haberto Mommoro		i–ii
Praefatio ad eundem	4	i–viii
Liber primus. De natura, & proprietatibus sonorum	25	1–9
Tabula propositionum libri secondi		9
Liber secundus, de causis sonorum	43	10–34
Tabula propositionum libri tertii		35
Liber tertius. De fidibus, nervis et chordis	22	36–49
Tabula propositionum libri quarti		50
Liber quartus. De sonis consonis, seu consonantiis	29	51–67
Tabula propositionum libriquinti		68–69
Liber quintus. De musicae dissonantiis, de rationibus, et proportionibus	40	69–88
Tabula propositionum libri sexti		89
Liber sextus. De speciebus consonantiarum, deque modis, et generibus	26	90–112
Tabula propositionum libri septimi		113
Liber septimus. De cantibus, seu cantilenis	19	114–57
Liver octavus. De compositione musica	27	161–84
Liber novus praelusorius (only in the 1648 edition)		1–4
Harmonicorum instrumentorum libri IV		
Harmonicorum instrumentorum libri IV (only in the 1636 edition)		i–ii
Nobilissimo Viro Nicolae Claudio Fabry, Peirescii (dedication) (only in the 1636 edition)		i
Praefatio ad lectorem amicum (only in the 1636 edition)		i
Liber primus. De instrumentis harmonicis	44	1–72
Liber secundus. De instrumentis pneumaticis	21	73–112
Liber tertius. De organis, campanis, tympanis	30	113–44
Liber quartus. De campanis	20	145–68

Acknowledgments

I would first of all like to thank Sebastian Klotz for welcoming me into the Department of Musicology and Media Studies at the Humboldt University of Berlin and supporting my work ever since I set foot in his office. His curiosity and precision sharpened the book's argument and has helped me to keep my eye on the bigger picture. The colloquium at the chair of Transcultural Musicology and Historical Anthropology and the annual seminar in the picturesque Rheinsberg were the start of many new ideas and valuable conversations, and I will continue to cherish my time there. Viktoria Tkaczyk made it possible for me to come to Berlin and pursue my doctorate, something for which I will be forever grateful. Starting in our days together in Amsterdam, she has stimulated my work with generosity and wisdom. This book would not have been possible without her continuous support, and I would like to thank her for her enthusiasm and careful feedback over the years.

The institutional framework of the Collaborative Research Centre (CRC) 980 "Episteme in Motion: Transfer of Knowledge from the Ancient World to the Early Modern Period" at the Freie Universität Berlin and of the Max Planck Institute for History of Science was an exhilarating context in which to begin this project. From the CRC, I would especially like to thank Kristiane Hasselman and my colleagues from the working group "Medium and Materiality." I would also like to thank Dagmar Schäfer and all of Department III at the Max Planck Institute for the History of Science for their generous feedback on my project and the time and resources to finish my dissertation project. Work on the finishing stages were made possible by a fellowship at the Descartes Centre for the History and Philosophy of the Sciences and the Humanities at Utrecht University in 2021, for which I must thank Sven Dupré and Bert Theunissen. In the last two years, I was able to continue refining this work in the research group "Visualizing Science in Media Revolutions" at the Bibliotheca Hertziana – Max Planck Institute for Art History in the context of my project BH-P-22-17.

As soon as I arrived in Berlin, Viktoria Tkaczyk's Max Planck Research Group "Epistemes of Modern Acoustics" became an intellectual home filled with good company and inspiring researchers. I would like to thank Joeri Bruyninckx, Leon Chisholm, Fanny Gribenski, Anna Laqua, Birgitta von Mallinckrodt, João Romão, Lotte Schüßler, Kate Sturge, and Hannah Wiemer and for their attentive comments, support, and general good spirits over all these years. Sharing office V006 with João and Lotte was a delight. At the

Humboldt University of Berlin, I similarly found a thoughtful and generous community through the colloquiums at the chair of Transcultural Musicology and Media and Knowledge. I would like especially thank Christopher Klauke, Mona Mirkamali, Razieh-Sadat Mousavi, and Sebastian Schwesinger for their feedback and encouragement.

From Berlin I took the project to Rome to the Bibliotheca Hertziana. I would like to thank Sietske Fransen for welcoming me to her research group in Rome and the support she has demonstrated. Having her as an interlocutor, boss, and an all-around kind colleague has been invaluable. I would also like to thank Marie-Pierre Dabadie and Mathis Goddet from the Emmanuel Community at Trinità dei Monti for their hospitality. Visiting Trinità made the subject matter of my research far more palpable than I could ever hope.

Many people provided critical feedback at various stages in conversations and writing. Most of the chapters of this book were presented at some stage at the colloquiums of the Max Planck Institute for the History of Science and Humboldt University, and I would like to thank all participants for their time and active support. Rebekah Ahrendt, Eric Bianchi, Emily Kent, and Carmel Raz generously commented upon previous versions of some of the chapters. Sonja Brentjes, Floris Cohen, Flora Dennis, Judith Haug, Noa Hegesh, Julia Kursell, Jan Lazardzig, Karsten Mackensen, Alejandro Nodarse, Théodora Psychoyou, Jacomien Prins, Rose Pruiksma, Rudolf Rasch, Ivana Rentsch, and Skúli Sigurdsson have all stimulated ideas, revisions, and new rabbit-holes. Their conversations, suggestions, and own work have been a perpetual source of inspiration.

This work would not have been possible without the help of many librarians at the various institutions where I had the privilege to conduct my research. I would especially like to thank the librarians of the Bibliothèque nationale de France in Paris and the Bibliothèque Inguimbertine in Carpentras for their help locating sources and their generosity in allowing me to reuse some of their materials. At the Max Planck Institute for the History of Science, I would especially like to thank Ruth Kessentini and her team, whose skill in tracking down difficult-to-find books was immensely helpful. Julia Steinmetz and Rafaela Arregui provided further assistance with locating references and helping to transcribe sources. Their help was invaluable. I would also like to thank Andreas Freytag-Hill for transcribing and translating the Arabic material in Peiresc's music manuscripts and Robbert-Jan Sebregts for his help with creating the music examples.

In the final stages of this project, I had the pleasure to co-organize a conference on Marin Mersenne with Emily Kent. I would like to thank all the participants for their openness and willingness to share ideas and sources. In 1934 the composer Harry Partch asked the early music performer Arnold Dolmetsch about a particular Mersenne edition, upon which Dolmetsch exclaimed: "For twenty years I have been talking about Mersenne, and nobody even knows who I am talking about."[1] Although my preoccupation with Mersenne at times felt like a lonely endeavor, Dolmetsch's problem was not my problem and I have encountered a wonderful and generous community interested in his work. I hope that future studies can correct the faults in mine.

Stefano Gulizia, Scientiae Studies series editor, and Erika Gaffney at Amsterdam University Press have expertly supported and guided the project throughout the entire process. I would like to thank them, the series board, and the anonymous reader for the helpful criticism and suggestions for improvement.

Since my first steps into musicology in Amsterdam, Katty, Kelsey, Kjell, Lydia, Mari, Pieter, and Robbert-Jan have accompanied me with insights and, crucially, distractions. Tom and Editha made Berlin a second home. My piano teachers Cathelijne and Andor showed me the joy not only in playing music but also talking about it. My sisters and my parents have stimulated my studies since the beginning, and I would like to dedicate the book this book to them, Jan, Heleen, Hanna, and Tessa.

And finally, Lea Greenberg has read every word and made it better. More than that, her companionship and joy have kept me afloat. I cannot thank her enough.

1 Harry Partch, *Bitter Music: Collected Journals, Essays, Introductions, and Librettos*, edited by Thomas McGeary (Urbana: University of Illinois Press, 1991), 27.

Bibliography

Manuscript sources

Carpentras, Bibliothèque Inguimbertine

MS 1774. Papers of Nicolas-Claude Fabri de Peiresc.

MS 1831. Papers of Nicolas-Claude Fabri de Peiresc.

MS 1874. Letters of Nicolas-Claude Fabri de Peiresc.

Oxford, Bodleian Library

MS Tanner 204. Edmund Chilmead. "An Examination."

Paris, Bibliothèque de l'Arsenal

MS 2884. Marin Mersenne. "Livre de la nature des sons."

Paris, Bibliothèque nationale de France

MS Dupuy 668. A volume of seventeenth-century letters by various authors.

MS. fr. 6205. Letters of Marin Mersenne.

MS fr. 9119. Papers of Gilles Personne de Roberval.

MS fr. 9530. Papers of Nicolas-Claude Fabri de Peiresc.

MS fr. 9531. Papers of Nicolas-Claude Fabri de Peiresc.

MS fr. 12357. Marin Mersenne's annotations in his *Harmonie universelle.*

MS fr. 24256. Manuscript with melodies of eight notes, attributed to Marin Mersenne.

MS NAF 28730. Fonds Michel Foucault.

Paris, Bibliothèque Mazarine

MS 4146. Catalogue of the library of the Minim convent in Paris from the mid-seventeenth century.

Paris, Observatoire de Paris

MS a, b, 5, 11, 351–64. Ismael Boulliau's manuscript on music.

Printed primary sources

Agricola, Martin. *The "Musica instrumentalis deudsch" of Martin Agricola: A Treatise on Musical Instruments, 1529 and 1545*. Translated and edited by William E. Hettrick. Cambridge: Cambridge University Press, 1994.

Alsted, Johannes Heinrich. *Encyclopaedia septem tomis distincta.* Herborn: n.p., 1630.

Aristotle. *Politics: With an English Translation by H. Rackman.* London: Heinemann, 1959.

Aristotle. *Posterior Analytics.* Translated by Jonathan Barnes. Oxford: Oxford University Press, 1993.

Aristotle. *On the Soul and Other Psychological Works.* Translated by Fred D. Miller, Jr. Oxford: Oxford University Press, 2018.

Augustine. *Confessions.* Translated by Henry Chadwick. Oxford: Oxford University Press, 1991.

Augustine. *Earlier Writings.* Edited and translated by J. H. S. Burleigh. Louisville: Westminster John Knox Press, 2006.

Augustine. *On the Free Choice of the Will, On Grace, and Free Choice.* Edited and translated by Peter King. Cambridge: Cambridge University Press, 2010.

Augustine. *De musica liber VI. A Critical Edition with a Translation and an Introduction by Martin Jacobsson.* Stockholm: Almqvist & Wiksell, 2002.

Augustine. *The Trinity.* Translated by Stephen McKenna. Washington, DC: Catholic University of America Press, 1963.

Bacon, Francis. *Advancement of Learning and New Atlantis.* Edited by Arthur Johnston. Oxford: Clarendon Press, 1980.

Baillet, Adrien. *Vie de Monsieur Des-Cartes, seconde partie.* Paris: Daniel Horthemels, 1691.

Ban, Joan Albert. *Zangh-bloemzel & kort sangh-bericht.* With an introduction by Frits Noske. Amsterdam: Knuf, 1973.

Barker, Andrew. *Greek Musical Writings II: Harmonic and Acoustic Theory.* Vol. 2 of 2 vols. Cambridge: Cambridge University Press, 1989.

Beeckman, Isaac. *Journal tenu par Isaac Beeckman de 1603 à 1634.* 4 Vols. Edited by Cornelis de Waard. The Hague: Martinus Nijhoff, 1939–53.

Belon, Pierre. *Les observations de plusieurs singularitez et choses memorables trouvées en Grèce, Asie, Judée, Egypte, Arabie et autres pays étrangèrs.* Antwerp: Steelsius, 1555.

Bergier, Nicolas. *La musique speculative.* Edited and translated by Ekkehard Jost. Cologne: Arno Volk Verlag, 1970.

Bettini, Mario. *Apiaria universæ philosophiæ mathematicæ. In quibus paradoxa et nova pleraque machinamenta ad usus eximios traducta, & facillimis demonstrationibus confirmata. Tomus Secundus.* Vol. 2 of 2 vols. Bologna: Ferroni, 1642.

Binet, Étienne [Rene Francois]. *Essay des merveilles de nature, et des plus nobles artifices.* Rouen: Osmon, 1623.

Bodin, Jean. *The Six Bookes of a Common-Weale. Out of the French and Latine copies, Done into English, by Richard Knolles.* London: Islip, 1606.

Boethius, Anicius Manlius Severinus. *Fundamentals of Music.* Translated by Calvin M. Bower, edited by Claude V. Palisca. New Haven: Yale University Press, 1989.

Bottrigari, Hercole. *Il Desiderio, or Concerning the Playing Together of Various Musical Instruments*. Translated by Carol MacClintock. Rome: American Institute of Musicology, 1962.

Bourgoing, Francis. *Le David françois, ou traité de la saincte psalmodie, contenant plusieurs advis ou enseignements méthodiques, pour ceux qui sont appliquez par estat à chanter les loüanges de Dieu. Avec un petit traité touchant la saincte messe*. Paris: Huré, 1641.

Braithwaite, William. *Methodus nova docendi discendique veterem (ut ita dicam) musicam* [...]. London: Norton, 1639.

Cassiodorus. *Institutions of Divine and Secular Learning and on the Soul*. Translated by James W. Halporn and Mark Vessey. Liverpool: Liverpool University Press, 2004.

Chapard, François. *Fabriciani cimeliarchii promptuarium triceps*. Aix: Roize, 1647.

Charleton, Walter. *Physiologia Epicuro-Gassendo-Charltoniana* [...]. London: Heath, 1654.

Cicero, Marcus Tullius. *On the Republic and The Laws*. Translated with an introduction, notes, and indexes by David Fott. Ithaca/London: Cornell University Press, 2014.

Colonna, Fabio. *La sambuca lincea overo dell'istromento musico perfetto. Con annotazioni critiche manoscritte di Scipione Stella (1618–1624) (rist. anast. Napoli, 1618)*. Edited by Patrizio Barbieri. Lucca: Libreria musicale italiana, 1991.

De Caus, Salomon. *Institution harmonique*. Frankfurt: Jan Norton, 1615. Reprint, Geneva: Minkoff, 1980.

De Coste, Hilarion. *La vie du R.P. Marin Mersenne*. Paris: Cramoisy, 1649.

De Cousu, Antoine. *La musique universelle, contenant toute la pratique, et toute la theorie*. Paris: Ballard, 1658. Reprint, Geneva: Minkoff, 1972.

De Gouy, Jacques. *Airs à quatre parties, sur la paraphrase des pseaumes de Messire Antoine Godeau, évesque de Grasse*. Paris: Ballard, 1650.

De Lery, Jean. *Histoire d'un voyage faict en la terre du Brésil, autrement dite Amerique*. Geneva: s.n., 1585.

De Paule, François. *Les regles des freres, des soeurs, et des fideles de l'un & de l'autre sexe de l'Ordre des Minimes*. Paris: Martin, 1632.

De Paule, François. *Regulae fratrum et sororum ac fidelium utriusque sexus Ordinis Minimorum*. Paris: Dehors, 1631.

Della Valle, Pietro. "Della musica dell'età nostra." In Giovanni Battista Doni, *Lyra Barberini*, vol 2 of 2 vols., 249–64. Florence: Stamperia imperiale, 1763.

Descartes, René. *Compendium musicae*. Utrecht: Van Zijl, 1650.

Descartes, René. *Musicae compendium/Leitfaden der Musik*. Edited by Johannes Brockt. Darmstadt: Wissenschaftliche Buchgesellschaft, 1992.

Descartes, René. *Oeuvres completes III: Correspondance, janvier 1640–juin 1643*. Vol. 3 of 11 vols. Edited by Charles Adam and Paul Tannery. Paris: Vrin, 1975.

Descartes, René. *The Philosophical Writings of Descartes*. Vol. 1 of 2 vols. Translated by John Cottingham, Robert Stoothoff, and Dugald Murdach. Cambridge: Cambridge University Press, 1985.

Dony-d'Attichy, Louis. *Histoire generale de l'ordre sacre des Minimes. Divisee en huict livres*. Paris: Cramoisy, 1624.

Du Caurroy, Eustache. *Fantasies a III. IIII. V. et VI. parties*. Paris: Ballard, 1610.

Du Caurroy, Eustache. *Preces ecclesticae ad numeros musices redactae*. Paris: Ballard, 1609.

Du Vivier, Claude. *Vie et miracles de Sainct Francois de Paule Instituteur de Lordre des Freres Minimes*. Paris: Cramoisy, 1609.

Evelyn, John. *The Diary of John Evelyn*. Oxford: Oxford University Press, 1960.

Faret, Nicolas. *L'honneste-homme ou, l'art de plaire a la court*. Paris: Du Bray, 1630.

Fludd, Robert. *The Temple of Music*. Translated by Peter Hauge. London/New York: Routledge, 2017.

Furetière, Antoine. *Dictionnaire universel contenant généralement tous les mots françois tant vieux que modernes* [...]. The Hague/Rotterdam: Arnout & Reinier Leers, 1690.

Gaffarel, Jacques. *Curiositez inouyes sur la sculpture talismanique des Persans, horoscope des Patriarches, et lecture des estoilles*. Paris: Masnil, 1629.

Gaffurius, Franchinus. *Theorica musicae*. Milan: Mantegatium, 1492.

Galilei, Vincenzo. *Dialogue on Ancient and Modern Music*, translated by Claude V. Palisca. New Haven/London: Yale University Press, 2003.

Galilei, Vincenzo. "Discorso di Vicentio Galilei intorno all'uso dell enharmonio." In *Die Kontrapunkttraktate Vincenzo Galileis*, edited by Frieder Rempp, 163–80. Cologne: Gerig, 1980.

Galilei, Vincenzo. *Discorso intorno all'opera di* [...] *Zarlino*. Florence: Mareschotti, 1589.

Galilei, Vincenzo. "A Special Discourse Concerning the Diversity of the Ratios of the Diapason." In *The Florentine Camerata: Documentary Studies and Translations*, edited and translated by Claude V. Palisca, 180–97. New Haven: Yale University Press, 1989.

Gantez, Annibal. *L'entretien des musiciens*. Auxerre, 1643. Reprint, Geneva: Minkoff, 1971.

Gassendi, Pierre. *The Mirrour of True Nobility and Gentility Being the Life of the Renowned Nicolaus Claudius Fabricius, Lord of Pieresk, Englished by W. Rand*. London: Streater for Moseley, 1657.

Gassendi, Pierre. *Iniation à la théorie ou partie speculative de la musique*. Introduction, translation, and notes by Patrice Bailhache. Turnhout: Brepols, 2005.

Gassendi, Pierre. *Opera omnia*. 6 Vols. Lyon: Anisson, 1658.

Gay, Godefroy. *Ode sur les raretés du cabinet de Pierre Trichet, advocat bourdelois*. N.p.: s.n., ca. 1635.

Haranger, René. *Articles, statuts, ordonnances et réglemens de la Communauté des Vergetiers, Raquetiers et Brossiers* [...]. Paris: Grou, 1659.

Hobbes, Thomas. *Angli Malmesburiensis philosophi vita*. London: Cooke, 1682.

Hobbes, Thomas. *Leviathan*. Edited by J. C. A. Gaskin. Oxford: Oxford University Press, 2008.

Huygens, Christiaan. *Oeuvres complètes. Tome XIII: Dioptrique*. Edited by Diederik Johannes Korteweg. The Hague: Nijhoff, 1916.

Huygens, Christiaan. "Origine de chant." In *Oeuvres complètes: Musique et Mathématique*, vol. 20 of 22 vols., edited by Johan Adriaan Volgraff, 30–37. The Hague: Nijhoff, 1940.

Huygens, Constantijn. *Driehonderd brieven over muziek, van, aan en rond Constantijn Huygens*. Edited by Rudolf Rasch. 2 vols. Hilversum: Verloren, 2007.

Jurgens, Madeleine. *Documents du minutier central concernant l'histoire de la musique (1600–1650), Volume 1*. Vol. 1 of 2 vols. Paris: S.E.V.P.E.N., 1967.

Kepler, Johannes. *The Harmony of the World*. Translated by E. J. Aiton, A. M. Duncan, and J. V. Field. Philadelphia: American Philosophical Society, 1997.

Kircher, Athanasius. *Musurgia universalis sive ars magna consoni et dissoni in X libros digesta*. Rome: Corbelletti, 1650.

La Mothe le Vayer, François, de. "Discours sceptique sur la musique." In Marin Mersenne, *Questions harmoniques. Dans lesquelles sont contenuës plusieurs choses remarquables pour la physique, pour la morale, et pour les autres sciences*, 84–165. Paris: Villery, 1634.

La Mothe le Vayer, François de. "Des doutes raisonnes." In *Oeuvres des François de la Mothe le Vayer, conseiller d'estat ordinaire*, vol 12 of 15 vols., 309–317. Paris: Billaine, 1669.

La Mothe le Vayer, François de. *Opuscule ou petit traité sceptique sur cette commune façon de parler: N'avoir pas le sens commun*. Paris: Sommaville, 1646.

La Mothe le Vayer, François de. *Petit discours de l'immortalité de l'ame. Avec le corollaire, & un discours sceptique sur la musique. Seconde edition*. Paris: Camusat, 1640.

La Mothe le Vayer, François de. "La physique du Prince." In *Oeuvres des François de la Mothe le Vayer, Tome Premier*, vol. 1 of 2 vols., 933–1015. Paris: Courbe, 1662.

La Primaudaye, Pierre de. *The French Academie Fully Discoursed and Finished in Foure Bookes*. London: Legat for Adams, 1618.

Le Jeune, Claude. *Dodecacorde contenant douze pseaumes de David, mis en musique selon les douze modes, approuvez des meilleurs autheurs anciens & moderns*. La Rochelle: Haultin, 1598.

Le Jeune, Claude. *Le Printemps*. Paris: Ballard, 1603.

Lescarbot, Marc. *Histoire de la Nouvelle France. Contenant les navigations, découvertes, & habitations faites par les François és Indes Occidentales & Nouvelles-France ...* . Paris: Milot, 1609.

Lippius, Johannes. *Synopsis musicae novae omnino verae atque methodociae universae.* Strasbourg: Ledertz, 1612. Reprint, Olms: Hildesheim, 2004.

Locke, John. *The Locke Reader: Selections from the Works of John Locke with a General Introduction and Commentary.* Edited by John W. Yolton. Cambridge: Cambridge University Press, 1977.

Lodewijcksz, Willem. *Prima pars descriptionis itineris navalis in Indiam Orientalem ….* Amsterdam: Nicolai, 1598.

Masson, Charles. *Nouveau traité des règles pour la composition de la musique.* Paris: Ballard, 1705.

Maugars, André. *Response faite à un curieux, sur le sentiment de la musique d'Italie.* Paris: s.n., 1639–40.

Mersenne, Marin. *Cogitata physico-mathematica.* Paris: Bertier, 1644.

Mersenne, Marin. *Correspondance du P. M. Mersenne.* Edited by Paul Tannery, Cornelis de Waard, Bernard Rochot, Armand Beaulieu, and René Pintard. Paris: Centre National de la Recherche Scientifique, et al., 1932–88.

Mersenne, Marin. *Harmonicorum libri XII.* Paris: Baudry, 1648. Reprint, Geneva: Minkoff, 1972.

Mersenne, Marin. *Harmonie universelle contenant la théorie et la pratique de la musique (Paris, 1636). Édition fascimilé de l'exemplaire conserve à la Bibliothèque des Arts et Métiers et annoté par l'auteur.* Introduction by François Lesure. Paris: Éditions du Centre National de la Recherche Scientifique, 1963.

Mersenne, Marin. *L'impiété des deistes, et des plus subtils libertins découverte, & refute par raisons de theologie, & de philosophie.* Paris: Billaine, 1624.

Mersenne, Marin. *Les mechaniques de Galilée: mathematicien & ingenieur du duc de Florence.* Paris: Guenon, 1634.

Mersenne, Marin. *Observationes et emendationes ad Francisci Giorgii Veneti problemata.* Paris: Cramoisy, 1623.

Mersenne, Marin. *Les preludes de l'harmonie universelle ou questions curieuses. Utiles aux predicateurs, aux theologiens, aux astrologues, aux medicins & aux philosophes.* Paris: Guenon, 1634.

Mersenne, Marin. *Quaestiones celeberrimae in Genesim, cum accurata textus explicatione: in hoc volumine athei, et deistae impugnantur & expugnantur, & vulgata edition ab hereticorum calumnijs vindicator. Graecorum, & Hebraeorum musica instauratur.* Paris: Cramoisy, 1623.

Mersenne, Marin. *Questions harmoniques. Dans lesquelles sont contenuës plusieurs choses remarquables pour la physique, pour la morale, et pour les autres sciences.* Paris: Villery, 1634.

Mersenne, Marin. *Questions inouyes, ou recreation des scavans.* Paris: Villery, 1634.

Mersenne, Marin. *Les questions theologiques, physiques, morales, et mathematiques. Où chacun trouvera du contentement ou de l'exercice.* Paris: Guenon, 1634.

Mersenne, Marin. *Traité de l'harmonie universelle*. Paris: Baudry, 1627.
Mersenne, Marin. *L'usage de la raison, texte revu par Claudio Buccolini*. Paris: Fayard, 2002.
Mersenne, Marin. *La verité des sciences, contre les sceptiques ou pyrrhoniens*. Paris: Du Bray, 1625.
Montaigne, Michel de. *The Complete Works of Montaigne*. Translated by Donald Frame. Stanford: Stanford University Press, 1958.
Moryson, Fynes. *An Itenerary written by Fynes Moryson Gent. First in the Latine tongue, and then translated by him into English* [...]. London: Beale, 1617.
Niceron, Jean-François. *La perspective curieuse ou magie artificielle des effets merveilleux*. Paris: Billaine, 1638.
Nicot, Jean. *Thresor de la langue françoyse, tant ancienne que moderne*. Paris: Douceur, 1606.
Nivers, Guillaume Gabriel. *Traité de la composition de musique*. Paris: Ballard, 1667.
Parran, Antoine. *Traité de la musique théorique et pratique, contenant les préceptes de la composition*. Paris: Ballard, 1639. Reprint, Geneva: Minkoff, 1972.
Pascal, Blaise. *Histoire de la roulette, appellée autrement la trochoïde, ou la cycloïde* [...]. N.p.: s.n., 1658.
Peiresc, Nicolas Claude-Fabri de. *Les correspondants de Peiresc, XV: Lettres inédites écrites de Tunis à Peiresc*. Vol. 15 of 21 vols. Edited by Philippe Tamizay de Larroque. Algiers: Adolphe Jourdain, 1889.
Peiresc, Nicolas-Claude Fabri de. *Lettres a Claude Saumaise et a son entourage (1620–1637)*. Edited by Agnès Bresson. Florence: Olschki, 1992.
Peiresc, Nicolas-Claude Fabri de Peiresc. *Lettres de Peiresc*. Vol. 4 of 7 vols.: *Lettres de Peiresc à Borrilly, à Bouchard et à Gassendi. Lettres de Gassendi à Peiresc, 1626–1637*. Edited by Philippe Tamizey de Larroque. Paris: Imprimerie Nationale, 1893.
Peiresc, Nicolas-Claude Fabri de. *Lettres de Peiresc aux frères Dupuy. Tome troisième: janvier 1634–juin 1637*. Vol. 3 of 7 vols. Edited by Philippe Tamizay de Larroque. Paris: Imprimerie Nationale, 1892.
Peiresc, Nicolas-Claude Fabri de. *Lettres de Peiresc. Tome septième: lettres à divers, 1602–1637*. Vol. 7 of 7 vols. Edited by Philippe Tamizey de Larroque. Paris: Imprimerie National, 1898.
Piccini, Alessandro. *Intavolatura di liuto e di chitiarrone libro primo*. Bologna: Heredi di Gio, 1623.
Picot, Claude. *Les principes de la philosophie escrits en Latin, par René Descartes. Et traduits en François par un des Amis*. Paris: Pierre Des-Hayes, 1647.
Plato. *Republic*. Translated by C. D. C. Reeve. Indianapolis/Cambridge: Hackett, 2004.
Plautus. *Stichus; Three-Dollar Day; Truculentus; The Tale of a Travelling-Bag; Fragments*. Edited and translated by Wolfgang de Melo. Cambridge: Harvard University Press, 2013.

Postel, Guillaume. *De la republique des Turcs*. Poitier: Enquibert de Marnef, 1560.

Praetorius, Michael. *Syntagmatis musici Michaelis Praetorii tomus secundus De Organographia*. Vol. 2 of 3 vols. Wolfenbüttel: Holwein, 1618.

Quiccheberg, Samuel. *Inscriptiones vel tituli theatri amplissimi* [...]. Munich: Berg, 1565.

Renaudot, Théophraste. *Troisieme centurie des questions traictees aux conferences du Bureau d'Adresse, depuis le 18. fevrier 1636. jusques au 17. janvier 1639*. Paris: Bureau D'Adresse, 1639.

Ripa, Cesare. *Iconologia, or, Moral Emblems*. London: Motte, 1709.

Rivault, David de Flurance. *L'art d'embellir tiré du sens de ce sacré Pradoxe. Estendu en toute sorte de beauté, et és moyens de faire que le corps retire en effect son embellissement des belles qualitez de l'ame*. Paris: Bertault, 1608.

Ronsard, Pierre de. *Abbrège de l'art poëtique françois*. Paris: Buon, 1565.

Sagard, Gabriel. *Histoire du Canada et voyages que les Freres Mineurs Recollects y ont faicts pour la conversion des infidelles*. Paris: Sonnius, 1636.

Sanches, Francisco. *That Nothing Is Known* (*Qvod Nihil Scitur*). Translated by Douglas F. S. Thomson. Cambridge: Cambridge University Press, 1988.

Sauveur, Joseph. *Collected Writings on Musical Acoustics* (*1700–1713*). Edited by Rudolf Rasch. Utrecht: The Diapason Press, 1984.

Schott, Gaspar. *Magiae universalis naturae et artis, pars II: Acustica, in vii libros digesta* [...]. Würzburg: Hertz, 1657.

Seville, Isidore of. *The Etymologies*. Translated and edited by Stephen A. Barney, W. J. Lewis, J. A. Beach, and Oliver Berghof. Cambridge: Cambridge University Press, 2006.

Sextus Empiricus. *Against the Musicians* (*Adversus musicos*). *A New Critical Text and Translation on Facing Pages, with an Introduction, Annotations, and Indices Verborum and Nominum et Rerum by Denise Davidson Greaves*. Lincoln/London: University of Nebraska Press, 1986.

Sextus Empiricus. *Outlines of Scepticism*. Edited by Julia Annas and Jonathan Barnes. Cambridge: Cambridge University Press, 2000.

Stevin, Simon. "Vande Spiegheling der Singconst." In *Principal Works of Simon Stevin*. Vol. 5 of 5 vols., *Engineering, Music, Civic Life*, edited by Adriaan Daniel Fokker, 413–64. Amsterdam: Swets & Zeitlinger, 1966.

Stelluti, Francesco. *Persio tradotto in verso sciolto e dichiarato*. Rome: Mascardi, 1630.

Suárez, Francesco. *Complectens tractatum secundum de opera sex dierum, ac tertium De anima*. Lyon: Boissat, 1635.

Tamizey de Larroque, Philippe. *Notice inédite sur le livre de raison du muet de Laincel d'après les manuscrits de Peiresc*. Digne: Chaspoul et Ve Barbaroux, 1895.

Taveau, Laurence. "Le manuscrit 2884 du P. Mersenne à la bibliothèque de l'Arsenal: Étude et édition critique." PhD diss., Université de Paris-Sorbonne, Laboratoire IREMUS, 2018.

Terzago, Paolo Maria. *Museo ò galeria adunata dal sapere, e dallo studio del Sig. Canonico Manfredo Settala nobile Milanese*. Tortona: Viola, 1666.

Thévenot, Jean. *Relation d'un voyage fait au Levant: dans laquelle il est curieusement traité des estats sujets au Grand Seigneur* [...]. Paris: Joly, 1664.

Thuillier, René. *Traduction nouvelle des regles, du correctoire, et du ceremonial de l'Ordre des Minimes*. Paris: Giffard, 1703.

Trichet, Pierre. *Synopsis rerum variarum tam naturalium quam artificialium, quae in Musaeo Petri Tricheti Burdigalae reperiuntur/Denombrement de diverses et curieuses choses du Cabinet de Pierre Trichet Bourdelois.* Bordeaux: s.n., 1631.

Trichet, Pierre. *Traite des instruments de musique (vers 1640)*. Edited by François Lesure. Geneva: Minkoff, 1978.

Trigault, Nicholas. *Histoire de l'expédition chrestienne au royaume de la Chine ...* . Lyon: Horace Cardon, 1616.

Tyard, Pontus de. *Les discours philosophiques*. Paris: L'Angelier, 1587.

Tyard, Pontus de. *Solitaire second.* Edited by Cathy M. Yandell. Geneva: Droz, 1980.

Van Nierop, Dirck Rembrandtsz. *Wis-konstige musyka: vertoonende de oorsaecke van 't geluyt*. Amsterdam: Goedesbergen, 1659.

Vanneo, Stephano. *Recanetum de musica aurea*. Rome: Apud Valerium Doricum Brixiensem, 1533.

Virdung, Sebastian. *Musica getutscht: A Treatise on Musical Instruments (1511) by Sebastian Virdung*. Translated and edited by Beth Bullard. Cambridge: Cambridge University Press, 1993.

Zarlino, Gioseffo. *Dimostrazioni harmoniche del R. M. Gioseffo Zarlino da Chioggia*. Venice: Francesco Franceschi, 1571.

Zarlino, Gioseffo. *Sopplimenti musicali del Rev. M. Gioseffo Zarlino da Chioggia: Terzo volume*. Venice: Sanese, 1588.

Secondary literature

Adkins, Cecil. *A Trumpet by Any Other Name: A History of the Trumpet Marine*. Buren: Frits Knuf, 1991.

Ammann, Peter J. "The Musical Theory and Philosophy of Robert Fludd." *Journal of the Warburg and Courtauld Institutes* 31, no. 1 (1967): 198–227.

Atkinson, Niall. *The Noisy Renaissance: Sound, Architecture, and Florentine Urban Life*. University Park: Pennsylvania State University Press, 2016.

Alexis-Collins, Marie."Eustache du Caurroy: Un compositeur français aux confins du XVIe et du XVIIe siècle." *Acta Musicologica* 73, no. 2 (2001): 189–258.

Anthony, James R. *French Baroque Music: From Beaujoyeulx to Rameau*. Portland: Amadeus Press, 1997.

Ashworth, William B. "Catholicism and Early Modern Science." In *God and Nature: Historical Essays on the Encounter between Christianity and Science*, edited by David C. Lindberg and Ronald L. Numbers, 136–66. Berkeley: University of California Press, 1986.

Austern, Linda Phyllis. "'Alluring the auditorie to effeminacie': Music and the Idea of the Feminine in Early Modern England." *Music & Letters* 74, no. 3 (1993): 343–354.

Austern, Linda Phyllis. *Both from the Ears and Mind: Thinking about Music in Early Modern England*. Chicago: University of Chicago University Press, 2020.

Bank, Katie. *Knowledge Building in Early Modern English Music*. New York: Routledge, 2021.

Barker, Andrew. *Scientific Method in Ptolemy's Harmonics*. Cambridge: Cambridge University Press, 2001.

Behar, Cem. "The Ottoman Musical Tradition." In *The Cambridge History of Turkey*, edited by Suraiya N. Faroqhi, 319–407. Cambridge, Cambridge University Press, 2006.

Belin, Christian. "La notion de collection scientifique dans les *Questions inouïes* de Marin Mersenne." In *La Tradition rassemblée: Journées d'études de l'Université de Fribourg*, edited by Guy Bedouelle, Christian Belin, and Simone de Reyff, 169–82. Fribourg: Academic Press Fribourg, 2007.

Beltrán, José "Nature *au natural* in Late Seventeenth-Century France." In *Ad vivum? Visual Materials and the Vocabulary of Life-Likeness in Europe before 1800*. Edited by Thomas Balfe, Joanna Wooddall, and Claus Zittel, 272–93. Leiden: Brill, 2019.

Benedict, Barbara M. *Curiosity: A Cultural History of Early Modern Inquiry*. Chicago: University of Chicago Press, 2002.

Bennett, Peter. *Music and Power at the Court of Louis XIII: Sounding the Liturgy in Early Modern France*. Cambridge: Cambridge University Press, 2021.

Bennett, Peter. *Sacred Repertories in Paris under Louis XIII: Bibliothèque nationale de France MS vm res.571*. Farnham: Ashgate, 2009.

Benoit, Marcelle. "L'apprentissage chez les facteurs d'instruments de Musique a Paris, 1600–1661, 1715–1774." *Recherches sur la Musique Française Classique* 24 (1986): 5–106.

Ben-Zaken, Avner. *Cross-Cultural Scientific Exchanges in the Eastern Mediterranean, 1560–1660*. Baltimore: The Johns Hopkins University Press, 2010.

Berrada. Tarek. "Music at Home: Spaces for Music in French Seventeenth-Century Residential Architecture." In *The Music Room in Early Modern France and Italy*, edited by Deborah Howard and Laura Moretti, 290–307. Oxford: Oxford University Press, 2012.

Bertoglio, Chiara. *Reforming Music: Music and the Religious Reformations of the Sixteenth Century*. Berlin: De Gruyter, 2017.

Bertoloni Meli, Domenico. "The Role of Numerical Tables in Galileo and Mersenne." *Perspectives on Science* 12, no. 2 (2004): 164–90.

Bertoloni Meli, Domenico. *Thinking with Objects: The Transformation of Mechanics in the Seventeenth Century.* Baltimore: The Johns Hopkins University Press, 2006.

Bevilacqua, Alexander. *The Republic of Arabic Letters: Islam and the European Enlightenment.* Cambridge, MA: The Belknap Press of Harvard, 2018.

Bianchi, Eric. "Asses at the Lyre: Latin as Musical Language, and the Benefits of Exclusion." In *Baroque Latinity: Studies on "Baroque" European Neo-Latin Literature*, edited by Jacqueline Glomski, Gesine Manuwald, and Andrew Taylor, 171–93. London: Bloomsbury, 2023.

Bianchi, Eric. "Bad Latin, Bad Manners: Giovanni Battista Doni, Marin Mersenne, and Literary Style in Seventeenth-Century Music Theory." *Music & Letters* 96, no. 2 (2015): 167–84.

Bianchi, Eric. "Scholars, Friends, Plagiarists: The Musician as Author in the Seventeenth Century." *Journal of the American Musicological Society* 70, no. 1 (2017): 61–128.

Bianconi, Lorenzo. *Music in the Seventeenth Century.* Translated by David Bryant. Cambridge: Cambridge University Press, 1987.

Biddle, Ian, and Kristen Gibson. *Cultural Histories of Noise, Sound and Listening in Europe.* London/New York: Routledge, 2017.

Blasius, Leslie. "Mapping the Terrain." In *The Cambridge History of Western Music Theory*, edited by Thomas Christensen, 25–45. Cambridge: Cambridge University Press, 2002.

Bleichmar, Daniela, and Peter C. Mancall. *Collecting across Cultures: Material Exchanges in the Early Modern Atlantic World.* Philadelphia: University of Pennsylvania Press, 2011.

Bloechl, Olivia. *Native American Song at the Frontiers of Early Modern Music.* Cambridge: Cambridge University Press, 2008.

Bloechl, Olivia. "The Pedagogy of Polyphony in Gabriel Sagard's *Histoire du Canada*." *The Journal of Musicology* 22, no. 3 (2005): 365–411.

Bloechl, Olivia. "Wendat Song and Carnival Noise in the Jesuit Relations." In *Native Acts: Indian Performance, 1603–1832*, edited by Laura L. Mielke and Joshua David Bellin, 117–43. Lincoln, NE: University of Nebraska Press, 2012.

Boiteux, Martine. "Fêtes et identité nationale à la Trinité-des-Monts." In *La chiesa e il convento della Trinità dei Monti: ricerche, nuove letture, restauri*, edited by Colette Di Matteo and Sebastiano Roberto, 68–75. Rome: De Luca Editori d'Arte, 2016.

Bolzoni, Lina. *The Gallery of Memory: Literary and Iconographic Models in the Age of the Printing Press.* Toronto: University of Toronto Press, 2001.

Bredekamp, Horst. *Antikensehnsucht und Maschinenglauben: Die Geschichte der Kunstkammer und die Zukunft der Kunstgeschichte.* Berlin: Wagenbach, 1993.

Brentjes, Sonja. *Peiresc's Interests in the Middle East and Northern Africa in Respect to Geography and Cartography.* Berlin: Max Planck Institute for the History of Science, 2004.

Brooks, Jeanice. *Courtly Song in Late Sixteenth-Century France.* Chicago: University of Chicago Press, 2000.

Brown, Howard Mayer. "*Ut musica poesis:* Music and Poetry in France in the Late Sixteenth Century." Early Music History 13 (1994): 1–63.

Brunet, Pierre. "Peiresc et le microscope." *Revue d'histoire des Sciences* 4, no. 2 (1951): 154–58.

Buccolini, Claudio. L'exégèse du dernier Mersenne et le cartésianisme." In *The Philosophers and the Bible: The Debate on Sacred Scripture in Early Modern Thought*, 233–47. Leiden: Brill, 2021.

Buccolini, Claudio. "Mersenne et la philosophie baconienne en France à l'époque de Descartes." In *Bacon et Descartes: genèse de la modernité philosophique*, edited by Élodie Cassan, 115–34. Lyon: ENS Éditions, 2014.

Buccolini, Claudio. "Mersenne Translator of Bacon?" *Journal of Early Modern Studies* 2, no. 1 (2013): 33–59.

Burnett. Charles. "European Knowledge of Arabic Texts Referring to Music: Some New Material." *Early Music History* 12 (1993): 1–17.

Burnett, Charles. "Musical Instruments as Conveyors of Meaning from One Culture to Another: The Example of the Lute." In *The Power of Things and the Flow of Cultural Transformations: Art and Culture between Europe and Asia*, edited by Lieselotte E. Saurma-Jeltsch and Anja Eisenbeiß, 156–69. Berlin: Deutscher Kunstverlag, 2010.

Buzon, Frédéric de. "Harmonie et métaphysique: Mersenne face a Kepler." *Les Études philosophiques* 1, no. 2 (1994): 119–28.

Cavicchi, Camilla, and Philippe Vendrix. "L'érudit et l'amateur: Collectionner la musique à la Renaissance." In *Collectionner la musique*. Vol. 1 of 3 vols.: *Histoires d'une passion*, edited by Denis Herlin, Catherine Massip, Jean Duron, and Dinko Fabris, 25–54. Turnhout: Brepols, 2010.

Champion, Matthew S., and Miranda Stanyon. "Musicalising History." *Transactions of the Royal Historical Society* 29 (2019): 79–103.

Christensen, Thomas. "Mersenne and the Mechanics of Musical Proportion." In *Proportions: science, musique, peinture & architecture*, edited by Sabine Rommevaux, Philippe Vendrix, and Vasco Zara, 247–60. Turnhout: Brepols, 2011.

Christensen, Thomas. "Mersenne's Sound World." In *Structures of Feeling in Seventeenth-Century Cultural Expression*, edited by Susan McClary, 60–89. Toronto: University of Toronto Press, 2013.

Christensen, Thomas. "Rules, License and Taste in 17th-Century French Music Theory: From Mersenne to Rameau." In *Musikalische Norm um 1700*, edited by Rainer Bayreuther, 81–95. Berlin/New York: Walter de Gruyter, 2010.

Christensen, Thomas. "Sensus, Ratio, and Phtongos: Mattheson's Theory of Tone Perception." In *The Work of Music Theory: Selected Essays*, 289–310. London/New York: Routledge, 2016.

Clark, Suzannah, and Alexander Rehding. "Introduction." In *Music Theory and Natural Order from the Renaissance to the Early Twentieth Century*, edited by Suzannah Clark and Alexander Rehding, 1–13. Cambridge: Cambridge University Press, 2001.

Clarke, Desmond. *French Philosophy, 1572–1675*. Oxford: Oxford University Press, 2016.

Coelho, Victor, ed. *Music and Science in the Age of Galileo*. Dordrecht: Kluwer, 1992.

Coelho, Victor. "Preface." In *Music and Science in the Age of Galileo*, edited by Victor Coelho, ix–xii. Dordrecht: Kluwer, 1992.

Cohen, Albert. "The *Fantaisie* for Instrumental Ensemble in 17th-Century France: Its Origin and Significance." *The Musical Quarterly* 48, no. 2 (1962): 234–43.

Cohen, Albert. "Jean le Maire and La Musique Almérique." *Acta Musicologica* 35, no. 4 (1963): 175–81.

Cohen, Albert. "Jehan Titelouze as Music Theorist." In *Festa Musicologica: Essays in Honor of George J. Buelow*, edited by Thomas J. Mathiesen and Benito V. Rivera, 391–406. Stuyvesant, NY: Pendragon Press, 1995.

Cohen, Albert. *Music in the French Royal Academy of Sciences: A Study in the Evolution of Musical Thought*. Princeton: Princeton University Press, 1981.

Cohen, H. Floris. "Music as Science and as Art: The Sixteenth/Seventeenth-Century Destruction of World Harmony." In *The Making of the Humanities, Volume 1: Early Modern Europe*, edited by Rens Bod, Jaap Maat, and Thijs Weststeijn, 59–71. Amsterdam: Amsterdam University Press, 2010.

Cohen, H. Floris. *Quantifying Music: The Science of Music at the First Stage of the Scientific Revolution, 1580–1650*. Dordrecht: Reidel, 1984.

Copenhaver, Brian. "The Occultist Tradition and Its Critics." In *The Cambridge History of Seventeenth-Century Philosophy*, vol. 1 of 2 vols., edited by Daniel Garber and Michael Ayers, 454–512. Cambridge: Cambridge University Press, 1998.

Cook, Harold J. *Matters of Exchange: Commerce, Medicine, and Science in the Dutch Golden Age*. New Haven/London: Yale University Press, 2007.

Coumet, Ernest. "Mersenne: dénombrements, repertoires, numérotations de permutations." *Mathématiques et sciences humaines* 38 (1972): 5–37.

Coumet, Ernest. *Oeuvres d'Ernest Coumet*, vol. 2 of 2 vols., edited by Catherine Goldstein. Besançon: Presses universitaires de Franche-Comté, 2019.

Cowart, Georgia. *The Origins of Modern Music Criticism: French and Italian Music, 1600–1750*. Ann Arbor: UMI Research Press, 1981.

Crombie, Alistair C. "Marin Mersenne and Scientific Acceptability." In *Science, Optics and Music in Medieval and Early Modern Thought*, 399–417. London: Hambledon Press, 1990.

Crombie, Alistair. "Music, Mathematics and Medical Science." In *Science, Optics and Music in Medieval and Early Modern Thought*, 363–77. London: Hambledon Press, 1990.

Crombie, Alistair. *Styles of Scientific Thinking in the European Tradition: The History of Argument and Explanation Especially in the Mathematical and Biomedical Sciences and Arts*. Vol. 2 of 2 vols. London: Duckwort, 1994.

Cypess, Rebecca. *Curious and Modern Inventions: Instrumental Music as Discovery in Galileo's Italy*. Chicago: University of Chicago Press, 2016.

Cypess, Rebecca. "Giovanni Battista Della Porta's Experiments with Musical Instruments." *Journal of Musicological Research* 35, no. 5 (2016): 159–75.

Dakhlia, Jocelyne. "Une archéologie du même et de l'autre: Thomas-Osman d'Arcos dans la Méditerranée du XVIIe siècle." In *Les Musulmans dans l'histoire de l'Europe*. Vol. 2 of 2 vols.: *Passages et contacts en Méditerranée*, edited by Jocelyne Dakhlia and Wolfgang Kaiser, 61–163. Paris: Albin Michel, 2013.

Dandrey, Patrick. "La phoniscopie, c'est-à-dire la science de la voix." *Littératures classiques* 12 (1990): 13–76.

Daston, Lorraine. "The Nature of Nature in Early Modern Europe." *Configurations* 7, no. 2 (1998): 149–72.

Daston, Lorraine, and Katharine Park. *Wonders and the Order of Nature, 1150–1750*. New York: Zone Books, 1998.

Davies, Surekha. *Renaissance Ethnography and the Invention of the Human: New Worlds, Maps, and Monsters*. Cambridge: Cambridge University Press, 2016.

De Buzon, Frédéric. "The *Compendium Musicae* and Descartes's Aesthetics." In *The Oxford Handbook of Descartes and Cartesianism*, edited by Steven Nadler, Tad M. Schmaltz, and Delphine Antoine-Mahut, 254–68. Oxford: Oxford University Press, 2019.

De Buzon, Frédéric. "Harmonie et métaphysique: Mersenne face a Kepler." *Les Études philosophiques* 1, no. 2 (1994): 119–28.

De Rosa, Agostino and Alessio Bortot. "Anamorphosis: Between Perspective and Catoptrics." *Handbook of the Mathematics of the Arts and Sciences*, edited by Bharath Sriraman, 243–289. Cham: Springer, 2021.

De Rosa, Agostino, Alessio Bortot, and Luca Ruggiero D'Elia. "The Digital Reconstruction of the Minim's Convent in Place Royal and its Optical-Perspective Games." *Disegnarecon* 14, no. 27 (2021): 1–14.

Dear, Peter. *Discipline and Experience: The Mathematical Way in the Scientific Revolution*. Chicago: University of Chicago Press, 1995.

Dear, Peter. "Marin Mersenne and the Probabilistic Roots of 'Mitigated Scepticism.'" *Journal of the History of Philosophy* 22, no. 2 (1984): 173–205.

Dear, Peter. *Mersenne and the Learning of the Schools*. Ithaca: Cornell University Press, 1988.

Dear, Peter. "Method and the Study of Nature." In *The Cambridge History of Seventeenth-Century Philosophy*, edited by Daniel Garber and Michael Ayers, vol. 1 of 2 vols., 147–77. Cambridge: Cambridge University Press, 2000.

Dell'Antonio, Andrew. *Listening as Spiritual Practice in Early Modern Italy*. Berkeley: University of California Press, 2011.

Dennis, Flora. "When is a Room a Music Room? Sounds, Spaces, and Objects in Non-Courtly Italian Interiors." In *The Music Room in Early Modern France and Italy: Sound, Space and Object*, edited by Deborah Howard and Laura Moretti, 37–49. Oxford: Oxford University Press, 2012.

Dew, Nicholas. *Orientalism in Louis XIV's France*. Oxford: Oxford University Press, 2009.

Dostrovsky, Sigalia. "Early Vibration Theory: Physics and Music in the Seventeenth Century." *Archive for History of Exact Sciences* 14, no. 3 (1975): 169–218.

Downey, Peter. "Fantini and Mersenne: Some Additions to Recent Controversies." *Historic Brass Society Journal* 6 (1994): 355–62.

Drake, Stillman. "Renaissance Music and Experimental Science." *Journal of the History of Ideas* 31, no. 4 (1970): 483–500.

Dufourcq, Norbert. "Une dynastie française: Les Denis." *Revue de Musicologie* 38, no. 114 (1956): 151–55

Duncan, David Allen. "Persuading the Affections: Rhetorical Theory and Mersenne's Advice to Harmonics Orators." In *French Musical Thought, 1600–1800*, edited by Georgia Cowart, 149–75. Ann Arbor: UMI Research Press, 1989.

Duncan, David Allen. "Mersenne and Modern Learning: The Debate over Music." In *The Rise of Modern Philosophy: The Tension between the New and Traditional*, edited by Tom Sorell, 89–106. Oxford: Clarendon Press, 1993.

Eamon, William. "Markets, Piazzas and Villages." In *The Cambridge History of Science. Volume 3: Early Modern Science*, edited by Katharine Park and Lorraine Daston, 206–23. Cambridge: Cambridge University Press, 2006.

Erlmann, Veit. *Reason and Resonance: A History of Modern Aurality*. New York: Zone Books, 2010.

Evans, R. J. W., and Alexander Marr, eds. *Curiosity and Wonder from the Renaissance to the Enlightenment*. Aldershot: Ashgate, 2006.

Fabbri, Natacha. "Genesis of Mersenne's *Harmonie universelle*: The Manuscript *Livre de la Nature des Sons*." *Nuncius: Journal of the History of Science* 22, no. 2 (2007): 287–308.

Fabbri, Natacha. "Hérésie et orthodoxie dans la pensée mersennienne: vers la 'concordia' de la République chrétienne." *Historia philosophica* 2 (2004): 13–34.

Fabbri, Natacha. "Questioning Fludd, Kepler and Galileo: Mersenne's Harmonious Universe." In *Space, Imagination, and the Cosmos from Antiquity to the Early Modern Period*, edited by Frederik A. Bakker, Delphine Bellis, and Carla Rita Palmerino, 201–32. Cham: Springer, 2018.

Fabbri, Natacha. *De l'utilité de l'harmonie: filosofia, scienza e musica in Mersenne, Descartes e Galileo*. Pisa: Edizioni della Normale, 2008.

Falletti, Franca, Renato Meucci, and Gabriele Rossi-Rognoni, eds. *Marvels of Sound and Beauty: Italian Baroque Musical Instruments.* Florence: Giunti Editore, 2007.

Feingold, Mordechai, and Penelope Gouk. "An Early Critique of Bacon's *Sylva Sylvarum*: Edmund Chilmead's Treatise on Sound." *Annals of Science* 40, no. 2 (1983), 139–57.

Feld, Steven. "Acoustemology." In *Keywords on Sound*, edited by David Novak and Matt Sakaneeny, 12–21. Durham, NC: Duke University Press, 2015.

Fend, Michael. "The Changing Functions of 'Senso' and 'Ragione' in Italian Music Theory of the Late Sixteenth Century." In *The Second Sense: Studies in Hearing and Musical Judgement from Antiquity to the Seventeenth Century*, edited by Charles Burnett, Penelope Gouk, and Michael Fend, 199–221. London: Warburg Institute/University of London, 1991.

Fend, Michael. "Probleme mit der Idee der 'harmonia universalis' in der frühen Neuzeit." *Archiv für Musikwissenschaft* 71, no. 4 (2014): 307–34.

Filippi, Daniel V. "'Their sound hath gone forth into all the earth': Music and Sound in the Ministries of Early Modern Jesuits: Introduction." *Journal of Jesuit Studies* 3 (2016): 357–64.

Findlen, Paula. "The Janus Faces of Science in the Seventeenth Century: Athanasius Kircher and Isaac Newton." In *Rethinking the Scientific Revolution*, edited by Margaret J. Osler, 221–46. Cambridge: Cambridge University Press, 2000.

Findlen, Paula. *Possessing Nature: Museums, Collecting, and Scientific Culture in Early Modern Italy.* Berkeley: University of California Press, 1996.

Fix, Adam. "The Acoustical Paradox: How Music Was Unmoored and Set Adrift from the Sciences in the Seventeenth Century." In *Music and Science from Leonardo to Galileo*, edited by Rudolf Rasch, 377–420. Turnhout: Brepols, 2022.

Fix, Adam. "Esperienza, Teacher of All Things: Vincenzo Galilei's Music as Artisanal Epistemology." *Nuncius* 34 (2019): 535–74.

Flandrin, Jean-Louis. "Distinction through Taste." In *A History of Private Life, Volume 3: Passions of the Renaissance*, edited by Roger Chartier, translated by Arthur Goldhammer, 265–307. Cambridge, MA: Belknap Press, 1989.

Foucault, Michel. *The Order of Things: An Archaeology of the Human Sciences.* London/New York: Routledge, 2002.

Frisch, Andrea. *The Invention of the Eyewitness: Witnessing and Testimony in Early Modern France.* Chapel Hill: University of North Carolina Press, 2004.

Garber, Daniel. "On the Frontlines of the Scientific Revolution: How Mersenne Learned to Love Galileo." *Perspectives on Science* 12, no. 2 (2004): 135–63.

Garber, Daniel. "Philosophia, Historia, Mathematica: Shifting Sands in the Disciplinary Geography of the Seventeenth Century." In *Scientia in Early Modern Philosophy Seventeenth-Century Thinkers on Demonstrative Knowledge from First Principles*, edited by Tom Sorell, G. A. Rogers, and Jill Kraye, 1–17. Dordrecht: Springer, 2010.

Gaukroger, Stephen. *Descartes: An Intellectual Biography.* Oxford: Clarendon Press, 1995.

Gauvin, Jean François. "Instruments of Knowledge." In *The Oxford Handbook of Philosophy in Early Modern Europe*, edited by Desmond M. Clarke and Catherine Wilson, 315–37. Oxford: Oxford University Press, 2011.

Gauvin, Jean François. *Instruments of Knowledge: Finding Meaning in Objects, Habits, and Museums.* Leiden: Brill, 2023.

Gerbino, Giuseppe. "Skeptics and Believers: Music, Warfare, and the Political of Renaissance Italy according to Francesco Bocchi." *The Musical Quarterly* 90, no. 3–4 (2007): 578–603.

Gétreau, Florence. "Entre l'oral et l'écrit: Pratique, transmission et théorie du métier de facteur d'instruments de musique." *Ethnologie française* 26, no. 3 (1996): 504–19.

Ghrab, Anas. "Livre de la générosite dans la connaisance des modes: Edition et traduction." Diplome d'etudes approfondies, Université Lumiere-Lyon, 2002.

Godwin, Joscelyn. *The Harmony of the Spheres: The Pythagorean Tradition in Music.* Rochester: Inners Traditions, 1993.

Goldberg, Randall. *Where Nature and Art Adjoin: Investigations into the Zarlino-Galilei Dispute, Including an Annotated Translation of Vincenzo Galilei's Discorso intorno all'opera Di Messer Gioseffo Zarlino.* Ann Arbor: UMI Research Press, 2011.

Gominet-Brun, Julien. "Une curieuse invention du Père Mersenne. La psophologie et la musique du coeur." *Revue Bossuet* 7 (2016): 87–102.

Goodman, Glenda. "Sounds Heard, Meaning Deferred: Music Transcription as Imperial Technology." *Eighteenth-Century Studies* 52, no. 1 (2018): 39–45.

Gordon, Bonnie. *Monteverdi's Unruly Women: The Power of Song in Early Modern Italy.* Cambridge: Cambridge University Press, 2004.

Gordon, Bonnie. "Orfeo's Machines." *The Opera Quarterly* 24, no. 3–4 (2008): 200–222.

Gouk, Penelope. *Music, Science, and Natural Magic in Seventeenth-Century England.* New Haven: Yale University Press, 1999.

Gouk, Penelope. "The Role of Harmonics in the Scientific Revolution." In *The Cambridge History of Western Music Theory*, edited by Thomas Christensen, 223–45. Cambridge: Cambridge University Press, 2002.

Gouk, Penelope. "The Role of Harmonics in the Scientific Revolution." In *The Cambridge History of Western Music Theory*, edited by Thomas Christensen, 223–45. Cambridge: Cambridge University Press, 2002.

Gozza, Paolo, ed. *Number to Sound: The Musical Way to the Scientific Revolution.* Dordrecht: Kluwer, 2000.

Gozza, Paolo. "A Renaissance Mathematics: The Music of Descartes." In *Number to Sound: The Musical Way to the Scientific Revolution*, edited by Paola Gozza, 155–72. Dordrecht: Kluwer, 2002.

Greenblatt, Stephen. *Marvelous Possessions: The Wonder of the New World*. Chicago: University of Chicago Press, 1991.

Grosslight, Justin. "Small Skills, Big Networks: Marin Mersenne as Mathematical Intelligencer." *History of Science* 51, no. 3 (2013): 337–74.

Guillo, Laurent. *Pierre I Ballard et Robert III Ballard. Imprimeurs du roy pour la musique (1599–1673)*, 2 vols. Sprimont: Mardaga, 2003.

Guillo, Laurent. "Sous la main du père Mersenne: la bibliothèque de musique des Minimes de la place Royal (Paris, XVIe–XVII siècles)." In *Collectionner la musique*. Vol. 3 of 3 vols.: *Erudits collectionneurs*, edited by Denis Herlin, Catherine Massip, and Valérie De Wispelaere, 29–80. Turnhout: Brepols, 2015.

Hallyn, Fernard. *The Poetic Structure of the World: Copernicus and Kepler*. New York: Zone Books, 1990.

Hamilton, Alistair. "'To divest the east of all its manuscripts and all its rarities': The Unfortunate Embassy of Henri Gournay de Marcheville." In *The Republic of Letters and the Levant*, edited by Alastair Hamilton, Maurits H. van den Boogert, and Bart Westerweel, 123–50. Leiden/Boston: Brill, 2005.

Hamlin, William. "On Continuities between Skepticism and Early Ethnography: Or, Montaigne's Providential Diversity." *Sixteenth Century Journal* 31 (2000): 361–80.

Hammond, Nicholas. *The Powers of Sound and Song in Early Modern Paris*. University Park: Pennsylvania State University Press, 2020.

Harne, George A. "Unstable Embodiments of Musical Theory and Practice in the *Speculum musicae*." *Plainsong and Medieval Music* 21, no. 2 (2012): 113–36.

Haug, Judith I. *Ottoman and European Music in ʿAli Ufuḳī's Compendium, MS Turc 292: Analysis, Interpretation, Cultural Context*. Münster: Universitäts- und Landesbibliothek Münster, 2019.

Heider, Daniel. "Suárez on Sound and Hearing." *Anuario Filosófico* 50, no. 2 (2017): 323–44.

Heilman, Anja. *Boethius' Musiktheorie und das Quadrivium: Eine Einführung in den neuplatonischen Hintergrund von "De institutione musica."* Göttingen: Vandenhoeck & Ruprecht, 2007.

Heller, Henry. *Labour, Science and Technology in France, 1500–1620*. Cambridge: Cambridge University Press, 1996.

Heller-Roazen, Daniel. *The Fifth Hammer: Pythagoras and the Disharmony of the World*. New York: Zone Books, 2011.

Hine, William. "Marin Mersenne: Renaissance Naturalism and Renaissance Magic." In *Occult and Scientific Mentalities in the Renaissance*, edited by Brian Vickers, 165–76. Cambridge: Cambridge University Press, 1984.

Hine, William. "Mersenne and Vanini." *Renaissance Quarterly* 29, no. 1 (1976): 52–65.

Hicks, Andrew. *Composing the World: Harmony in the Medieval Platonic Cosmos*. Oxford: Oxford University Press, 2017.

Hollander, John. *Untuning of the Sky: Ideas of Music in English Poetry, 1500–1700*. Hamden: Archon Books, 1993.

Honisch, Erika Supria. "Encounters with Music in Rudolf II's Prague." *Austrian History Yearbook* 52 (2021): 64–80.

Honn Hoegberg, Elisabeth. "From Theory to Practice: Composition and Analysis in Marin Mersenne's Harmonie universelle." PhD diss., Indiana University, 2005.

Hubbard, Frank. *Three Centuries of Harpsichord Making*. Cambridge: Harvard University Press, 1965.

Hunt, John N. "Jurors of the Guild of Musical Instrument Makers of Paris." *The Galpin Society Journal* 51 (1998): 110–13.

Huppert, George. "Divinatio et Eruditio: Thoughts on Foucault." *History and Theory* 13, no. 3 (1974): 191–207.

Irving, David. "Ancient Greeks, World Music, and Early Modern Constructions of Western European Identity." In *Studies on a Global History of Music: A Balzan Musicology Project*, edited by Reinhard Strohm, 21–41. London/New York: Routledge, 2018.

Irving, David. "Comparative Organography in Early Modern Empires." *Music and Letters* 90, no. 3 (2009): 372–98.

Irving, David. "The Dissemination and Use of European Music Books in Early Modern Asia." *Early Music History* 28 (2009): 39–59.

Irving, David. *The Making of European Music in the Long Eighteenth Century*. Oxford: Oxford University Press, 2024.

Isherwood, Robert M. *Music in Service of the King: France in the Seventeenth Century*. Ithaca: Cornell University Press, 1973.

Jackson, Myles. "Music and Science during the Scientific Revolution." *Perspectives on Science* 9, no. 1 (2001): 106–15.

Jaffé, David. "Peiresc: Wissenschaftlicher Betrieb in einem Raritäten-Kabinett." In *Macrocosmos in Microcosmo: Die Welt in der Stube: Zur Geschichte des Sammelns 1450 bis 1800*, edited by Andreas Grote, 301–22. Opladen: Leske + Budrich, 1994.

Jeserich, Philipp. *Musica Naturalis: Speculative Music Theory and Poetics, from Saint Augustine to the Late Middle Ages in France*. Baltimore: Johns Hopkins University Press, 2013.

Kassler, Jamie C. *The Beginnings of the Modern Philosophy of Music in England: Francis North's A Philosophical Essay of Musick (1677) with Comments of Isaac Newton, Roger North and in the Philosophical Transactions*. London/New York: Routledge, 2004.

Kassler, Jamie C. *Inner Music: Hobbes, Hooke, and North on Internal Character*. Madison/London: Fairleigh Dickinson University Press, 1995.

Keller, Vera. *The Interlopers: Early Stuart Projects and the Undisciplining of Knowledge*. Baltimore: Johns Hopkins University Press.

Kellman, Jordan. "Mendicants, Minimalism, and Method: Franciscan Scientific Travel in the Early Modern French Atlantic." *Journal of Early Modern History* 26 (2022): 10–37.

Kent, Emily. "'Minimizing' Marin Mersenne: Tracing the Intellectual Culture of a Seventeenth-Century Parisian Convent." *History of Universities* 36, no. 2 (2023): 58–79.

Kirnbauer, Martin. *Vieltönige Musik: Spielarten chromatischer und enharmonischer Musik in Rome in der ersten Hälfte des 17. Jahrhunderts.* Basel: Schwabe, 2013.

Kittler, Friedrich. "Wie der Sound von heute vor 300 Jahren enstand." In *Weltbürgertum und Globalisierung*, edited by Norbert Bolz, Friedrich Kittler, and Raimar Zons, 145–55. München: Fink, 2000.

Klotz, Sebastian. *Kombinatorik und die Verbindungskünste der Zeichen zwischen 1630 und 1780.* Berlin: Akademie Verlag, 2006.

Klotz, Sebastian. "Vibration und Vernunft: Zur experimentellen Agenda in Marin Mersennes *Harmonie universelle* (Paris, 1636)." In *Spektakuläre Experimente: Praktiken der Evidenzproduktion im 17. Jahrhundert*, edited by Helmar Schramm, Ludger Schwarte, and Jan Lazardzig, 279–94. Berlin: De Gruyter, 2006.

Knighton, Tess, and Ascensión Mazuela-Anguita. *Hearing the City in Early Modern Europe.* Turnhout: Brepols, 2018.

Knowlson, James R. "Jean le Maire, the Almérie, and the "musique almérique": A Set of Unpublished Documents." *Acta Musicologica* 40, no. 1 (1968): 86–89.

Koch, Erec. *The Aesthetic Body: Passion, Sensibility and Corporeality in Seventeenth-Century France.* Newark: University of Delaware, 2008.

Konoval, Brandon. "Pythagorean Pipe Dreams? Vincenzo Galilei, Marin Mersenne, and the Pneumatic Mysteries of the Pipe Organ." *Perspectives on Science* 26, no. 1 (2018): 1–51.

Krakovitch, Odile. "L'architecture des trois couvents minimes de Paris." In *Saint François de Paule et les Minimes en France de la fin du XVe au XVIIIe siècle*, edited by André Vauchez and Pierre Benoist, 229–48. Tours: Presses universitaires François-Rabelais, 2011.

Krakovitch, Odile. "La vie intellectuelle dans les trois couvents minimes de la Place Royale, de Nigeon et de Vincennes." *Bulletin de la Société de l'Histoire de Paris et de L'Ile-de-France* 109 (1983): 23–175.

Kvíčalová, Anna. "Hearing Difference in Calvin's Geneva: From Margins to Center." *Sixteenth Century Journal* 49, no. 1 (2018): 25–47.

Latour, Melinda. *The Voice of Virtue: Moral Song and the Practice of French Stoicism, 1574–1652.* Oxford: Oxford University Press, 2023.

Launay, Denise. "La musique à Venise vers 1645: Ismaël Boulliau, astronome français méloname et voyageur." *Revue de Musicologie* 77, no. 2 (1991): 269–77.

Lazardzig, Jan. "Die Maschine als Spektakel: Funktion und Admiration im Maschinendenken des 17. Jahrhunderts." In *Instrumente in Wissenschaft und Kunst – Zu Architektonik kultureller Grenzen im 17. Jahrhundert*, edited by Helmar Schramm, Ludger Schwarte, and Jan Lazardzig, 167–93. Berlin: De Gruyter, 2006.

Lehr, André. "Middeleeuwse klokkengietkunst." In *Klokken en Klokkengieters: Bijdragen tot de Campanologie*, 13–158. Culemborg: Historische Commissie van Nederlandse Klokkenspel-Vereniging, 1963.

Leinkauf, Thomas. "Kirchers Musikverständnis im Kontext der barocken Universalwissenschaft." In *Ars magna musices: Athanasius Kircher und die Universalität der Musik*, edited by Markus Engelhardt and Michael Heinemann, 1–21. Laaber: Laaber Verlag, 2007.

Lenoble, Robert. *Mersenne ou la naissance du mécanisme*. Paris: Vrin, 1943.

Lesure, François. *Musique et musiciens français du XVI siècle*. Geneva: Minkoff, 1976.

Levenberg, Jeffrey. "A Diagram from the Mubarak Shah Commentary." In *The Museum of Renaissance Music: A History in 100 Exhibits*, edited by Vincenzo Borghetti and Tim Shephard, 392–95. Turnhout: Brepols, 2022.

Licoppe, Christian. *La formation de la pratique scientifique: Le discourse de l'expérience en France et en Angleterre (1630–1820)*. Paris: La Découverte, 1996

Lindley, Mark. "Mersenne on Keyboard Tuning." *Journal of Music Theory* 24, no. 2 (1980): 166–203.

Linhart Wood, Jennifer. *Sounding Otherness in Early Modern Drama and Travel: Uncanny Vibrations in the English Archive*. Cham: Palgrave Macmillan, 2019.

Locke, Ralph P. *Music and the Exotic from the Renaissance to Mozart*. Cambridge: Cambridge University Press, 2015.

Long, Pamela O. *Openness, Secrecy, Authorship. Technical Arts and the Culture of Knowledge from Antiquity to the Renaissance*. Baltimore: The Johns Hopkins University Press, 2001.

Loubet de Sceaury, Paul. *Musiciens et facteurs d'instruments de musique sous l'Ancien Régime: Status corporatifs*. Paris: Pedone, 1949.

Luppi, Andrea. "Musiche immaginarie e critica scettica nel Discours di François de la Mothe." *Studi musicali* 20, no. 1 (1991): 111–40.

Mace, Dean T. "Marin Mersenne on Language and Music." *Journal of Music Theory* 14, no. 1 (1970): 2–34.

Mackensen, Karsten. *Musik und die Ordnung der Dinge im ausgehende Mittelalter und in der Frühen Neuzeit*. Frankfurt: Peter Lang, 2017.

Mackensen, Karsten. "System und Kritik. Zur Krise des Musikwissens bei Johann Mattheson." *International Review of the Aesthetics and Sociology of Music* 40, no. 2 (2009): 179–205.

Mahoney, Michael Sean. *The Mathematical Career of Pierre de Fermat, 1601–1665*. 2nd ed. Princeton: Princeton University Press, 1994.

Mai Groote, Inga. "Studying Music and Arithmetic with Glarean: Contextualizing the Epitomes and Annotationes among the Sources for Glarean's Teaching." In *Heinrich Glarean's Books: The Intellectual World of a Sixteenth-Century Musical Humanist*, edited by Iain Fenlon and Inga Mai Groote, 195–222. Cambridge: Cambridge University Press, 2013.

Malcolm, Noel. *Aspects of Hobbes*. Oxford: Clarendon Press, 2002.

Mancosu, Paolo. *Philosophy of Mathematics and Mathematical Practice in the Seventeenth Century*. Oxford: Oxford University Press, 1999.

Margócsy, Daniel. *Commercial Visions: Science, Trade, and Visual Culture in the Dutch Golden Age*. Chicago: University of Chicago Press, 2014.

Marr, Alexander. "Introduction." In *Curiosity and Wonder from the Renaissance to the Enlightenment*, edited by R. J. W. Evans and Alexander Marr, 1–20. Aldershot: Ashgate, 2006.

Martin, Craig. *Subverting Aristotle: Religion, History, and Philosophy in Early Modern Science*. Baltimore: Johns Hopkins University Press, 2014.

Massip, Catherine. *La vie des Musiciens de Paris au temps de Mazarin: 1643–1661: essai d'études sociale*. Paris: Picard, 1976.

Massip, Catherine. "Facteurs d'instruments et maîtres à danser parisiens au XVIIe siècle." In *Instrumentistes et luthiers parisiens, XVIIe–XIX siècles*, edited by Florence Gétreau, 17–34. Alençon: Impremerie Alençonnaise, 1988.

Mehl, Éduouard. "Gassendi in the Philosophical Debate: Stakes of the Essay Concerning the Principles of Robert Fludd's Philosophy (1630)." In *Pierre Gassendi: Humanism, Science, and the Birth of Modern Philosophy*, edited by Delphine Bellis, Daniel Garber, Carla Rita Palmerino, 169–92. New York: Routledge, 2023.

Meyer, Andreas. "In Sound Similar to the Harps: Early Descriptions of African Musical Instruments." In *Instruments in Art and Science: On the Architectonics of Cultural Boundaries in the 17th Century*, edited by Helmar Schramm, Ludger Schwarte, and Jan Lazardzig, 269–79. Berlin: De Gruyter, 2008.

Miller, Peter. "Looking at the Past, Nature, and Peoples in Peiresc's Archive." In *Historia: Empiricism and Erudition in Early Modern Europe*, edited by Gianna Pomata and Nancy Siraisi, 355–97. Cambridge, MA: The MIT Press, 2005.

Miller, Peter. "Making the Paris Polyglot Bible: Humanism and Orientalism in the Early Seventeenth Century." In *Die europäische Gelehrtenrepublik im Zeitalter des Konfessionalismus/The European Republic of Letters in the Age of Confessionalism*, edited by Herbert Jaumann, 59–85. Wiesbaden: Harrassowitz, 2001.

Miller, Peter. "Peiresc in the Parisian 'Jewel House.'" In *Knowledge and Discernment in the Early Modern Arts*, edited by Sven Dupré and Christine Göttler, 213–35. London/New York: Routledge, 2017.

Miller, Peter. *Peiresc's Europe: Leaning and Virtue in the Seventeenth Century*. New Haven: Yale University Press, 2000.

Miller, Peter. *Peiresc's Mediterranean World*. Cambridge, MA/London: Harvard University Press, 2015.

Moreno, Jairo. *Musical Representations, Subjects, and Objects: The Construction of Musical Thought in Zarlino, Descartes, Rameau and Weber*. Bloomington: Indiana University Press, 2004.

Moyer, Ann E. *Musica Scientia: Musical Scholarship in the Italian* Renaissance. Ithaca: Cornell University Press, 1992.

Moyer, Ann. E. "Musical Scholarship in Italy and the End of the Renaissance, 1500–1650: From Veritas to Verisimilitude." In *History and the Disciplines: The Reclassification of Knowledge in Early Modern Europe*, edited by Donald R. Kelley, 185–202. Rochester: University of Rochester Press, 1997.

Mueller, Paul R. "Marin Mersenne's 'Questions theologiques, Physiques, Morales, et Mathématiques': Agenda and Structure, with an Annotated Translation." PhD diss., University of Chicago, 2006.

Ndiaye, Noémie. *Scripts of Blackness: Early Modern Performance Culture and the Making of Race*. Philadelphia: University of Pennsylvania Press, 2022.

Nelson, Eric. "Remembering the Martyrdom of Saint Francis of Paola: History, Memory and Minim Identity in Seventeenth-Century France." *History and Memory* 26, no. 2 (2014): 76–105.

Norton-Wright, Jenny. "Innovations in Music Notation in Late Medieval Syria: British Library Manuscript Or. 13019." *British Library* (music blog), June 29, 2020. https://blogs.bl.uk/music/2020/06/innovations-in-music-notation-in-late-medieval-syria-british-library-manuscript-or-13019.html.

Oates, Rosamund. "Speaking in Hands: Early Modern Preaching and Signed Languages for the Deaf." *Past & Present* 256, no. 1 (2022): 49–85.

Obelkevich, Mary Romen. "Turkish Affect in the Land of the Sun King." *The Musical Quarterly* 63, no. 3 (1977): 367–38

Oosterhoff, Richard J. *Making Mathematical Culture: University and Print in the Circle of Lefèvre d'Étaples*. Oxford: Oxford University Press, 2018.

Ossa-Richardson, Anthony. "Nicolas Peiresc and the Delphic Tripod in the Republic of Letters." *Journal of the Warburg and Courtauld Institutes* 74, no. 1 (2011): 263–79.

Owens, Jessie Ann. *Composers at Work: The Craft of Musical Composition 1450–1600*. Oxford: Oxford University Press, 1997.

Paganini, Gianni. "Gassendi's Interplay between Skepticism and Empiricism." In *Pierre Gassendi: Humanism, Science, and the Birth of Modern Philosophy*, edited by Delphine Bellis, Daniel Garber, and Carla Rita Palmerino, 75–105. New York: Routledge, 2023.

Palisca, Claude V. *G.B. Doni's Lyra Barberina: Commentary and Iconographical Study: Facsimile Edition with Critical Notes*. Bologna: A.M.I.S., 1981.

Palisca, Claude V. *Humanism in Italian Renaissance Musical Thought.* New Haven: Yale University Press, 1985.

Palisca, Claude V. "Mersenne pro Galilei contra Zarlino." In *Essays in honour of David Evatt Tunley*, edited by Frank Callaway, 61–72. Nedlands: Callaway International Resource Center for Music Education, 1995.

Palisca, Claude V. *Music and Ideas in the Sixteenth and Seventeenth Centuries.* Urbana: University of Illinois Press, 2017.

Palisca, Claude V. "Scientific Empiricism in Musical Thought." In *Seventeenth-Century Science and the Arts*, edited by Hedley H. Reys, 91–137. Princeton: Princeton University Press, 1961.

Partch, Harry. *Bitter Music: Collected Journals, Essays, Introductions, and Librettos*, edited by Thomas McGeary. Urbana: University of Illinois Press, 1991.

Pasnau, Robert. "Sensible Qualities: The Case of Sounds." *Journal of the History of Philosophy* 38, no. 1 (2000): 27–40.

Peil, Dietmar. "Concordia discors: Anmerkungen zu einem politischen Harmoniemodell von der Antike bis in die Neuzeit." In *Geistliche Denkformen in der Literatur des Mittelalter*, edited by Klaus Grübmüller, Ruth Schmidt-Wiegand, and Klaus Speckenbach, 401–34. Munich: Wilhelm Fink, 1984.

Perler, Dominik. "Was There a Pyrrhonian Crisis in Early Modern Philosophy? A Critical Notice of Richard H. Popkin." *Archiv für Geschichte der Philosophie* 86, no. 2 (2006): 209–20.

Phillips, Henry. *Church and Culture in Seventeenth-Century France.* Cambridge: Cambridge University Press, 1997.

Pierre, Benoist, and Andre Vauchez, eds. *Saint François de Paule et les Minimes en France de la fin du XVe au XVIIIe siècle.* Tours: Presse Universitaire François Rabelais, 2011.

Pomata, Gianna. "Observation Rising: Birth of an Epistemic Genre, 1500–1650." In *Histories of Scientific Observation*, edited by Lorraine Daston and Elizabeth Lunbeck, 45–80. Chicago: University of Chicago Press, 2011.

Pomian, Krysztof. *Collections and Curiosities: Paris and Venice, 1500–1800.* Translated by Elizabeth Wiles-Portier. Cambridge: Polity Press, 1990.

Popkin, Richard. *The History of Scepticism: From Savonarola to Bayle.* Oxford: Oxford University Press, 2003.

Popplow, Marcus. "Why Draw Pictures of Maschines? The Social Contexts of Early Modern Machine Drawings." In *Picturing Maschines 1400–1700*, edited by Wolfang Lefèvre, 17–48. Cambridge, MA: MIT Press, 2004.

Pons, André. *Droit ecclésiastique et musique sacrée. Tome III: Décadence et réforme du Chant liturgique.* Vol. 3 of 5 vols. St-Maurice: Editions St-Augustine, 1958.

Prins, Jacomien. *Echoes of an Invisible World: Marsilio Ficino and Francesco Patrizi on Cosmic Order and Music Theory.* Leiden/Boston: Brill, 2012.

Prins, Jacomien. "Patrizi's and Mersenne's Critiques of Ficino's Interpretation of the Harmony of the Spheres." In *Perfect Harmony and Melting Strains*, edited by Cornelia Wilde and Wolfram R. Keller, 59–80. Berlin: De Gruyter, 2021.

Psychoyou, Théodora. "De la musique universelle à l'oreille des autres: altérités culturelles dans le discours sur la musique en France au XVIIe siècle." In *La musicologie francophone et la circulation des savoirs dans un contexte multiculturel: interpénétration et résilience*, edited by Achille Davy-Rigaux, Catherine Deutsch, Hamdi Makhlouf, and Anas Ghrab, 43–65. Tunis: Centre des Musiques Arabes et Méditerranéennes, 2022.

Psychoyou, Théodora. "Plaisirs de l'esprit, plaisirs de l'oreilles: anatomie et paradoxes d'un nouveau critère théorique." In *Le plaisir musical en France au XVIIe siècle*, edited by Thierry Fabrier and Manuel Couvreur, 61–78. Liège: Mardaga, 2006.

Psychoyou, Théodora. "Trajectories of Musical Acoustics in Seventeenth-Century French Theory and Experiment: Sound, Noise, and the Authorities of the Past." In *Music and Science from Leonardo to Galileo*, edited by Rudolf Rasch, 361–75. Turnhout: Brepols, 2022.

Quint, David. "A Reconsideration of Montaigne's *Des cannibales*." *Modern Language Quarterly* 51, no. 4 (1990): 459–90.

Randel, Don M. "Al- Fārābī and the Role of Arabic Music Theory in the Latin Middle Ages." *Journal of the American Musicological Society* 29/2 (1976): 173–188.

Raphael, Renée. *Reading Galileo: Scribal Technologies and the Two New Sciences*. Baltimore: John Hopkins University Press, 2017.

Rasch, Rudolf. "René Descartes and Isaac Beeckman: The Philosopher and the Schoolmaster." In *Music and Science from Leonardo to Galileo*, edited by Rudolf Rasch, 277–304. Turnhout: Brepols, 2022.

Rasch, Rudolf. "*Me veux-tu voir mourir*: Joan Albert Ban versus Antoine Boësset (1640–1641)." In *Music and Science from Leonardo to Galileo*, edited by Rudolf Rasch, 211–49. Turnhout: Brepols, 2022.

Rasch, Rudolf. "Simon Stevin and the Calculation of Equal Temperament." In *Music and Mathematics in Late Medieval and Early Modern Europe*, edited by Philippe Vendrix, 253–320. Turnhout: Brepols, 2008.

Redwood, André. "Combinatorics, Composition, Copia: Mersenne's Permutations as Rhetoric of Abundance." *Music Theory Spectrum* 45, no. 1 (2023): 71–88.

Redwood, André. "Mersenne and the Art of Delivery." *Journal of Music Theory* 59, no. 1 (2015): 99–119.

Reiss, Timothy. *Knowledge, Discovery and Imagination in Early Modern Europe: The Rise of Aesthetic Rationalism*. Cambridge: Cambridge University Press, 1997.

Reiss, Timothy. "Montaigne, the New World, and Precolonialisms." In *The Oxford Handbook of Montaigne*, edited by Philippe Desan, 196–214. Oxford: Oxford University Press, 2016.

Rentsch, Ivana. *Die Höflichkeit musikalischer Form: Tänzerische und anthropologische Grundlagen der frühen Instrumentalmusik.* Kassel: Bärenreiter, 2012.

Rentsch, Ivana. "Sinn statt Verstand: Johann Mattheson und das Geschmacksurteil im 18. Jahrhundert." In *Gut oder schlecht? Urteil und Werturteil in der Musik*, edited by Claudia Maurer Zenck and Ivana Rentsch, 31–43. Frankfurt am Main: Peter Lang, 2015.

Rheinberger, Hans-Jörg. *Spalt und Fuge: Eine Phänomenologie des Experiments.* Berlin: Suhrkamp, 2021.

Rivera, Benito V. *German Music Theory in the Early 17th Century: The Treatises of Johannes Lippius.* Ann Arbor: UMI Research Press, 1980.

Roberts, Lissa, and Simon Schaffer. "Preface." In *The Mindful Hand: Inquiry and Invention from the Late Renaissance to Early Industrialisation*, edited by Lissa Roberts, Simon Schaffer, and Peter Dear, xiii–xxvii. Amsterdam: Royal Netherlands Academy of Arts and Sciences, 2007.

Rochot, Bernard. *La correspondance scientifique du Père Mersenne.* N.p.: Conférence donnée au Palais de la Découverte le 4 juin, 1966.

Rochemonteix, Camille de. *Un college de jésuites aux XVIIe & XVIIIe siècles: le Collége Henri IV de La Flèche.* Le Mans: Leguicheux, 1889.

Rothman, Aviva. *The Pursuit of Harmony: Kepler on Cosmos, Confession, and Community.* Chicago: University of Chicago Press, 2017.

Ruellet, Aurélien. "'Le grand négociant des lettres': contours, évolution, et usages du réseau épistolaires de Mersenne." In *Saint François de Paule et les Minimes en France de la fin du XVe au XVIIIe siècle*, edited by André Vauchez and Pierre Benoist, 277–89. Tours: Presses universitaire François Rabelais, 2011.

Ruellet, Aurélien. "'Voir cette merveille et la faire voir': Chains of Circulation and Recommendations around the Microscope in the Early 1620s." *Nuncius* (2024): 1–26.

Scherpereel, Joseph. "Peiresc et la musique." In *Peiresc ou la passion de connaître: colloque de Carpentras, novembre 1987*, edited by Anne Reinbold, 153–204. Paris: Vrin, 1990.

Schnapper, Antoine. *Le géant, la licorne, la tulipe: collections françaises au XVII siècle.* Paris: Flammarion, 1988.

Schneider, Herbert, *Die Französische Kompositionslehre in der ersten Hälfte des 17. Jahrhundert.* Tutzing: Schneider, 1972

Seidel, Wilhelm. "Französische Musiktheorie im 16. und 17 Jahrhundert." In *Enstehung nationaler Traditionen: Frankreich, England* (=*Geschichte der Musiktheorie*, vol. 9), edited by Frieder Zaminer, 56–82. Darmstadt: Wissenschaftliche Buchgesellschaft, 1986.

Seipel, Wilfried, ed. *Für Aug' und Ohr: Musik in Kunst- und Wunderkammern.* Milan: Skira, 1999.

Shapin, Steven. *The Scientific Revolution.* Chicago: University of Chicago Press, 1996.

Shapiro, Barbara. *A Culture of Fact: England, 1550–1720.* Ithaca: Cornell University Press, 2000.

Shaw, Robert L. J. "The Minims, Their Rule, and French Religious Society at the Turn of the Sixteenth Century." *Viator* 51, no. 2 (2020): 335–87.

Sherr, Richard. "Plainsong in the Age of Polyphony." In *The Cambridge History of Fifteenth-Century Music*, edited by Anna Maria Busse Berger and Jesse Rodin, 771–84. Cambridge: Cambridge University Press, 2015.

Shiloah, Amnon. *Music in the World of Islam: A Socio-Cultural Study.* Farnham: Ashgate, 1995.

Shiloah, Amnon, and Annie Berthier. "A propos d'un 'petit livre arabe de musique.'" *Revue de Musicologie* 71, no. 1–2 (1985): 164–77.

Singer, Gerhard. "Lodovico Zacconi's Treatment of the Suitability and Classification of All Musical Instruments in the *Prattica di musica* of 1592." PhD diss., University of Southern California, 1969.

Smith, Justin E. H. *Nature, Human Nature, and Human Difference: Race in Early Modern Philosophy.* Princeton: Princeton University Press, 2015.

Smith, Pamela H. *From Lived Experience to the Written Word: Reconstructing Practical Knowledge in the Early Modern World.* Chicago: University of Chicago Press, 2022.

Smith, Pamela H. "Science on the Move: Recent Trends in the History of Early Modern Science." *Renaissance Quarterly* 62, no. 2 (2009): 345–75.

Smith, Pamela H., and Paula Findlen. *Merchants and Marvels: Commerce, Science, and Art in Early Modern Europe.* New York: Routledge, 2002.

Smith, Pamela H., and Isabella Lores-Chavez. "Counterfeiting Materials, Imitating Nature." In *The Matter of Mimesis: Studies of Mimesis and Materials in Nature, Art and Science*, edited by Marjolijn Bol and Emma C. Spary, 27–53. Leiden: Brill, 2023.

Stoessel, Jason, and Denis Collins. "'La grandezza del numero sonoro': Canonic Techniques, Combinatorics, and Early Scientific Thought in Seventeenth-Century Rome." In *Music and Science from Leonardo to* Galileo, edited by Rudolf Rasch, 155–86. Turnhout: Brepols, 2022.

Swann, Marjorie. *Curiosities and Texts: The Culture of Collecting in Early Modern England.* Philadelphia: University of Pennsylvania Press, 2001.

Taussig, Sylvie. *L'examen de la philosophie de Fludd de Pierre Gassendi par ses hors-texte.* Pisa/Roma: Fabrizio Serra, 2009.

Taussig, Sylvie. "La réception des curiositez inouyes par Gassendi et Mersenne." In *Jacques Gaffarel Between Magic and Science*, edited by Hiro Hirai, 57–71. Pisa/Rome: Fabrizio, 2014.

Taveau, Laurence. "Le manuscrit 2884 du P. Mersenne à la bibliothèque de l'Arsenal: Étude et édition critique." PhD diss., Université de Paris-Sorbonne, Laboratoire IREMUS, 2018.

Taylor, Kathryn. "Ancients and Moderns in Sixteenth-Century Ethnography." *History of European Ideas* 46, no. 2 (2020): 113–130

Tolbert, Jane. "Ambiguity and Conversion in the Correspondence of Nicolas-Claude Fabri de Peiresc and Thomas d'Arcos, 1630–1637." *Journal of Early Modern History* 13, no. 1 (2009): 1–24.

Tomlinson, Gary. *Music in Renaissance Magic: Toward a Historiography of Others.* Chicago/London: University of Chicago Press, 1993.

Tomlinson, Gary. *The Singing of the New World: Indigenous Voice in the Era of European Contact.* Cambridge: Cambridge University Press, 2007.

Tsien, Jennifer. *The Bad Taste of Others: Judging Literary Value in Eighteenth-Century France.* Philadelphia: University of Pennsylvania Press, 2012.

Vaast, Corrine. "L'esthétique musicale du père Mersenne et l'austérité musicale des Minimes." In *Saint François de Paule et les Minimes en France de la fin du XVe au XVIIIe siècle,* edited by Pierre Benoist and André Vauchez, 249–71. Tours: Presses universitaires François-Rabelais, 2011.

Valleriani, Matteo. "Galileo's Abandoned Project on Acoustic Instruments at the Medici Court." *History of Science* 50, no. 1 (2012): 1–31.

Van der Miesen, Leendert. "Songs at the Docks: Nicolas-Fabri de Peiresc and Musical Entanglements in Seventeenth-Century Marseille." *Music and Letters* 105, no. 1 (2024): 27–51.

Van der Miesen, Leendert. "Studying the Echo in the Early Modern Period: Between the Academy and the Natural World." *Sound Studies* 6, no. 2 (2020): 196–214.

Van der Miesen, Leendert. "'Unbelievably Hard Work.' Marin Mersenne's *Harmonie universelle* at the Printers." In *Early Printed Music and Material Culture in Central and Western Europe,* edited by Andrea Lindmayr-Brandl and Grantley McDonald, 231–244. London: Routledge, 2021.

Van Orden, Kate. "Descartes on Musical Training and the Body." In *Music, Sensation, and Sensuality,* edited by Linda Phyllis Austern, 17–38. New York/London: Routledge, 2002.

Van Orden, Kate. "Hearing Franco-Ottoman Relations circa 1600: The Chansons Turcquesques of Charles Tessier (1604)." In *Seachanges: Music in the Mediterranean and Colonial Worlds, 1550–1800,* edited by Kate van Orden, 33–68. Florence: Officina, 2021.

Van Orden, Kate. *Music, Discipline, and Arms in Early Modern France.* Chicago: University of Chicago Press, 2005.

Varwig, Bettina. *Music in the Flesh: An Early Modern Musical Physiology.* Chicago: University of Chicago Press, 2023.

Vendrix, Philippe. "L'augustinisme musical en France au XVII siècle." *Revue de musicologie* 78, no. 2 (1992): 237–55.

Vendrix, Philippe. "Music and Model in the Renaissance." In *Music and Mathematics in Late Medieval and Early Modern Europe*, edited by Philippe Vendrix, 9–21. Turnhout: Brepols, 2008.

Vendrix, Philippe. "Pour les grands et les autres: la réforme oratorienne du plain-chant." In *Plain-chant et liturgie en France au xviie siècle*, edited by Jean Duron, 87–96. Versailles: Éditions du Centre de musique baroque de Versailles, Klincksieck, Fondation Royaumont, 1997.

Vérin, Hélène. "Rédiger et réduire en art: un projet de rationalisation des pratiques." In *Réduire en art. La technologie de la Renaissance aux Lumières*, edited by Pascal Dubourg Glatigny and Hélène Vérin, 17–58. Paris: Éditions de la Maison des sciences de l'homme, 2008.

Verwaal, Ruben E. "Fluid Deafness: Earwax and Hardness of Hearing in Early Modern Europe." *Medical History* 65, no. 4 (2021): 366–83.

Wald-Fuhrmann, Melanie. "'Discors Concordia': Diskursverflechtungen in Athanasius Kirchers musikalischem Staatsmodell." *Zeitschrift für Ästhetik und allgemeine Kunstwissenschaft* 53 (2008): 285–321.

Wald-Fuhrmann, Melanie. *Welterkenntnis aus Musik: Athanasius Kircher's "Musurgia Universalis" und die Universalwissenschaft im 17. Jahrhundert.* Kassel: Bärenreiter, 2006.

Walker, Daniel Pickering. "Mersenne's Musical Competition of 1640 and Joan Albert Ban." In *Studies in Musical Science in the Late Renaissance*, 81–110. London: Brill, 1978.

Walker, Daniel Pickering. *Studies in Musical Science in the Late Renaissance.* London/Leiden: Warburg Institute/Brill, 1978.

Wallace, Johan, and Alexander McGrattan. *The Trumpet.* New Haven: Yale University Press, 2011.

Wardhaugh, Benjamin. *Music, Science, Experiment in England 1653–1705.* Farnham: Ashgate, 2008.

Weber, Joseph G. "Pascal and Music: World Harmony in Early Seventeenth-Century France." *Symposium: A Quarterly Journal in Modern Literatures* 30, no. 1 (1976): 75–91.

Wegman, Rob. *The Crisis of Music in Early Modern Europe, 1470–1530.* New York/London: Routledge, 2005.

Wegman, Rob C. "Sense and Sensibility in Late-Medieval Music: Thoughts on Aesthetics and 'Authenticity.'" *Early Music* 23, no. 2 (1995): 298–312.

Weiss, Gillian. "Infidels at the Oar: A Mediterranean Exception to France's Free Soil Principle." *Slavery & Abolition*, 32, no. 3 (2011): 397–412.

Welch, Ellen. "The Specter of the Turk in Early Modern French Court Entertainments." *L'Esprit Créateur* 53, no. 4 (2013): 84–97.

Whatley, Janet. "Savage Hierarchies: French Catholic Observers of the New World." *The Sixteenth Century Journal* 17, no. 3 (1986): 319–30.

Whitmore, P. J. S. *The Order of Minims in Seventeenth-Century France.* The Hague: Nijhoff, 1967.

Wilbourne, Emily, and Suzanne G. Cusick. *Acoustemologies in Contact: Sounding Subjects and Modes of Listening in Early Modernity.* Cambridge, UK: Open Book Publishers, 2021.

Wolfe, Jessica. *Humanism, Machinery, and Renaissance Literature.* Cambridge: Cambridge University Press, 2004.

Wright, Owen. "The Sight of Sound." *Muqarnas: An Annual on the Visual Culture of the Islamic World* 21 (2004): 359–71.

Wright, Owen. "Turning a Deaf Ear." In *The Renaissance and the Ottoman World,* edited by Anna Contadini and Claire Norton, 143–65. Farnham: Ashgate, 2013.

Wymeersch, Brigitte van. *Descartes et l'évolution de l'esthétique musicale.* Liège: Mardaga, 1999.

Wymeersch, Brigitte van. "Gabriel de la Charlonye (1588–1646): portrait d'un intellectuel de province à l'aube de la première modernité." *Albineana* 33 (2021): 181–201.

Wymeersch, Brigitte van. "Peiresc et la musique." In *Peiresc (1604–2004): actes du colloque international 26–30 août 2004*, edited by Jean Dhombres and Agnès Bresson, 111–34. Turnhout: Brepols, 2006.

Yates, Frances. *The French Academies of the Sixteenth Century.* London: Warburg Institute, 1947.

Zecher, Carla. *Sounding Objects: Musical Instruments, Poetry, and Art in Renaissance France.* Toronto: University of Toronto Press, 2007.

Zittel, Claus. *Theatrum philosophicum: Descartes und die Rolle ästhetischer Formen in der Wissenschaft.* Berlin: Akademie Verlag, 2009.

Index

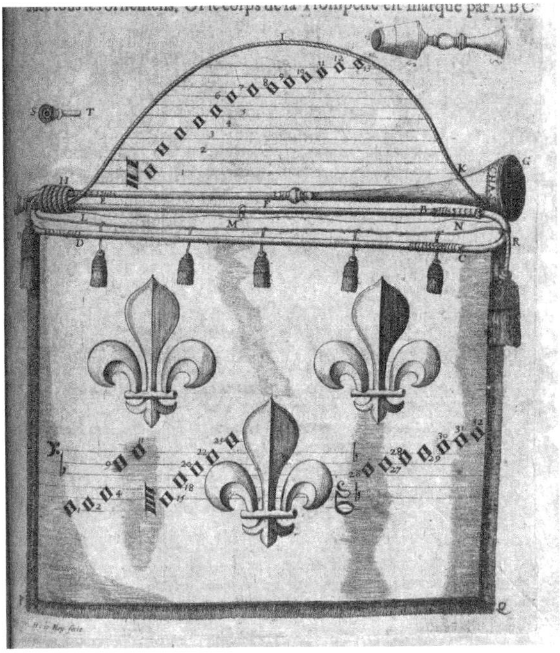

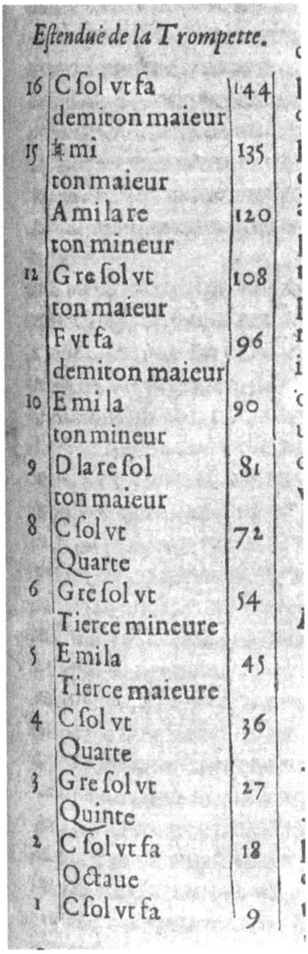

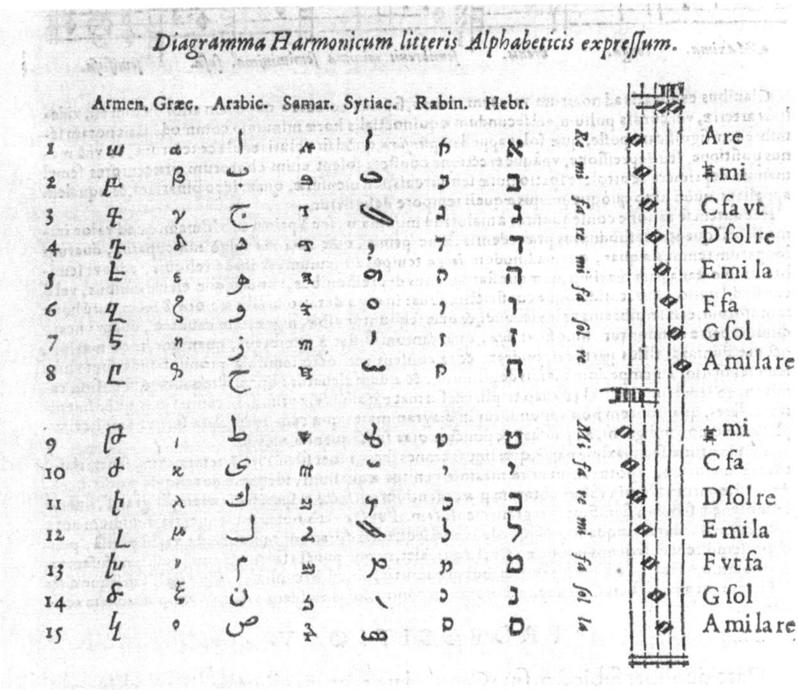

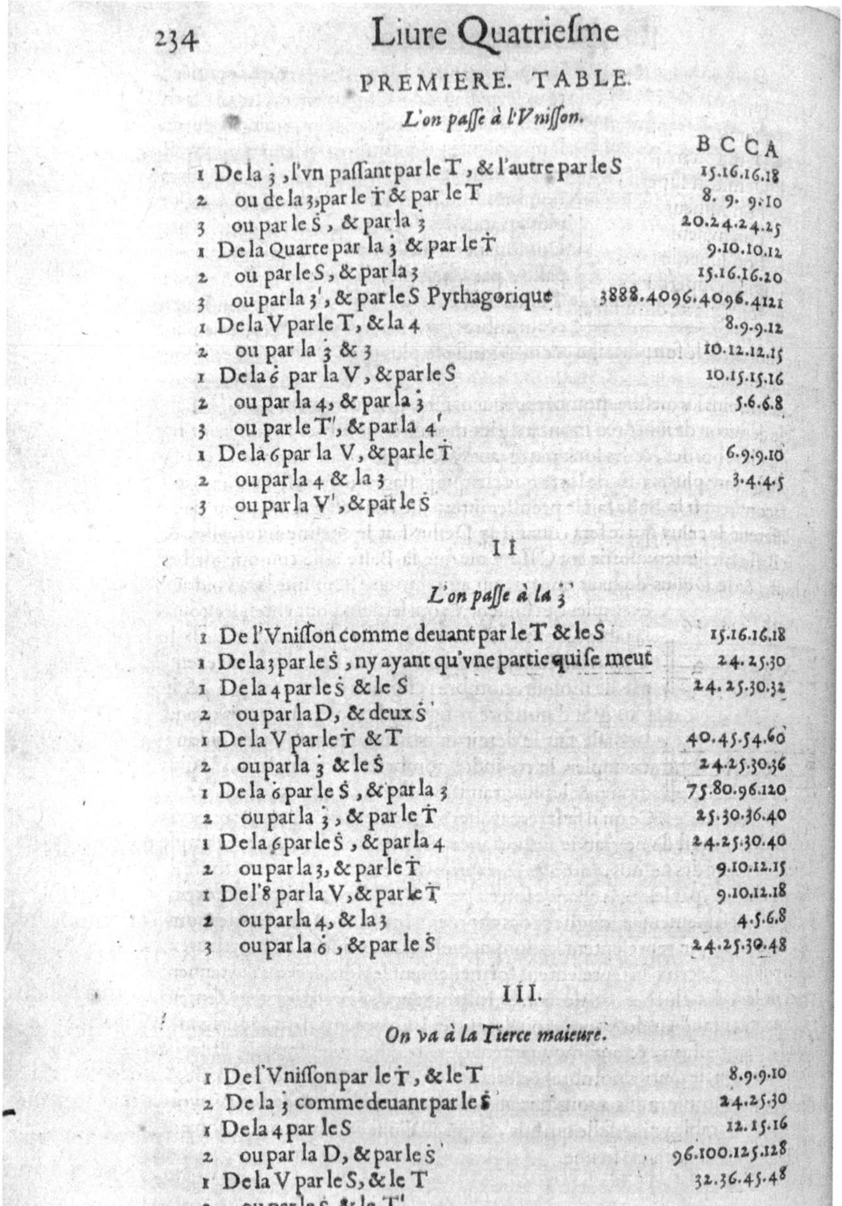

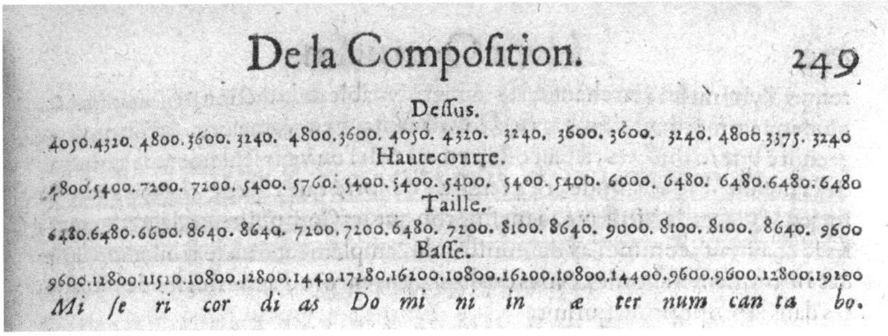

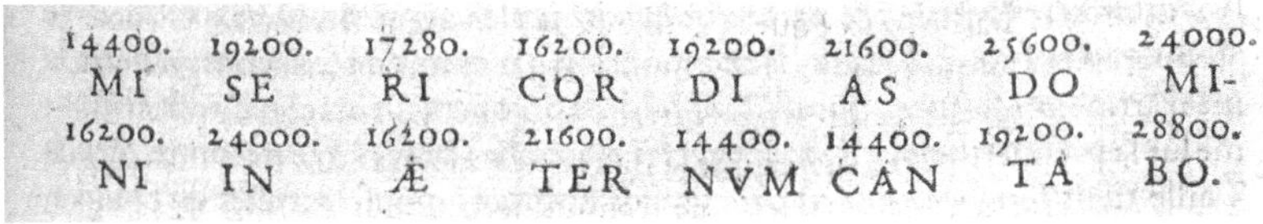

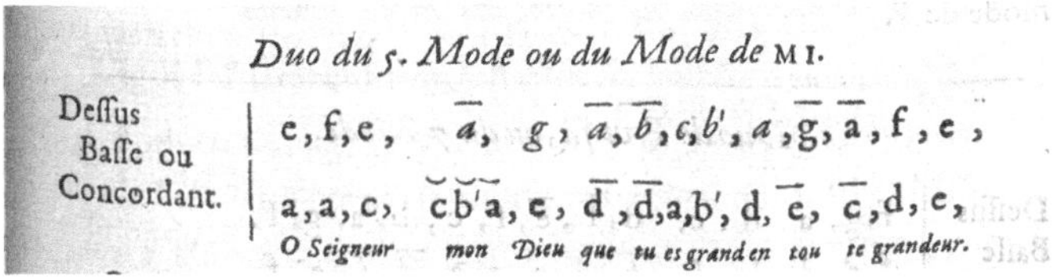

For Product Safety Concerns and Information please contact our
EU representative GPSR@taylorandfrancis.com Taylor & Francis
Verlag GmbH, Kaufingerstraße 24, 80331 München, Germany